Making Environmental Law

The Politics of Protecting the Earth

Nancy E. Marion

AN IMPRINT OF ABC-CLIO, LLC
Santa Barbara, California • Denver, Colorado • Oxford, England

Library of Congress Cataloging-in-Publication Data

Marion, Nancy E.
Making environmental law : the politics of protecting the Earth / Nancy E. Marion.
p. cm.
Includes bibliographical references and index.
ISBN 978-0-313-39362-4 (hardcopy : alk. paper) — ISBN 978-0-313-39363-1 (ebook)
1. Environmental law—United States. 2. Environmental law—Political aspects—United States. I. Title.
KF3775.M3737 2011
344.7304'6—dc22 2011013550

ISBN: 978-0-313-39362-4
EISBN: 978-0-313-39363-1

15 14 13 12 11 1 2 3 4 5

This book is also available on the World Wide Web as an eBook.
Visit www.abc-clio.com for details.

Praeger
An Imprint of ABC-CLIO, LLC

ABC-CLIO, LLC
130 Cremona Drive, P.O. Box 1911
Santa Barbara, California 93116-1911

This book is printed on acid-free paper ∞

Manufactured in the United States of America

Contents

Preface

Protecting the environment has become a concern to both citizens and politicians over the past century. Realizing that a poor environment affects the health of citizens as well as the health of the nation, presidents and the Congress have suggested and passed many different policies to maintain the quality of the air, water, and other natural resources.

The current book is an analysis of the federal response to environmental issues. It provides a look into the political process and explores the presidential support, legislative actions, and court decisions that have influenced environmental policy. Each chapter seeks to explain key developments in a specific environmental policy area. The book will be relevant for the environmental community and for others who are concerned with finding solutions to environmental problems. The purpose of this book is to provide a systematic account of how American environmental policy has developed and to answer the question, What is the state of environmental policy, and how has it changed over the years?

Each chapter focuses on a different aspect of pollution and the action taken to reduce the impact it might have on the environment and human health. Chapter 2 provides an analysis of policy created to diminish air pollution, and water pollution is the topic of chapter 3. The fourth chapter focuses on the congressional response to ocean pollution. Chapter 5 demonstrates congressional activity regarding pesticides, while chapter 6 looks at the problems and solutions revolving around solid waste. Toxic and hazardous waste is the focus of chapter 7. Chapter 8 examines public policy to protect wilderness land, and chapter 9 focuses on policies to protect endangered species.

Overall, the book provides a summary of environmental policy in the United States over time. It shows readers what issues were addressed by elected officials, as well as what issues were ignored. In the end, readers will have a greater understanding of how environmental policy came to be and where it is headed for the future.

Chapter 1

Introduction

On April 20, 2010, an explosion off the coast of Louisiana on the *Deepwater Horizon* drilling rig owned by oil giant BP left 11 platform workers dead and another 17 injured. Oil gushed out of the underwater well for three months afterward, releasing 4.9 million barrels of crude oil into the Gulf of Mexico.[1] The leaking of crude oil continued until September 19, 2010, when the federal government declared the well "effectively dead."[2] During that time, the spilled oil caused extensive damage to marine and wildlife habitats. Even though government officials, representatives from BP, environmentalists, and residents of the Gulf tried many methods to protect the marine environment, such as anchored barriers, skimmer ships, dispersal, dispersants, floating containment booms, and sand-filled barricades along the shorelines, the risk to wildlife remained. Over 400 species of animals were threatened, including many kinds of turtles (such as the endangered Kemp's ridley turtle, the loggerhead turtle, and the leatherback turtle), pelicans, herons, fish, mollusks, and even dolphins.[3]

The oil spill quickly became a political issue, with many politicians getting involved in some way. Many members of Congress attacked the existing law pertaining to cleaning up oil spills, originally enacted in 1989 after an oil rig called the *Exxon Valdez* went aground. That law limits an oil company's economic damages to $75 million, not nearly enough to compensate victims for the losses brought on by the oil spill in the Gulf. Some senators proposed new legislation that would increase the cap on what companies would have to spend to clean up spills from $75 million to $10 billion.

Investigating who was to blame for the spill became a little easier under HR 5481, which gave the Oil Spill Commission (the National Commission on the BP Deepwater Horizon Oil Spill and Offshore Drilling) the power to issue subpoenas. The commission could then require that witnesses give testimony. The release of books, records, correspondence, memoranda, and other documents related to the oil spill investigation would also be required.

Under S 3305, if found guilty, BP would be required to pay the entire bill for cleaning up the Gulf oil spill and compensating victims. Another bill, HR 5629, would change the liability cap for the cleanup costs of oil spills and for harm caused to victims. The minimum level of financial responsibility would rise from $10 million to $1.5 billion. The amount could be raised by the president under some circumstances. Additionally, the bill would require all vessels drilling in the 200-mile area known as the U.S. exclusive economic zone to be U.S.-flag vessels owned by American citizens, so that they would be required to follow U.S. safety regulations, including inspection of oil rigs by the Coast Guard.[4]

Many other new bills were proposed in Congress to help mitigate the harm from the oil spill. One of those was S 3473, a proposal to increase the president's access to the Oil Spill Liability Trust Fund. This bill would allow President Obama to withdraw more money from the fund to pay for efforts to clean up the environment. The president would be allowed to take multiple withdrawals of up to $100 million each from the fund, whereas without the law he was limited to a total of $150 million per fiscal year.

Other proposals focused on protecting victims harmed by the spill. S 3466 was a proposal to require mandatory restitution for victims, which is currently discretionary.[5] This bill would also require the U.S. Sentencing Commission to strengthen federal sentencing guidelines for violating the Clean Water Act (PL 92-500). The sentences would account for the actual harm to both the public and the environment in a more appropriate manner and could include incarceration and monetary fines for law violations. Additionally, a proposal in the House (HR 5503) would eliminate barriers to compensation for spill victims by revising the maritime liability law (PL 66-165, PL 66-2610) and allowing the families of those involved in the initial explosion to be compensated for injuries and loss of life. The victims could be compensated for nonmonetary losses such as pain and suffering or loss of care, comfort, and companionship. The family members of any workers who died could file lawsuits under the Death on the High Seas Act and receive compensation for noneconomic damages. At the present time, families of deceased workers can recover damages only for lost income or funeral expenses.

Proposals were made to decrease the chances of a similar explosion and spill occurring again. For example, S 3516 was a proposal to overhaul the federal management of offshore oil rigs and impose new safety requirements on drilling operations. The proposed legislation would require oil and gas companies to show their ability to cover damages from any oil spill that occurred. It would increase both civil and criminal penalties for wrongdoing. Additionally, drilling leases would be sold only to companies that demonstrated a strong safety and environmental record. In another bill (HR 5716), the deep-water drilling research program, which had originally been targeted for termination by the president, would be overhauled. If the bill passed, the research program would continue and would support initiatives to develop safer deepwater drilling technologies. Finally, a third proposal was HR 3534, an effort to overhaul offshore drilling requirements and give the Department of the Interior the responsibility to increase inspections of offshore drilling rigs.[6]

President Obama also took action after the oil spill. He made it clear that he expected BP to cover the entire cost of cleaning up the Gulf as well as pay for the economic damages the spill caused.[7] He temporarily suspended additional offshore drilling until the practice could be more fully discussed. In doing so, President Obama delayed the Dutch Royal Shell company's plans to start drilling exploratory wells in the waters off the Chukchi and Beaufort seas north of Alaska,[8] which had been approved by the Department of the Interior only three weeks prior to the disaster in the Gulf.[9]

Since the spill, there have been numerous legal, congressional, and scientific inquiries into the cause of the oil rig explosion. Some of the inquiries have revolved around determining civil and criminal responsibilities, while others have focused on recommendations to prevent such disasters in the future. One investigation was initiated by the Department of Justice. Attorney General Eric Holder Jr. announced an investigation into the spill and promised to enforce all of the relevant laws against those responsible. The Departments of the Interior and Homeland Security were also interested in investigating the events. These agencies established a joint investigation into the cause of the oil spill. The inquiry was overseen by the Minerals Management Service, housed within the Department of the Interior, which oversees offshore oil leases and enforces regulations, along with the Coast Guard, which was managing the cleanup in the Gulf. If their investigation resulted in evidence of criminal misconduct, the Coast Guard would decide whether to refer it to the Department of Justice for possible prosecution. The House Energy and Commerce Committee, which has jurisdiction over energy policy and some environmental issues, also investigated the spill.

Another investigation was completed by the National Commission, an organization established by President Obama through an executive order. This was a bipartisan commission set up to investigate the oil spill and make recommendations for preventing similar accidents in the future. The president also asked the National Academy of Engineering to investigate the causes of the explosion. This is a scientific panel of experts with the technical knowledge to understand the reasons for the accident.[10]

Acting as a political insider, the CEO of London-based BP, Tony Hayward, met with members of Congress after the spill to answer questions about the explosion and to try to control the potential damage to the company and its reputation. A trained geologist, Hayward had been at the Capitol many times before, reaching out to Congress in an attempt to develop positive, ongoing relationships with members.[11]

This scenario shows how an environmental accident like the *Deepwater Horizon* explosion had major political and policy implications in both the short and the long term. Members of Congress and the president took immediate action to ensure the cleanup of the spill and to reduce the harm to the wildlife in the Gulf. They made numerous policy proposals to ensure that the victims and their families were financially compensated for their losses and that future accidents (and the potential harm from accidents) would be limited to the greatest extent possible.

Action by presidents and the Congress to protect the environment and victims of environmental crime is not new. Over the past 70 years, government officials have taken a variety of steps to protect the quality of the environment in which we live and our natural resources,[12] including the air, water, land, vegetation, and wildlife. These actions make up our *environmental policy.* Environmental policy includes all laws, rules, policies, and regulations that elected officials have passed or implemented that are designed to protect environmental features and natural resources. It also includes what they choose not to do.[13] In short, environmental policy involves any action the government has taken that affects or impacts the natural environment in some way,[14] such as any laws to establish standards regarding the use and preservation of the environment or its assets and penalties for ignoring those laws.

For the most part, environmental policy is not criminal policy but rather can be categorized as regulatory policy. This is because most environmental policy is designed to regulate the behavior of those industries that might have an impact on the environment.[15] The intent of environmental policy is to deter acts that could cause harm to the natural environment.[16] That means that any violations of criminal and civil laws that are intended to protect the environment from polluters are environmental crimes.[17]

The need for environmental policy stems from problems associated with pollution. Pollution occurs when society creates too much solid or liquid waste, noise, congestion, or other things that negatively affect the environment or produce unpleasantness such as obnoxious odors, unsightly views, and loss of wildlife, plants, trees, and water and air quality. Pollution is a major concern because it can cause illnesses such as cancer, allergies, infertility, birth defects, and heart disease in people who are exposed to it on a regular basis.

Pollution is the by-product of a modern, consumer-based economy in which there is a demand for goods and services that are readily produced and then easily consumed. As consumers demand more products at low prices, businesses produce goods quickly in order to maximize their profits. The production process often causes innumerable tons of waste that can harm the environment. At this point, our society does not have the technology to produce goods and services in a way that does not create any pollution. Unfortunately, products that are manufactured in an environmentally safe manner are oftentimes more expensive than those that are not. Additionally, many daily activities in a modern society cause pollution, both directly and indirectly. Activities such as driving cars, buying clothing made from synthetic fibers, and using herbicides and insecticides on gardens and lawns all lead to pollution of the environment. The government itself often causes pollution with many of its activities, like testing military aircraft or nuclear weaponry.[18]

Not surprisingly, the pollution problem across the nation has increased as the population has grown. The number of people living in the United States rose sharply from 1939 to 1956, then leveled off for a short time. There was another increase in the population in the 1980s. Along with that came a rise in consumer spending and consumption. This then required considerable energy for manufacturing goods and transporting them to the market, and then for consumers' purchase and use of the goods.[19] In this sense, modern society often clashes with the environment, causing pollution and ill effects on resources.

The control of pollution is one element of environmental policy. This entails the prevention, management, and cleanup of waste discharges, accidental spills, and dispersion of toxic materials. Another element of environmental policy is the management of sustainable natural resources and the maintenance of naturally renewable resources such as water, wildlife, and soil. The third element of environmental policy is the preservation of our natural and cultural heritage and areas of beauty, historical and cultural significance, and ecological importance.[20]

Today's environmental policy in the United States is not to be found in any single criminal or civil statute on the federal, state, or local level. Instead, legislation to protect the environment is a patchwork of hundreds of different laws, statutes, regulations, policies, and court decisions that have been passed down over the years.[21] Environmental law is based on the Constitution, but it is also created by Congress, state legislatures, and administrative agencies. It is also the result of international actions such as treaties that have been signed by the president and ratified by the Senate.[22]

This means that today there is an elaborate set of policies geared toward protecting the environment and reducing future health problems that stem from pollution.[23] To make the issue more complex, environmental policy is constantly changing and growing.[24] It is reliant on public moods, on resources such as money and personnel, and on changes in political parties' control of government (in particular the Congress and the White House) as well as changes in state governments,[25] scientific research, and current events. It is also influenced by precedents and the attitudes of officials responsible for making, implementing, and enforcing it.[26] It is impacted by a wide variety of social, political, and economic forces that each attempt to shape the final product during the legislative process. As a result, environmental policy in the United States is a disparate collection of policies that have been enacted at different times, to different extents, and for different purposes.[27]

It must be recognized that the federal role in making environmental policy is a relatively new one. For many years, environmental policy was recognized as a state responsibility. During that time, state officials developed different strategies for dealing with environmental issues that were specific to their state. Consequently, states differed widely as to how much support they gave to environmental laws. Even today, in the House of Representatives, states differ as to their support of environmental policy. The states that typically have the highest support for environmental legislation include the New England states, New York, New Jersey, the Upper Great Lakes states, Florida, and the West Coast states. Those with the least support are the Gulf coast states, the Plains states from North Dakota to Oklahoma, and the Mountain states.[28]

States are involved in making federal environmental policy through the Environmental Council of the States, the top environmental protection agency of all states. The National Conference of State Legislatures conducts extensive research on a wide range of environmental, energy, and natural resource issues for state legislators and citizens and has collected an extensive set of publications, specialized reports, and books on the environment. Finally, the National Caucus of Environmental Legislators represents state legislators with a strong interest in environmental policy.

Even though the states at one time had the primary responsibility for environmental policy, common sense indicates that the federal government should be involved in making and implementing environmental policy as well. To begin with, the federal government has traditionally assigned and enforced property rights and determined who has which rights to use or alter the environment. For example, they have determined what land will become a national park, who can use it, and under what circumstances. They have also protected and decided mining rights. Second, the federal government has traditionally defined and enforced the rules of the market system through its power to enforce contracts that are aimed at protecting both buyers and sellers. Today, government officials make policies about selling land contaminated with toxic waste that could harm residents and others using that property. Third, the federal government is responsible for protecting the health and safety of the public. This means that it needs to protect citizens against polluted air and water. Fourth, the federal government has the duty to protect environmental assets from "tragedies of the commons," or the cumulative effects of individual actions that can ruin "open-access" areas such as oceans and parks. Under this argument, public officials should protect against the cumulative effects of overfishing by individuals, or the common resource will be destroyed. Fifth, the federal government is to provide collective goods that markets do not. Many environmental conditions are collective goods, including clean air and public parks, which benefit many. People using those goods cannot be forced to pay for them. Sixth, the federal government provides environmental services that people prefer to have provided collectively, such as public water supplies and trash services. Seventh, the federal government's actions have environmental impacts themselves. In some cases, government actions both cause environmental problems (e.g., through military weapons experiments) and correct them.[29]

Additionally, federal policies that are geared toward protecting the environment are appropriate because many types of pollution do not respect state boundaries. Emissions allowed into the air do not remain in one place. At the same time, states depend on the federal government for scientific research, expertise, and funding for the implementation of effective antipollution policies.

Today, federal environmental policy in the United States is overseen by the Environmental Protection Agency (EPA), created in 1970 during the Nixon administration.[30] Later, during the early 1990s, there was a proposal to elevate the EPA to cabinet status. The proposal was originally made at the end of the George H. W. Bush administration, and the idea was strongly endorsed by Clinton early in his first term. A proposal was passed in the Senate in May 1993 to create the U.S. Department of Environmental Protection, which would have elevated the EPA administrator to a cabinet secretary position without altering the EPA's duties or modifying the agency in any way, but this did not pass in the Senate.[31] Today, the EPA is not in the cabinet, but the administrator is given cabinet rank.

While the EPA oversees the implementation of environmental policy, the primary responsibility for making environmental policy lies with the members of Congress. Many members of that institution have a common interest in protecting the environment, yet there is no firm agreement on what bills should be passed to do that or to what extent action should be taken.

Some of the debate stems from the liberals and conservatives in Congress who have different perspectives on the direction that environmental policy should take. Party affiliation is a major factor in legislative decision making on environmental policy.[32] Voting records in the House of Representatives show that Democrats tend to vote about 2–1 in support of environmental policies, while the Republicans are more likely to vote about 2–1 against them.[33] In the 1970s and 1980s, about two-thirds of Democrats supported policies to improve the environment, whereas only about one-third of Republicans were in favor of policies to protect the environment.[34] Voting records also show that Democrats in the House did not support policies that were described as attempts to erase years of protecting the environment.[35]

Generally speaking, Republicans today tend to lean heavily toward increasing the use of environmental resources and economic development rather than conservation.[36] They have been accused of supporting policies that exploit environmental resources by making promises to deal with any consequences later. On the flip side, liberal Democrats typically are strong supporters of environmental legislation.[37] They tend to support legislation to increase protections on resources, even at the possible expense of business and the economy.

Beyond party affiliation, there are other characteristics that determine one's support or opposition to environmental policy. The first of those is membership in the farming, lumber, or mining industries or in the manufacturing industries that process and refine materials such as iron, steel, and wood. For those people, new policies to protect the environment are sometimes viewed as a threat to their established methods. They oppose environmental policy because they are sometimes reluctant to embrace new science, technology, and production practices. Instead, they view environmental policies as possible restraints on their economic activities.[38] Opposition to environmental policy can also come from land developers, who may be attempting to turn natural areas into developed areas and oppose regulations preventing that from happening.

History

Today's environmental policies have been strongly shaped by past policies and by their historical contexts.[39] Federal involvement in pollution control goes back to the 1940s, but Congress expanded its role in the early 1970s. At that time there was public demand for legislation that would protect land, mountains, lakes, and rivers from development. About that same time, concerns were raised about the role a polluted environment plays in human health and the realization that poor environmental conditions could be a serious threat to human health.[40]

When compared to other policy areas, environmental policy is a relatively new part of the federal legislative agenda. But chances are it will continue to be a crucial policy concern of the federal government. It is important to examine the policies established in previous years to have a better understanding of why the existing policies are the way they are and where we are headed for the future.

Prior to the 1970s

The founding fathers had little concern for protecting the environment. In their eyes, natural resources were plentiful. Human waste was thrown into streams without concern for the impact it might have on others or the environment.[41] Between 1850 and 1950, the country faced an increase in population with a resulting increase in consumption and industrial production. In most industries, coal replaced wood as an energy source for manufacture and transportation, which in turn produced pollution in areas downwind of factories. The increased population also led to changes in housing and food, transportation, and recreation. The number of factories increased, as did waste, including household waste, human waste, and consumer waste. This all had major impacts on the environment.[42]

The federal government's role in environmental policy making was sharply limited during this period.[43] Federal action until that time had been geared toward public land management by setting aside portions of the remaining public domain for preservation as national parks, forests, grazing lands, recreation areas, and wildlife refuges.[44] In 1812, Congress established a General Land Office within the Department of the Treasury that was responsible for public land functions.[45] But because the Treasury was becoming a catch-all for many unrelated functions about that time, the federal government decided in 1849 to create the Department of the Interior. It was believed that a "home department" to oversee the country's internal affairs was needed. The department was formed by reorganizing many offices, including the General Land Office, the Patent Office, Indian Affairs, and the Census Bureau. These agencies were all moved into a single new department.[46]

Although no federal legislation at the time dealt directly with environmental protection,[47] there was some related legislation that impacted the environment indirectly. For example, the Mining Law of 1866 stated that "the mineral lands of the public domain... are hereby declared to be free and open to exploration and occupation" subject to the customs of the individual local mining districts. The law authorized lode mining claims in the lands that contained valuable metal ores at a cost of $5 per acre. The person claiming the land would have to occupy it and spend over $1,000 in labor and improvements. The law was amended in 1872 to change the term *mineral lands* to *valuable* deposits, and it required that those claiming the land only spend $100 per year to develop the land in order to retain their claims.[48] The amended mining law applied to hard-rock minerals and gave mining first priority over all other uses for that land. This meant that if someone staked a claim to land that contained minerals, there was no way to prevent him from mining it except by buying up his claim, which was often quite expensive. Thus, the miner not only got the minerals but also received the land itself.[49] In 1920 the Mineral Leasing Act passed, which altered the government's policy from selling to leasing of fuel mineral lands.[50]

President Theodore Roosevelt and his administration held a strong belief that there could be multiple-use federal management for the national forest reserves. He knew that there was concern on the part of the public and in the scientific community that commercial logging could have some long-term effects, including severe deforestation, which in turn could lead to diminishing water supplies from the forested watersheds. Another concern was the risk of a "timber famine" in the future. Because of these concerns, there was support for legislation to protect land reserves. Presidents Harrison and Cleveland each reserved millions of acres of land and forests in the 1890s. Opponents called their actions simply a way to "lock up" valuable resources. In 1897, the National Forestry Committee of the National Academy of Sciences recommended that Congress pass a Forest Management Act that would authorize the secretary of the interior to manage the reserved forestlands. The law protected the reserves and guaranteed they would remain under federal management for years to come.[51]

By 1900, Congress had created some national parks, including Hot Springs (1832), Yellowstone (1872), Mackinac Island (1885), and Sequoia and General Grant (1890). With the support of Presidents Roosevelt and

Taft, Congress created five more parks in the early 1900s, including Crater Lake (1902), Wind Cave (1903), Platt/Sulfur Springs (1906), Mesa Verde (1906), and Glacier (1910). In 1903, Roosevelt established the first national wildlife refuge at Pelican Island, Florida, and, in 1908, the National Bison Range in Montana.[52] From 1906 to 1910, Presidents Roosevelt and Taft proclaimed 23 areas as national monuments, many of which became national parks. Those included the Grand Canyon, Death Valley, Katmai, and Glacier Bay. To do this, they used their powers under the American Antiquities Act that allowed them to set aside land as national monuments.[53] Then, in 1911, the Weeks Act authorized the federal government to purchase lands for reforestation.

The first major federal law in the United States that focused on protecting the environment was the Rivers and Harbors Act, otherwise known as the Refuse Act, passed in 1899. This law was intended to protect water quality by preventing discharges of refuse into navigable waterways, thereby keeping the waters free from blockages created by excessive dumping that might block ships.[54] The law made it a federal offense to put any refuse of any kind into U.S. navigable waters without a permit from the U.S. Army Corps of Engineers. Anyone who did so was subject to criminal sanctions of up to $2,500 in fines and up to one year in jail per offense.[55] The law applied only to solid matter that might be hazardous to navigation and specifically exempted liquid wastes such as sewage.[56] Then in 1924 the federal Oil Pollution Act was passed. This legislation prohibited oil pollution from oceangoing vessels, but it did not address other sources of oil pollution and was not very effective in controlling the waste from vehicles.[57]

In the years following World War II, policies to control the most obvious forms of pollution were slowly developed at the local, state, and federal levels; however, the power to make environmental laws belonged almost entirely to the state and local governments. Throughout the 1940s, every state established an agency that was responsible for limiting water pollution, although the exact responsibilities of the agency varied widely from state to state.[58] The federal government's role was limited to assisting local authorities to build sewage treatment plants and playing a limited part in a program to research the effects of air pollution. Air and water pollution were considered to be a strictly local or state matter, and they were not high on the national agenda.[59]

By the late 1940s and throughout the 1950s, the federal role in environmental policy grew. A major legislative act that focused on the environment passed in 1948, called the Water Pollution Control Act (WPCA) of 1948. The law authorized the growth of federal water pollution research and planning grants and gave low-interest loans to local governments.[60] Not long afterward, Congress passed the Air Pollution Control Act of 1955 and the 1956 amendments to the Water Pollution Control Act.[61]

Many new environmental policies emerged in the 1960s, including the creation of a federal wilderness system and the expansion of federal concern over many issues, from pollution control to solid waste disposal problems. The laws passed in the 1960s included the Clean Water Act of 1960, the Clean Air Act of 1963, the Wilderness Act of 1964, the Water Quality Act of 1965, and the Solid Waste Disposal Act of 1965.[62] During this period, many federal environmental policies focused on water resources planning, air and water pollution control, and recreation. But it was soon discovered that the problem-to-problem approach did not address the lack of boundaries when it came to pollution.[63] As public awareness of environmental issues grew, there was a push for more laws to protect the environment, and environmental concerns appeared on policy agendas with greater frequency.[64]

Air pollution was the subject of the Air Quality Act of 1967 (PL 90-148), which was amended by the Clean Air Act Amendments of 1970 and 1977. With these laws, the administrator of the EPA was required to set national standards for air quality but left it up to the states to enforce the standards. However, the federal legislation set criminal sanctions for violations, with, among other things, misdemeanor penalties of up to one year of incarceration and up to $25,000 in fines.[65]

One significant law to protect the environment was passed in 1969. The National Environmental Policy Act (S 1075/PL 91-190), or NEPA, established a comprehensive national policy on "the environment."[66] The law stated that "the Congress (recognizes) the profound impact of man's activity on the interrelations of all components of the natural environment, particularly the profound influences of population growth, high-density urbanization, industrial expansion, resource exploitation, and new and expanding technological advances."[67] The goal of the law was to "promote efforts which will prevent or eliminate damage to the environment and biosphere and stimulate the health and welfare of man."[68] It required federal agencies to provide detailed environmental impact statements for all major federal actions. Prior to undertaking a project, government officials would be forced to consider the potential environmental consequences of that project and complete an environmental impact study. The reports would be submitted to other agencies for review. There would be public hearings and citizen participation to allow environmental and community groups to challenge decisions by filing legal suits concerning the impact statements. If necessary, the federal agency would be forced to alter the proposed project. The law also established the Council on Environmental Quality (CEQ) within the executive office of the president to advise the president and Congress on environmental issues.[69] The council issued its first annual report in August 1970 and called for stricter enforcement of existing regulations.[70] President Nixon signed NEPA as his first official act of 1970 and proclaimed the 1970s as the "environmental decade."[71]

The 1970s

Until the early 1970s, the public was not fully aware of the health hazards related to pollution.[72] But this began to change, partly because of the publication of Rachel Carson's book *Silent Spring.* In this book the author described a fictional town that was inundated with illnesses that resulted from chemical pollution. The chemicals caused the disappearance of the livestock, birds, and fish. Carson's book reflected emerging concerns about possible health threats from environmental hazards, particularly the potential harm from pesticides and insecticides such as DDT that did not break down readily and remained in the environment for many years. Carson reported that DDT could be passed through the food chain, first affecting fish and wildlife and then reaching humans. She reported that there was evidence that some insects had already developed resistance to DDT. *Silent Spring* informed many Americans, some for the first time, about the possible harmful effects that pesticides and chemicals could have.[73]

Although the town in Carson's book was fictional, there were real events that affected the public's concern with environmental issues. Most prominently, in 1978, 263 families in New York State's "Love Canal" were forced to leave their homes because of hazardous waste dumping that had occurred there years before. And just as Carson had predicted, many residents became seriously ill with ailments caused by chemicals.

To bring attention to the possible dangers of pollution, the first Earth Day was held on April 22, 1970. This event marked a turning point in the history of public understanding of nature and of humans' place in it. In the months leading up to Earth Day, the media focused on pollution and other harm done to the air, water, and land.

Partly because of these events, a public consciousness about the environment surfaced, along with a public consensus about the need for a wider federal role in protecting the quality of the environment.[74] During this time, the framework for legal, political, and institutional actions to protect the environment was established. Environmental issues rose to a more prominent place on the political agenda in the United States, and a more active federal role in environmental protection emerged.[75] Since this time, the environment has been one of the most visible and debated public policy topics.[76]

In the next few years, Congress enacted major changes in environmental policy. For example, Congress passed the Water Pollution Control Act of 1972, which imposed standards for discharges from industrial plants and municipal sewage treatment facilities. The penalties for violating the law were set at a possible 15 years of imprisonment and/or a fine of up to $250,000. Corporate offenders could be fined up to $1 million.[77]

By the late 1970s, the Congress and environmentalists were frustrated with the lack of progress. At the same time, industries were complaining about costs and regulations. As a result, there was more disagreement in Congress as to the direction that environmental policy should take. Throughout the 1980s, there was vigorous debate on the direction of environmental policy.[78]

The 1980s

After leveling off for a period, population rates began to rise again in the 1980s. The nation's income grew, and consumer spending increased. Even though industries and factories were located far away from homes, making it easy to ignore their environmental impact, there was a corresponding increase in waste, and officials were forced to look for new ways to dispose of it. At the same time, people were coming to a better understanding of air pollution. It was no longer believed that air pollution affected only those in the immediate community. New synthetic toxic chemicals, created because they did not degrade, were building up in the air, water, and soil and becoming serious health hazards. Urbanization placed demands not only on cities but also on rural areas, which experienced lower air and water quality and changes in the habitats of plants and animals as chemicals were deposited onto the land and water.[79]

The pattern of congressional action that developed in the 1970s did not continue in the 1980s. Congressional enthusiasm for creating environmental policy declined and was replaced with an apprehension about its possible impacts on the economy. Many of the policies enacted during the 1970s to protect the environment were reevaluated in light of President Reagan's desire to reduce the scope of the federal government, return responsibilities to the states, and rely more on the private sector for environmental programs.[80] Congress cooperated with President Reagan and approved budget cuts that limited existing environmental programs.

But Congress eventually switched gears and began to defend environmental policies once again. They criticized the president's tough approach to the environment and voted to override his objections to legislation on hazardous waste and water quality. The public continued to support policies that protected the environment. Membership in national environmental groups rose, and new grassroots organizations developed.[81]

Environmental concerns were even part of the vice presidential debates in 1988 between Vice President Dan Quayle and challenger Lloyd Bentsen. Quayle reported that he had a very strong record on the environment while in the Senate, stating that he voted against the president on the override of the Clean Water Act, supported major pieces of environmental legislation, and supported efforts to control carbon dioxide to protect the ozone layer. In response, Bentsen claimed that the Democrats were the authors of the Clean Air Act, Clean Water Act, and the Superfund legislation to clean up brownfields. He claimed to have played a significant role in getting that legislation passed.[82]

Among Congress's most notable achievements of the 1980s were legislation to strengthen the Resource Conservation and Recovery Act (1984), the Superfund Amendments and Reauthorization Act (1986), the Safe Drinking Water Act (1986), and the Clean Water Act (1987).[83]

The 1990s

Throughout the 1990s, Congress continued to show concern for environmental issues. In 1990, Congress passed the Pollution Prevention Act, which made it national policy that, whenever possible, pollution should be prevented or reduced at the source. If that was not possible, pollution should be recycled in an environmentally safe manner. The disposal or release of pollution into the environment should be used as a last resort. The most significant part of the law was that it did not focus on one aspect of environmental pollution (air, water, or waste) but instead created a more encompassing approach to protecting the environment.[84]

After the Democratic candidates (Bill Clinton and Al Gore) won the presidential election in 1992, environmental supporters anticipated the opportunity to propose new laws to protect the environment and enhance existing environmental policies. But they were disappointed when some of Clinton's appointees began to work closely with members of the Republican Party on environmental policies. After the Republicans won the majority of both chambers of Congress in 1994, the relationship between opponents of the Republican Party became stronger. The Republicans created the Contract with America, in which they promised a reduction in government regulations and a smaller role for the federal government. Those Republicans who supported environmental laws came under pressure to conform to the anti-environmental position held by most party members.

Throughout the mid- to late 1990s, the Republican Party membership continued to attack environmental policies. They were able to get industry leaders appointed to legislative positions, where they supported anti-environmental positions.[85] Some state and local government officials were becoming tired of the costly regulations of federal mandates and often criticized proposed policies that might incur more costs.[86] But the Clinton administration opposed the anti-environmental movement,[87] and important legislation was passed to protect the environment, including the Clean Air Act Amendments of 1990, the Safe Drinking Water Act, and regulations on pesticide residues in food.

The 2000s

In the election of 2000, George W. Bush was elected to the presidency, and the Republican majority in both the Senate and House increased. The federal environmental agenda that existed during the Clinton administration shifted. The Bush administration asserted that the election results gave him a clear mandate for conservative environmental policies. Many environmentalists feared the worst.[88] Consequently, there was political conflict over proposed environmental legislation. It was difficult to get agreement on issues, and Congress struggled to pass environmental legislation.

Attention to environmental issues decreased significantly following the terrorist attacks of September 11, 2001. Environmental issues that had played a significant role in the past virtually disappeared from the political agenda for a time.[89] But the public continued to be concerned about the quality of the environment. In March 2004, the Gallup organization asked people how they would rate the "overall quality of the environment in the country today." A majority of those polled described it as "only fair" (46%) or even "poor" (11%). When pressed further, 58 percent feared that things were "getting worse."[90]

In recent years, President Obama has not addressed environmental issues, as he has been forced to deal with other issues in the first few years of his administration. The war on terror and health care have dominated his first years in the White House, and environmental concerns have taken a back seat.

Summary

As we gather more scientific evidence about the health effects of pollution, public concern about protecting the environment has grown. As the public demands more laws to ensure their health, the issues surrounding the environment have emerged as a primary concern of Congress. Controversy remains over the desired federal action that should be taken. Debate continues over whether we should protect the environment and, if so, how to do it and to what extent. The following section details how that debate influences the final policy that is enacted and the people who can have an effect on the laws that are passed.

Policy Process

The term *public policy* can be thought of as "a course of government action or inaction in response to a social problem,"[91] or "what governments do, why they do it, and what difference it makes."[92] Another definition is "the impact of government activity."[93] In short, public policy is what government does in response to a problem—what policies, laws, and regulations are passed to solve a concern that is held by many citizens. When considering environment policy, the public process works to keep the environment safe for people who use it or live in it.

All public policy, whether related to the environment or not, is made in a political atmosphere and involves a struggle over whose approach should prevail and become law. Conservatives and liberals, Republicans and Democrats, presidents, members of Congress, and interest groups all may disagree about the correct course of action. Each has a different goal they are trying to accomplish and a different perspective. Environmental policy making revolves around the struggle of balancing the goal of environmental health with other competing values such as productivity, economic well-being, individual rights, and social justice.

Generally speaking, there are five steps to the policy process or the policy cycle. It begins with problem identification, then moves to the agenda-setting stage. After this are policy formulation, policy implementation, and policy evaluation.

Problem Identification

In the first stage of the policy process, a problem must be identified and recognized as a significant public issue that needs action by policy makers.[94] Legislators must recognize a concern as a priority issue to voters. During this stage, the problem must capture the attention of decision makers and be identified as a viable and significant issue that needs attention. Cobb and Elder describe this process as the "expanding public" and detail how a problem must capture the attention of one person, who then voices his or her perception of that problem, and then other individuals agree and strengthen that voice.[95] Through this process, a condition is identified by a person (and then a group of individuals) as a problem that needs action, and it is then identified by lawmakers as such. During the 1970s, environmental policy moved from being a nonissue that was rarely the topic of action to one of the most significant issues of our time.[96]

Agenda Setting

The second stage of the policy process is agenda setting, where decision makers select what issues will be given priority and acted on and what issues will be ignored. In other words, they make choices about what problems are significant enough to deserve action. In this step, issues must be placed on the legislative calendars. Some problems emerge as viable political issues and demand attention and action from the government.[97]

The agenda is typically set by the president and Congress, but it can also be set by an interest group, congressional committee member, or even a private citizen. Sometimes it is the result of new information, or an agenda can be the result of an event or disaster that occurs. Many major environmental laws and regulations are direct responses to environmental disasters. For example, after the *Exxon Valdez* oil spill, Congress created the Oil Pollution Act of 1990. After the Three Mile Island nuclear reactor accident of 1979, new regulations from the Nuclear Regulatory Commission increased the requirements for emergency planning at commercial nuclear power plants. Similarly, the 1984 chemical plant disaster in Bhopal, India, in which 5,000 residents and plant workers were killed, resulted in a community right-to-know provision in the Superfund Amendments and Reauthorization Act of 1986. This required industries that used dangerous chemicals to disclose the type and amount of these chemicals to people who lived within an area likely to be affected by an accident on the site.[98]

Policy Formulation

The third step of the policy process is policy formulation, where goals for policy are set and specific plans and proposals to meet the goals are considered. This is usually done by Congress rather than the president. Congress has to decide the extent to which a policy will address the problem. They will choose to make either incremental changes or a complete overhaul. Congress must also determine who will carry out the policy—what agency will be responsible for implementing the new law. They also need to determine penalties, if any, for those who violate the new law.

A formal process exists for a legislative proposal to become a law in our system. The process begins with the introduction of a proposed bill into either the House of Representatives or the Senate. Eventually, both chambers will need to approve the proposal for it to become a law. That can occur either simultaneously or consecutively. After a bill is introduced into one of the chambers, the bill is assigned to the most relevant committee, which then assigns it to a subcommittee. Here, the subcommittee "marks up" a bill, or edits it to make it more accurate or more acceptable to the members. They can hold hearings where witnesses can provide the members with additional information. After the markup process, the subcommittee members vote on the proposed bill. If the majority votes in favor of the legislation, it moves to the entire committee, where the same process takes place. The full committee can hold hearings and mark up the bill if they choose. The members of the entire committee must then also vote on the proposal. If they vote in favor of the bill, then it moves to the floor of the chamber in which it originated for a full vote. If both chambers vote to pass the bill, it goes to the president for his action.

If the two chambers pass different versions of the proposed bill, it must go to a conference committee where attempts are made to resolve the differences between the two bills. The committee is comprised of representatives from both the House and Senate who attempt to iron out differences. This is necessary because only one version of the bill can be sent to the president for his approval. If the conference committee version is agreed to, then the bill is then returned to the floor of the entire House and Senate for another vote. If the conference version of the bill is approved by the chambers, the legislation is sent to the president for action. The president can either sign the bill or veto it. If he chooses to veto the bill, it can return to both congressional chambers for a re-vote. The members of Congress can override a presidential veto if two-thirds of the members vote in favor of that. If a proposal becomes law, the next step is the policy implementation stage.

Policy Implementation

The fourth step in the policy process is implementation, where programs are put into effect. Some policies are carried out through a governmental agency, whereas others are carried out by private action. Implementation refers to those activities that occur when agencies respond to the mandates of legislators. The laws are translated into operational programs, or carried out.

Most of the time, the laws passed by Congress do not clearly define terms and leave that to the agencies. In doing so, Congress delegates a great deal of discretion to the agencies to interpret the policies enacted. As the agencies fill in the gaps in the legislation, or as they define important terms, they are making policy. So the bureaucracies shape the impact of all public policies, making implementation a very political process. As the jurisdiction of the federal government expanded over the years in making environmental policy, so did the jurisdiction of agencies. They now have more power to make choices. Some of the agency choices are strongly contested, even as much as the legislation itself.

Sometimes, presidents can have some influence on the implementation of policy. From their power to nominate personnel, they can determine the direction of policy. Congress can also affect agency actions through its budgetary decisions. Congress also has a role in choosing who is selected to fill positions once the person is nominated by the president.

Policy Evaluation

The final step in the policy process is policy evaluation, where the program is reviewed and the impact of policies is determined. This can be accomplished by the courts, Congress (through the General Accountability Office), or the agency itself, or even informally through the media. The purpose is to determine whether the program is doing what it said it would do and if it is reaching the goals identified by policy makers. Some might also want to know if the program is being accomplished in a cost-efficient manner. The evaluations may point out modifications that should be made in order to make the program either more efficient or more effective. Or, in some cases, another problem may be identified. Most policies are continually monitored to assess their impacts.

Policy Participants

There are many actors in the policy process, and each has its own role to play. Each plays a vital part in creating policy. They include the legislative, executive, and judicial branches, as well as public opinion, agencies, and the media.

Legislative Branch

The legislative branch plays a vital role in the policy process. This includes the Congress on the federal level, state legislatures in the individual states, and township trustees or the city council on the local level. These groups have the primary responsibility in the agenda-setting and policy-formulation stages. The legislative branches may also help with implementation by approving appropriations for different agencies and programs. Congress is also part of the evaluation stage through its power to conduct oversight hearings and investigations.[99]

The role of Congress in the policy process is most prominent in making new laws. Their power is divided between the House and Senate and among many committees. Most of the legislative work of Congress is done in committees. There are many committees in both the House and Senate that have some jurisdiction over environmental policies. In the House, these include the Committees on Energy and Commerce; Agriculture; Science, Space, and Technology; Appropriations; and Natural Resources. In the Senate the relevant committees are Environment and Public Works; Agriculture, Nutrition, and Forestry; Energy and Natural Resources; Appropriations; and Commerce, Science and Transportation. One study found that 13 committees and 31 subcommittees in Congress had some jurisdiction over EPA activities.[100]

The Congress also plays an oversight role, overseeing the behaviors of the president and the bureaucracies. In 1946 (during the Truman administration) and in 1956 (during the Eisenhower administration), the Congress overrode presidential vetoes of the first federal water pollution acts since the Refuse Act of 1899 and an act of 1924 (repealed in 1970) relating to oil pollution of coastal waters.[101]

In making policy, it is necessary for the members of Congress to come to a consensus. But this can be very difficult at times. There are times when the members cannot agree as to how to reconcile their conflicting views, leading to policy gridlock. This refers to the inability of policy makers to resolve conflicts and pass laws. When there is policy gridlock, no action is taken. There is no consensus on what to do and therefore no movement in any direction. It is caused by divergent policy views or the complexity of environmental problems. Policy gridlock often occurred during the Reagan administration, when the president and other Republican members of Congress sided more often with developers, who sought to lessen environmental controls, than with environmentalists, who sought greater protections.

Executive Branch

The president has many powers in the policy-making process. Those powers are significant in determining both the content and impact of that policy.[102]

First, a president can propose new legislation to Congress, which is typically done in the State of the Union address but can also be done in other speeches throughout the year. In the end, the president can also veto legislation if he feels it is not good for the country for some reason. Second, the national budget is proposed each year by the president. The president can have a key impact on a particular policy through appropriating more or less money to that program from one year to the next. Third, the president has the power to appoint the leadership of many federal bureaucracies and key agencies. In terms of the environment, those include the Department of the Interior and the EPA. Presidents can also appoint different policy advisers to their personal staff and to various White House advisory committees. These officials can be removed at the president's choice. Fourth, the president can appoint federal judges if any vacancies occur, but the nominations must be approved by the Senate. This is important because many environmental policies are reviewed by the court to determine if they are fair. Fifth, the president has the power to issue executive orders and proclamations, which require no congressional approval. Some presidents have used this power to create new environmental regulations that had major impacts. Sixth, presidents have the power to influence public awareness of issues. Although this may not be a constitutional power, presidents are able to influence public perceptions about issues through statements to the media or in other ways. Finally, presidents have the power to reorganize or create federal agencies.[103]

While the president has many opportunities to impact the policy process, he can choose to take an active role or be more passive in the process. Obviously, if a president chooses a more passive role, then little, if any, legislation will be proposed by him on that topic. Conversely, if a president feels strongly about a particular issue, he will probably take a more active role and propose more legislation, work toward its passage, and fund it more generously.

In the United States, the presidency (or the executive branch) is very fragmented, meaning there are many parts to it. There are 12 cabinet departments that each bear some responsibility for policy. The Departments of the Interior, Agriculture, Energy, and State are cabinet departments that have some impact on environmental policy. There are also non-cabinet-level agencies such as the EPA and the Nuclear Regulatory Commission that are important actors as well.[104]

Presidents have been concerned with the environment to different extents and have supported different environmental policies. One of our earliest presidents, Thomas Jefferson, tried different methods to halt soil erosion in his fields at his home in Virginia. Another early president, James Madison, also voiced his concern about soil erosion.

Theodore Roosevelt. In more recent times, Theodore Roosevelt was the first president who brought serious attention to environmental concerns, becoming known as the conservation president.[105] He loved the wilderness and was an original member of the Boone and Crockett Club, often expressing concerns about the loss of forests, wildlife, and other natural resources. He relied on research and supported using reliable scientific management to control the nation's natural resources. One of the major goals of his administration was to protect nature. His intent was to both use and conserve the nation's forests and natural resources. Roosevelt sought to establish an administrator to oversee the health of the nation's forests, a position that was created in February 1905. In doing so, he created the Forest Service out of the Department of the Interior's Bureau of Forestry. He also created a special fund that came from the sale of forest products.[106]

Roosevelt established the Inland Waterways Commission in 1907 to develop a comprehensive plan for river development. The following year, in 1908, Roosevelt called all state governors to a White House Conference on Conservation, where they discussed the preservation of forests, wildlife, and other resources and the protection of human health. Conservation was the topic of another conference, after which Roosevelt appointed a National Conservation Commission to inventory natural resources. With the U.S. Forest Service, Roosevelt was able to protect many acres of land and establish 50 federal wildlife refuges across the nation. He supported the Antiquities Act of 1906, which allowed presidents to create national monuments in order to protect historic landmarks or other areas of scientific interest. Although the law was originally meant to protect archaeological sites from thieves, Roosevelt was able to use the terms of the law to turn the land into national monuments. For example, he proclaimed that more than 800,000 acres of land should be protected from development, creating the Grand Canyon National Monument in 1908. In 1919, Congress made it a national park. He created 18 national monuments, more than any other president. About a quarter of all U.S. national parks can be traced back to Roosevelt's creative interpretation of the Antiquities Act.[107]

Overall, Roosevelt's record on conservation was one of the strongest for any president.[108] He increased the acreage of the nation's forest reserves and created many wild bird reservations.[109] Knowing that the United States could not continue to exploit its natural resources, he reshaped environmental policy and made managing the environment a major responsibility of the federal government.[110] Overall, he successfully raised the problems related to the use of natural resources into a national issue, making the environment an important concept for many.

Harry S. Truman and Dwight D. Eisenhower. Environmental issues were not major elements of the Truman and Eisenhower administrations. President Eisenhower considered air and water pollution to be local problems, so he placed environmental concerns lower on the federal agenda.[111] Eisenhower was also dealing with other significant issues such as handling the Cold War, ending the Korean War, and rebuilding the nation after World War II. In the end, he often supported development over protection for the environment.[112] He opened up large areas for timber harvesting and turned offshore oil reserves over to the states. However, in 1956, President Eisenhower expressed support for "Mission 66," a program that maintained, reconstructed, and developed park roads and facilities, most of which had been neglected since the beginning of World War II.[113] He also supported the Fish and Wildlife Coordination Act of 1958, which authorized the inclusion of fish- and wildlife-enhancement features in water resource development projects.[114] In 1958, Eisenhower supported the Outdoor Recreation Resources Review Commission (ORRRC).[115] There was also a White House Conference on Natural Beauty in May 1965.

John F. Kennedy. President Kennedy also did not define a strong environmental agenda, although he recognized the public's growing environmental concerns. In his second State of the Union speech, Kennedy supported expansion of national parks and forests.[116] In another speech on conservation on March 1, 1962, Kennedy focused on seven areas: outdoor recreation; water resources; public lands; soil; watershed and range resources; timber resources; minerals; power; and research and technology. In that speech, Kennedy said that "we must reaffirm our dedication to the sound practices of conservation which can be defined as the wise use of our natural environment...the prevention of waste and despoilment while preserving, improving and renewing the quality and usefulness of all our resources." He proposed a White House Conference on Conservation to consider ways to protect the environment.[117]

Lyndon B. Johnson. President Johnson incorporated many of Kennedy's proposals into his own environmental agenda. Johnson believed that action should have been taken to protect the environment earlier, and he made many proposals to the Congress about the environment. These included proposals for 12 national seashore, lakeshore, and recreation areas. He proposed highway-beautification proposals that included banning billboards and junkyards and sought bills that would create a national wild rivers system and a national system of trails. New legislation to protect air and water and to deal with solid waste pollution, pesticides, and toxic chemicals was supported by Johnson. He announced a national center for environmental health and a White House Conference on Natural Beauty.[118]

President Johnson often publicly recognized the need to address environmental concerns. On March 8, 1968, he gave a message to Congress on conservation, called "To Renew a Nation." In it, he outlined his agenda for controlling pollution. His program included ideas to assure that people had pure and plentiful water. To do this, he suggested constructing community waste treatment plans and creating a National Water Commission. He also urged Congress to pass the Safe Drinking Water Act to reach that goal. Johnson also sought to protect waterways from oil spillage and other hazardous substances, to prevent the harm done by strip mining through the Surface Mining Reclamation Act of 1968, and to discover more efficient methods to dispose of millions of tons of trash in cities through the extension of the Solid Waste Disposal Act. To protect clean air, Johnson suggested finding new pollution abatement methods and investing in pollution control research. To protect outdoor recreation, Johnson proposed new national parks, adding thousands of acres of land to the wilderness system, increasing the number of scenic rivers and trails, and focusing on the problem of noise. Finally, Johnson had an agenda to protect the oceans that included cooperating with other nations.[119]

Richard M. Nixon. Nixon's general approach to pollution control was that it is not solely the responsibility of the Federal Government. Instead, heavy responsibilities fall on State and local governments, private industry and the general public as well.[120] Additionally, he believed that the costs of pollution should be more fully met in the free market, not in the federal budget.[121] However, according to Nixon, because there are no boundaries to the problems of our environment, the federal government must play an active, positive role.[122] He promised to continue vigorous enforcement of laws and federal regulations to protect the environment.[123]

Nixon did not have a personal stake in or commitment to environmental policy, and he personally tended to side with industrialists rather than environmental activists on environmental issues. However, he realized the political appeal of supporting the environment and jumped on the issue. He declared the 1970s as the "decade of the environment."[124] Although Nixon repeated that the basic responsibility for pollution control rests with state and local governments, industry, and the public, he understood that the federal government must provide leadership at the same time.[125]

During his administration, Nixon created many agencies to help further environmental issues. He created the EPA but also the National Industrial Pollution Control Council, which was housed in the Department of Commerce, with members from among the leading industrialists of the nation. About this time, Nixon also established a Quality of Life Committee housed in the Office of Management and Budget (OMB) that would review proposed environmental regulations to determine whether they gave balanced consideration to environmental quality and economic needs.[126] He also created the CEQ.

Nixon used his power of executive orders to protect the environment, most of which received support from environmental groups.[127] For example, he issued an executive order that required all projects or installations owned by or leased to the federal government to be designed, operated, and maintained so as to conform with air and water quality standards.[128] He also issued executive orders to implement NEPA (Executive Order 11514); to establish methods to prevent and abate pollution in federal government properties (Executive Order 11752); to increase the protection of historic and culturally important sites (Executive Order 11593); to ban giving federal contracts, grants, and loans to businesses convicted of Clean Air Act and Clean Water Act violations (Executive Order 11738); to delegate authority to the director of the OMB to implement provisions of the Water Resources Planning Act (Executive Order 11747); to delegate authority to the OMB to oversee the Rivers and Harbors Act (Executive Order 11592); to create the National Industrial Pollution Control Council (Executive Order 11592); and to establish a permit program for water quality control (Executive Order 11574).

In the various speeches Nixon gave, he brought up pollution and the environment often, mentioning many different topics and solutions. On February 8, 1970, Nixon gave a message to Congress on the environment. He said,

> The course of events in 1970 has intensified awareness of and concern about environmental problems. The news of more widespread mercury pollution, late summer smog alerts over much of the East Coast, repeated episodes of ocean dumping and oil spills, and unresolved controversy about important land use questions have dramatized with disturbing regularity the reality and extent of these problems. No part of the United States has been free from

> them, and all levels of government—federal, state and local—have joined in the search for solutions.... There can be no doubt about our growing national commitment to find solutions.

He called for strengthening pollution control programs with regard to air pollution, water quality, pesticides, toxic substances, and land use.[129]

In his State of the Union address in 1971, Nixon presented six goals to Congress, one of which related to the environment. Nixon told Americans that he was going to propose a strong new set of initiatives to clean up our air and water, to combat noise, and to preserve and restore our surroundings.[130] In another message to Congress on the environment, Nixon proposed setting authorizations for the federal share of water pollution programs administered by the EPA at $2 billion annually. Other proposals included revising laws on pesticide and noise control, adjusting programs on research and land use, and establishing international cooperation.[131] Nixon gave a third speech in which he suggested measures to strengthen pollution control programs, including measures to control toxic substances, noise pollution, and ocean dumping, as well as to expand international cooperation.[132] He outlined a plan for protecting the environment, with initiatives relating to tightening pollution control (i.e., toxic waste disposal control, emissions charge legislation, clean energy research), making technology an environmental ally, improving land use, protecting our natural heritage, expanding international cooperation on the environment, and protecting children from lead-based paint.[133]

Nixon took an international approach to controlling pollution. The president recognized that global environmental concerns transcend national boundaries, economic systems, and ideologies.[134] He said that "no Nation can keep its pollution to itself" and that "a broad international approach is therefore necessary."[135] He told the American public that the United States had broadened its discussions with our neighbors, Canada and Mexico, and with Japan, Argentina, Italy, and others, to solve certain basic environmental problems of particular concern. Further, he reiterated that it was the objective of his administration that the costs of pollution control be allocated in a uniform manner among different countries.[136]

Gerald R. Ford. President Ford continued Nixon's legacy, but his record on the environment was not as strong. He believed that the problem of protecting the environment had to incorporate international cooperation. In light of that, he signed a bill authorizing a $40 million contribution to the United Nations Environment Fund for five years. In doing so, he explained that the bill reflected broad agreement on the need for international cooperation to halt the degradation of the global environment. But Ford agreed that more action would be needed, including more knowledge about what the global problems are and how to cope with them.

Two of Ford's legislative goals included legislation on toxic substances and safe drinking water.[137] His budget included full funding of the Land and Water Conservation Fund for 1975.[138] During his presidency, Ford proposed action in eight areas that provided greater control over the sources of pollution: toxic substances, hazardous wastes, safe drinking water, sulfur oxides emissions charge, sediment control, environmental impacts of transportation, the United Nations Environmental Fund, and passage of the Ocean Dumping Convention. He also mentioned the need for legislation on a new Department of Energy and Natural Resources and adoption of the National Land Use Policy Act.[139] Ford vetoed legislation that would have created a program to regulate national surface mining,[140] but he proposed new legislation called the Comprehensive Oil Pollution Liability and Compensation Act of 1975, which would establish a system for assigning liability and settling claims for oil pollution damages in U.S. waters and coastlines.[141]

Jimmy Carter. During the 1976 presidential campaign, candidate Carter received the highest rating from environmental organizations. He promised to work for a cleaner, safer environment and careful management of natural resources.[142] Carter was an activist environmentalist who campaigned on environmental issues and, once in office, had an ambitious policy agenda for the environment.

Once he became president, Carter supported new energy sources, including hydroelectric, solar, and wind power. Carter gave a special message to Congress on the environment on May 23, 1977, in which he announced a "diverse but interrelated group of measures" designed to address pollution and expressed his concerns about preserving wilderness, wildlife, and other elements of America's national heritage. He also stressed his concern for the effects of pollution, toxic chemicals, and the damage caused by the demand for energy. He wanted to control pollution and protect health and said that he planned to improve enforcement of pollution control laws and make sure that regulation of a problem in one medium, such as water, did not create new environmental problems in another medium, such as air.[143] In that speech, Carter proposed actions that his administration could undertake to control pollution and protect health, to improve the urban environment, to protect our natural resources, to preserve our national heritage, to protect wildlife, to affirm our concern for the global environment, to improve implementation of environmental laws, and to deal with toxic chemicals, air pollution, and water quality.[144] The following year, in his State of the Union address, Carter promised to increase outlays for environmental concerns by more than 10 percent and to provide the new staff resources necessary to ensure that the nation's environmental laws were obeyed.[145]

Carter established an Interdepartmental Acid Rain Group to assist in discussions with Canada. He also supported the Surface Mining Control and Reclamation Act of 1977, the first national regulation of coal surface mining, and revisions of the Clean Air Act and Clean Water Act.[146] He proposed new legislation on water policy[147] that would achieve four basic objectives: improved planning and efficient management of federal water policy programs, a new national emphasis on water conservation, enhanced federal–state cooperation in water policy and in planning, and increased attention to environmental quality.[148]

Like Nixon, Carter used his powers of executive orders to help the environment. He issued Executive Order 11911 in 1977, which gave the CEQ the power to issue regulations and required federal agencies to abide by its directives.[149] He also issued Executive Order 12264, which required American vendors to notify importing nations of the risks of hazardous imported products;[150] Executive Order 11990, which dealt with the protection of wetlands; and Executive Order 11988, which protected floodplains. Finally, Executive Order 12113 provided technical review and tighter standards for the submission of water projects.[151]

However, many of Carter's environmental initiatives failed because he did not have the political skills and experience to convince Congress to act. When Carter got to Washington, the post-Watergate Congress was not favorable to his presidential leadership. Carter also faced a nationwide energy crisis that forced him and his advisors (and Congress) to deal with energy regulatory programs. Environmentalists were often disappointed by Carter's policies. After he lost the 1980 election, Carter preserved millions of acres of Alaskan wilderness and helped pass the Superfund bill to clean up toxic waste sites.

Ronald W. Reagan. When he first entered office, Reagan reversed much of the support that previous presidents had shown for the environment.[152] His agenda was avowedly anti-environmental and was considered by some to be even hostile to environmental protection.[153] During the Reagan administration, *deregulation* became the watchword as he sought to reduce the role of government in the private affairs of households and businesses.[154]

This approach was made clear when he said that when an environmental problem occurs, you must first ask if "there a private responsibility? Is some concern, a factory or something, responsible for this? And if so, then secure either help or turn this over completely to them and have them do it, if they're at a real fault. But then, other levels of government."[155]

When it came to the environment, Reagan sought to extend the principles of the market and privatization as far as possible,[156] and he wanted to loosen environmental regulations on businesses as a means to revitalize the economy.[157] Reagan clearly supported development over conservation and preservation and showed concern about what he considered to be increasingly burdensome regulations imposed on business and industry through environmental legislation.[158] The Reagan administration argued that the strict environmental regulations imposed on business by previous Democratic administrations were counterproductive. Instead, he supported weakening some laws and chose not to implement or enforce other environmental laws. He appointed personnel who were committed to rapid economic growth even if it exploited natural resources.[159] In some cases, he cut agency budgets and in other cases chose to reorganize the agencies, including the EPA, causing a loss of staff morale and credibility. Reagan criticized the Clean Air Act, claiming that trees were major air polluters.[160] He hampered efforts to solve acid deposition and the cleanup of hazardous waste sites and sought to abolish the CEQ. When that failed, he cut its staff drastically and ignored its members' advice. Finally, Reagan called for a reevaluation of all environmental programs based on their economic costs and benefits rather than effectiveness.[161]

In 1982, Reagan vetoed the Environmental Research Act (S 2577), which would have authorized the EPA's research and development programs. The law would have required the EPA Science Advisory Board to include representatives from states, industry, labor, academia, consumers, and the general public. However, Reagan said such a move would represent a major step backward in achieving the goal of assuring that environmental research programs reflect the best judgment of the scientific community, unhampered by partisan or interest group politics. He also argued that the law would authorize spending that was $46.4 million above that in the previously enacted appropriation bill. It also would be duplicative and wasteful effort to create another national environmental monitoring network.[162]

It did not take long for environmentalists to charge that, under Reagan, corporations were given a free pass to pollute the environment. Congress also did not support Reagan's environmental goals, nor was the public sympathetic to his assault on the environment.[163] By the spring of 1983, it was apparent to Reagan that his administration's approach to the environment was not what the public wanted and that he had seriously misjudged the public's commitment to protecting natural resources.[164] He understood that his approach to environmental policies could become a political liability for him, and he was forced to moderate his policies by 1984. During this time, the Democrats held the majority in the House of Representatives and prevented the administration from passing the extreme changes to existing environmental laws that Reagan had promised in the name of regulatory relief.[165]

Reagan then appointed an environmentally sensitive EPA administrator[166] and restored some funding for the EPA and the Department of the Interior. In his 1984 State of the Union address, Reagan said that the nation had improved the conditions of our natural resources but at the same time cited the need to preserve them. He described the need to clean up abandoned hazardous waste dumps, proposed a research program on acid rain, and promised action to restore lakes.[167]

Reagan also worked with other countries to solve pollution problems. He spoke to Canadian leaders on the problems related to acid rain.[168] He sent a treaty to the Senate for their approval that concerned an amendment to the Convention on International Trade in Endangered Species of Wild Fauna and Flora.[169] He also supported a treaty concerning protection of the ozone layer to protect the public from the potential adverse effects of depletion of stratospheric ozone.[170]

Overall, Reagan profoundly changed the course of environmental policy making.[171] He wanted to redefine environmental priorities and discredit the goals of environmental groups.[172] Environmental groups regarded Reagan as the most environmentally hostile president in a long time.[173] During his second term, he took few new initiatives, but the EPA was permanently damaged. During Reagan's two terms in office, his administration changed many of the pro-environmental stances of the Carter administration.[174]

Oddly, the Reagan years also had a positive impact on the environment. Reagan's opposition to environmental legislation grabbed the public's attention. Those opposed to Reagan's policies, and even many who supported him, became appalled by his lack of support of the environment. This led to a growth of the environmental movement in the 1980s. As Reagan's policies became public knowledge, more people joined environmental groups.[175]

George H. W. Bush. George H. W. Bush sought to distance himself from Reagan's policies. In his first two years as president, Bush was eager to adopt a more positive environmental policy agenda than Reagan had.[176] He promised he would be more sympathetic and active in environmental affairs and would act as the "environmental president." He demonstrated support for the environment by appointing people who advocated for environmental protection to high-level executive positions.[177] Bush's administration supported important new environmental policy initiatives and reform. He increased funding and staff for the EPA and promoted the Clean Air Act Amendments of 1990, describing them as among the "most important and urgently needed environmental policy initiatives since 1970."[178] He sought out advice from environmentalists and promised to restore the CEQ.

Some changes were made during the Bush administration. In 1989, Bush proposed legislation to amend the Clean Air Act so that every American would breathe clean air. He wanted to stop acid rain by the end of the 20th century and cut airborne toxic chemicals from major sources by at least 75 percent. He proposed legislation to reduce sulfur dioxide emissions and to provide tough sanctions for cities that did not make reasonable efforts and significant progress.[179] Overall, the proposed amendments were designed to curb three threats: acid rain, air pollution, and toxic air emissions.[180] As part of his clean air proposal, Bush called for the use of alternative fuels in 1 million vehicles by 1997.[181] As part of the fight for clean air, Bush promised to work with Canada.[182] However, the Bush White House was deeply divided on environmental issues for both ideological and economic reasons.[183] Some members of his administration were not as environmentally friendly as the president,[184] and he often disappointed most environmental advocates[185] by seeming to put the environment low on his agenda. Bush declared a moratorium on new environmental regulations and did not endorse international agreements on climate change and biodiversity at the 1992 Earth Summit in Rio de Janeiro, Brazil. He showed a reluctance to support global environmental concerns such as climate warming or the preservation of biodiversity. As his term went on, there seemed to be a resistance to domestic environmental regulation. He did not increase the EPA's budget in line with its increasing responsibilities. By the end of the Bush presidency, he had only partially restored resources that were needed to manage the environment and had enacted little legislation.[186]

William J. Clinton. Environmental issues received considerable attention during the 1992 presidential campaign. Bush, who was seeking reelection, criticized environmentalists as extremists who were putting Americans out of work. The Democratic candidate, Bill Clinton, took a more supportive stance toward environmental issues. He chose Senator Al Gore from Tennessee as his running mate. This appealed to environmentalists because Gore, the author of *Earth in the Balance*, had one of the strongest environmental records in Congress and was committed to improving the environment.[187] In his campaign, Clinton cultivated environmentalist votes.[188] He promised that he would be an environmental steward and that he would encourage mass transit programs, pass a new Clean Water Act, reform the Superfund program, and tighten enforcement of toxic waste laws. By doing all of this, he created high expectations among environmentalists.[189]

As president, Clinton appeared to have a strong environmental agenda. He appointed personnel who supported environmental protection to key positions. He and Gore pushed an extensive agenda of environmental policy reform and reversed many of the Reagan- and Bush-era actions that had been criticized by environmentalists. Clinton favored increased spending on environmental programs, alternative energy and conservation research, and international population policy.[190] He proposed a new series of policy initiatives intended to protect public lands.[191] He signed the Biodiversity Convention and the Law of the Sea Treaty, although neither has received formal ratification from the Senate. He issued executive orders creating or enlarging 22 national monuments and protecting millions of acres of forestlands. He passed many measures to protect public lands and endangered species, and he helped to broker agreements to protect and restore the Florida Everglades, Yellowstone National Park, and ancient redwood groves in California.

Like other presidents, Clinton believed that environmental policy should come from local governments. He did not want to "dictate from Washington" but rather to help communities develop their own plans to clean up their water supplies. He did not want a bureaucrat telling them that "water problems in Philadelphia are the same as they are in Phoenix."[192]

Clinton described his environmental program as based on three principles. First, he wanted to create a healthy environment, not only because people need a clean place to live, but also because a healthy economy and a healthy environment go hand in hand.[193] Second, he thought it important to protect the environment at home and abroad. In an era of global economics, global epidemics, and global environmental hazards, a central challenge of our time is to promote our national interest in the context of its connectedness with the rest of the world. And, third, he wanted to invest in more pollution prevention and water treatment.[194]

Clinton made many proposals to help the environment. He wanted to see more money invested in environmental cleanup, so he increased available funding for the fund to clean up toxic waste sites and proposed changes to make the process faster, cheaper, and more effective.[195] He asked Congress to pass stronger laws to protect lakes, rivers, beaches, and the water we drink. He asked for a new Safe Drinking Water Act and a new Clean Water Act,[196] because everyone wants "water we can drink and air we can breathe, food we can eat."[197] President Clinton proposed Project XL, Excellence and Leadership, as a new way to fight pollution. Through the program, the EPA would set certain pollution reduction goals, and companies could, on their own terms, figure out how to reach those goals in a cost-effective way.[198] Another policy supported by the Clinton administration was the Community Right to Know Act, which required manufacturers to tell the public how much they pollute.[199] He announced the creation of a White House Office on Environmental Policy that would replace the CEQ and would be responsible for coordinating environmental policy. He also proposed making the EPA a part of the cabinet.[200] Neither of these last two proposals were implemented.

Clinton understood the international aspects of pollution. He said, "In our era, the environment has moved to the top of the international agenda because how well a nation honors it will have an impact, for good or ill, not only on the people of that nation but all across the globe."[201] He told the American public that the United States agreed at the Kyoto summit to return to 1990 emissions levels between 2008 and 2012. Further, he wanted to commit to reduce emissions to below 1990 levels in the five-year period thereafter and believed we must work toward further reductions in the future.[202]

Despite this action, Clinton seemed to displease environmentalists as often as he gratified them.[203] It became clear that environmental protection was never Clinton's primary issue. His commitment to the environment was not as strong as believed, and he showed only occasional leadership on the environment.[204] He made only limited progress,[205] and his agenda was almost always limited to issues where deals could be made easily. He was faced with a Republican majority in Congress and often backed down.[206] Things moved slowly under Clinton, and environmentalists became frustrated with him.[207] In the end the Clinton administration seemed to be ambitious but had few actual accomplishments.

George W. Bush. During the 2000 campaign for the presidency, Bush strongly denied that he would return to Reagan's assault on environmental regulation and represented himself as a moderate environmentalist.[208] Most environmental organizations opposed the president's reelection in 2004 and supported the Democratic candidate, Al Gore. Overall, the campaign was dominated by terrorism, the Iraq War, and the economy rather than environmental issues.[209]

Although Bush seemed to recognize the political reality of the public's support for conservation, once in office, he and his cabinet focused more on the economic impact of environmental policy, and they put more emphasis on economic development than environmental protection or resource conservation. Bush seemed to support the core base of the Republican Party, particularly the interests of industry, corporations, timber, mining, agriculture, and oil companies. He drew from that base when filling major positions in the EPA and the Departments of the Interior, Agriculture, and Energy. Bush sought to reduce the burdens of environmental protection legislation by using voluntary, flexible, and cooperative programs and to transfer responsibility for enforcement of federal laws to the states.[210] Over time, Bush rescinded or amended many of the rules issued by the Clinton administration and replaced them with more industry-friendly approaches.[211]

Bush attempted to open federal lands, particularly the Arctic National Wildlife Refuge (ANWR), to energy exploration. Environmentalists charged that Bush rewrote scientific data to support his perspective on environmental issues.[212] The administration's proposal to rely more on the production of fossil fuels for energy angered many.[213] Bush continued to stress the importance of economic growth over environmental protection. These actions only confirmed environmentalists' concerns about Bush's policies regarding the environment, and because of this, Bush developed a confrontational relationship with environmental supporters in Congress.

Bush had many items on his environmental agenda, including the reauthorization of the Clean Water Act, the Clean Air Act, the Resource Conservation and Recovery Act, and the Endangered Species Act. Bush also proposed that the Congress create a Department of the Environment.[214] He proposed a budget with increased allocations for many environmental programs. His priorities included more money for the cleanup of facilities involved in manufacturing of nuclear weapons, accelerated construction of sewage treatment facilities in coastal cities, cleanup of Superfund toxic waste sites, and expansion of the world's largest global climate change research center.[215]

Like other presidents, Bush acknowledged that environmental problems are global and that every nation must help to solve them.[216] He repeated this idea when he said that "environmental protection requires international commitment and strategic American leadership."[217] In this light, he promised to work with other countries to help solve environmental problems. He worked with officials in Canada, Mexico, and Japan.[218] Along those same lines, Bush supported phasing out substances that deplete the Earth's ozone layer. He called on other nations to agree to an accelerated phase-out schedule.[219] Bush also sent the Senate the Protocol on Environmental Protection to the Antarctic Treaty that would designate Antarctica as a natural reserve devoted to peace and science.[220] However, he opposed the Kyoto treaty on global climate change. One controversial decision by Bush was to withdraw from the Kyoto Protocol on global climate change because, as he explained, it would place unfair burdens on the U.S. economy without requiring developing countries to control their emissions. Despite pleas from world leaders and much of the scientific community, Bush refused to change his mind.[221]

After September 11, 2001, the interest in the environmental movement was reduced.[222] At that time, environmental initiatives were overshadowed by antiterrorism and homeland security. The attacks also resulted in a change in thinking about the ability of state and local governments to continue to assume the responsibility for core public health functions, many of which involve environmental health and containment of communicable disease. Many state public health laws are more than a century old, and many state health departments are antiquated, yet they guide state policy in responding to public health threats such as environmentally borne diseases.[223]

Barack Obama. Although President Obama has been in office only a short time, he has shown a commitment to protecting the environment. Obama has said that one of his goals is to work toward curbing greenhouse gases and global warming.[224] He was a big supporter of a cap-and-trade system whereby emission standards would be set for companies. Those plants and refineries unable to meet the standards would be permitted to purchase permits from the federal government. Another way the companies could meet standards was to pay for carbon-offset projects, or activities that extract carbon from the atmosphere, including planting trees and building wind and solar electric plants.[225]

Obama has also shown concern for protecting oceans and waterways, saying that they are critical to supporting life.[226] He proclaimed June 2009 and June 2010 as National Oceans Months to help Americans learn more about the oceans and conservation of them.[227] He also formed the Interagency Ocean Policy Task Force to develop recommendations for a national policy to protect and maintain the health of oceans and other bodies of water such as the Great Lakes.[228]

When it came to protecting wilderness areas, Obama supported conservation efforts. He spoke in favor of strengthening the Endangered Species Act.[229] He also supported setting aside certain areas across the nation to protect them from development.[230] Obama proclaimed September 2009 and September 2010 as National Wilderness Months, bringing attention to those special places.[231]

Judicial Branch

The court system plays a major role in the policy process. Their role, called *judicial review*, is to interpret environmental laws to assure fairness and to ensure that agencies carry out their mandated responsibilities as required by legislation. Many times the laws from Congress are vague, and terms are left undefined. Other times, situations may arise that the law did not include. The courts review the law in relation to those situations and decide the appropriate course of action. When the courts choose a particular remedy, they are also shaping the law.[232] For example, in 1979, when the Supreme Court decided *Andrus v. Sierra Club* (442 U.S. 347), they decided that NEPA does not require federal agencies to prepare environmental impact statements to accompany their annual appropriations requests.[233] In essence, that decision created a rule, or policy, concerning the environmental impact statements made by agencies.

The majority of environmental cases begin in the federal court system rather than the state courts because, more than likely, the case will concern federal statutes or the Constitution. These cases begin in the federal district courts. If a decision made by a federal regulatory agency is contested or appealed, the case will go to the federal courts of appeals. Any unsatisfactory outcome in a lower court can be appealed to the U.S. Supreme Court. The decisions of appellate courts are considered precedents. This refers to "judge-made law" that guides and informs subsequent court decisions involving similar situations.[234] At this time there are more than 100 federal trial and appellate courts that potentially play a key role in interpreting environmental legislation and adjudicating disputes over administrative and regulatory actions.[235]

The use of the courts to determine the course of environmental policy is a recent pattern. The trend toward pursuing legal action on environmental issues began in the 1960s and has continued since then. In the early years, environmental agencies used legal action to require federal agencies to be more open. In more recent years, however, the courts have become a major battleground when it comes to environmental policy.[236] The number of environmental cases in the courts has increased, and judicial decisions have become significant determinants of policy. One of the reasons for this is that the EPA in 1971 "rediscovered" the 1899 Refuse Act and for the first time began to bring a significant number of enforcement actions against polluters.[237] Environmental litigation has been on the rise since 1970, both because of the number of environmental laws that were passed in the 1970s and because of the increasing level of information about these laws among all factions in environmental disputes.[238] Because of this, the courts are now playing a much greater role in controlling, defining, and supervising environmental administrators.[239]

Most environmental litigation comes from environmental interest groups, such as the Sierra Club and Environmental Defense Fund, or from business and property interests, including public interest law firms. Other cases stem from federal agencies such as the EPA and the Department of the Interior.[240] Some suits are by a private party (a citizen) against polluters (a business or private party).[241] But it can be difficult for a plaintiff to establish which polluter caused the harm, since multiple sources may contribute to the pollution. Additionally, there is the high cost of litigation.[242]

Cases are appealed to the court to provoke some sort of response from the EPA. Other policies end up in the courts for review or for clarification. The overriding goal of litigation is to force agencies to implement the law in a particular way. For example, in the past, the courts have authorized the EPA to begin legal action to force the abatement of hazardous waste pollution at Superfund sites that pose an "imminent and substantial" threat to public health or the environment.[243]

Since the presidency has been dominated by members of the Republican Party over the past 30 years, the judges these presidents have appointed have been increasingly conservative on environmental issues. Federal judges in the early 2000s are more likely to give priority to property rights over environmental goods, and less likely to grant discretion to federal regulatory agencies.[244]

Agencies. Agencies and bureaucracies are part of the policy process in significant ways. Most important, they are responsible for carrying out the laws passed by Congress. In implementing the laws, the agencies must sometimes interpret certain portions of the laws that were not fully defined or clarified by Congress. It is common for legislation to be written in broad and ambiguous terms by Congress, leaving agencies to define the specifics of the legislation. This means that the agency personnel are given great discretion to apply the laws to different situations, giving them a great amount of power in the process.

Agencies also play a role in other steps of the policy process. They are frequently involved in the agenda-setting stage because they are intimately aware of the issues and problems of their constituents. Since agency personnel work closely with those in the field, they understand the concerns that need to be addressed and changes that need to be made in existing policies. They are able to easily identify problems that need to be put on the agenda of Congress and the president.[245]

Agencies are sometimes involved in policy formulation because they sometimes help to design or write legislation to suggest improvements or changes in current laws.[246] They also occasionally lobby members of Congress in an attempt to win acceptance of a proposed bill they favor, or to kill one they oppose. They do this by meeting individually with members of Congress or their staff, or by testifying in front of congressional committees on proposed legislation.

For many years, agencies were seen as nonpolitical experts who could carry out the mandates of Congress effectively. But, with time, agencies have become political. They often are required to make choices among competing values that conflict with each other. They will make choices and implement policy based on their political ideology.

Many agencies are responsible for implementing policies dealing with pollution and the environment. These include the Departments of Agriculture, the Interior, Energy, Transportation, and Trade. They also include the CEQ and the EPA. The heads of these agencies have each been appointed by the president in a political process, and they generally hold the same political ideology as the president. The policies they implement will be the same.

Agencies sometimes take part in the evaluation stage of the policy process. Bureaucracies are sometimes the evaluators, determining if the program is working or if changes are necessary to alter a portion of the law. Sometimes an outside group will complete the evaluation, such as a university or private research center.

Agencies can be created by an administration. Nixon created the EPA through an executive branch reorganization (not by statute) in December 1970. He brought together people from the Departments of Health, Education and Welfare; the Interior; and Agriculture, as well as the Food and Drug Administration and the Atomic Energy Commission. Later, Clinton supported a bill that would make the EPA a cabinet agency; the Senate passed a bill to do so in May 1993, but the House did not consider the bill. The EPA acts as a regulator, educator, technical adviser, innovator, and leader. The EPA was given the responsibility to implement and then oversee and enforce environmental protection standards.[247] Nixon also created the CEQ, a cabinet-level advisory group that would review existing policies and suggest ways of improving them.[248] He also created the National Industrial

Pollution Control Council to allow businesspeople to communicate with the president about environmental issues.[249]

Interest Groups. Interest groups are groups of people who come together to support a particular policy approach. They are groups of individuals who organize themselves and attempt to influence legislation so that it reflects their policy goals.[250] They do this in many ways. Sometimes, members of an interest group may testify in committees and subcommittees in order to provide information to members of Congress. In doing so, they may also try to convince the members that their perspective is the right one. Sometimes, interest groups lobby members of Congress. This means that they meet with congressional members or their staff, provide information about a proposal, and try to convince members to vote a particular way. Interest groups also use litigation to influence policy choices.

Interest groups are involved in the policy process in many ways. They are involved in the problem-identification stage when they bring issues to the forefront. They are often more closely aware of issues that must be addressed. Groups often help identify significant problems and turn those problems into political issues that become part of the political agenda.

Agenda setting is also part of an interest group's role. Interest groups apply pressure on elected officials to have an issue placed on an agenda, or they can try to block legislation in a process called *negative blocking*. Through this, a policy alternative can be discredited so that it is not seriously considered. In this way, an existing policy can be protected.[251]

Interest groups can also be involved in the final stage of the policy process, program evaluation. Groups are often directly associated with program monitoring as they closely follow programs that affect their clients. If a policy is not working, a group may draw attention to its shortcomings and demand changes.

In the 1950s and 1960s, environmental organizations were primarily involved in lobbying Congress. They reached out to the public through their publications or through the media. Over time, the issues became more technical, and the groups added more scientific and technical experts to their staffs. Several new organizations were established that provided legal expertise, including the Environmental Defense Fund and the Sierra Club Legal Defense Fund (not connected with the Sierra Club). These groups brought together experts in the environmental field and attorneys who were trying to deal with evolving policies.[252]

The main strength of citizen environmental groups is found in the voting public, and this gives them more sway in Congress. Hence, much of the activity of these groups consists in mobilizing the public to influence the course of legislation. Over time, both the Congress and presidents have tended to become involved in specific decisions rather than general policy.[253]

The growth of the environmental movement has spawned many groups focused on the environment. The General Wildlife Federation was organized in 1937. The organization brought together thousands of local and state groups that could be considered conservationist organizations. They changed their name to the National Wildlife Federation in 1938. It convinced Congress to establish a tax on firearms, with the funds being used to support wildlife conservation efforts in every state. Other national nonprofits are dedicated to preserving the environment as well. One of the oldest and most powerful is the Sierra Club, founded in the 1890s by John Muir. Others are the Natural Resources Defense Council, the Environmental Defense Fund, and Friends of the Earth.

Opponents of environmental policy also became more visible during the 1970s, such as the Sagebrush Rebels, a group focused on reduction of federal authority. This gave way to the wise-use and county supremacy movements in the 1980s and 1990s. The wise-use movement proposes more development of public lands as well as increased construction of visitor accommodations in national parks. The county supremacy movement supports a reduction in the federal government's role in the management of natural resources.[254]

Public Opinion. The public's opinion about issues sometimes plays a role in creating public policy. It plays a role in problem identification when issues are recognized as important to the public and demanding action. When the public informs Congress of those issues, they are placed on the agenda. The laws Congress passes concerning the environment represent society's opinions about environmental goals and objectives and the methods needed to achieve them.

As more environmental legislation was passed by Congress, the public's interest in protecting nature grew alongside it. Public concern about protecting wetlands began in the 1960s and grew steadily, so that by the 1980s there was both federal and state action to protect the land. Public interest in nature also led to pressure to protect endangered species and wildlife.[255] There was also a concern for better health, as we recognized the role that pollution plays in a healthy lifestyle, especially for children and the elderly.[256] There have been public campaigns for waste reduction, and reuse and recycling have substantial popular appeal; companies acknowledge the popularity of environmental goals with a host of products labeled "green."[257] However, the nuclear industry has not been able to win public acceptance for new facilities.

Today, polls show that most people feel that the nation has not done enough to protect the environment and that environmental regulation hurts the economy.[258] Most opinion polls consistently report that substantial majorities in almost all major socioeconomic groups support the environmental movement and governmental programs to protect the environment and have supported them since Earth Day 1970.[259] In 1965 few Americans recognized environmental quality as an important problem. But by 1970 that had changed, and a consensus had emerged that the side effects of growth and prosperity were causing unacceptable damage to the environment and that the problems were so serious that they should be the subject of strong, albeit costly, federal regulation.[260] One Gallup poll taken around Earth Day 2003 reported that 79 percent of Americans stated they were active participants in or sympathetic to the environmental movement; 19 percent said they were active in environmental organizations.[261]

Gallup polls conducted in 1965 and 1970 showed that the percentage of the public selecting "reducing pollution of air and water" as a problem that should receive the attention of government more than tripled.[262] In 1971 a poll indicated that 64 percent of respondents thought that working to restore the natural environment was very important.[263] By 1976, 59 percent of respondents in a Harris survey indicated that it was very important for Congress to pass much stricter legislation to protect the environment and to curb air and water pollution.[264]

In 1994, only 1 percent of respondents to a poll indicated that Congress was doing an excellent job in handling environmental issues, with 33 percent reporting the congressional response as good and 37 percent saying it was "not so good."[265] Twenty-seven percent of respondents to a poll taken in 2001 believed that Congress and the administration should give protecting the environment the highest priority, 42 percent said it should get high priority but not the highest, 25 percent said it was middle priority, and only 5 percent rated it lower.[266]

In 2003, results of a Gallup poll showed that 25 percent of respondents believed that it was "extremely important" for the president and Congress to deal with the environment in the next year. Most of the respondents (38%) thought it was very important, 30 percent thought it was moderately important, and 6 percent reported that it was simply not that important.[267]

Two polls in 2007 indicated strong opinions about the congressional role in protecting the environment. One of those was a CBS News/*New York*

Times poll taken in April 2007. It showed that 63 percent of respondents believed that protecting the environment was so important that requirements and standards could not be too high and that continuing environmental improvements had to be made regardless of cost.[268] The second poll asked people, "During the next year, how much do you want...the U.S. Congress to do to help the natural environment?" To that, 43 percent reported "a great deal," 30 percent reported "a lot," 17 percent reported "a moderate amount," 5 percent reported "a little," and 5 percent reported "nothing."[269]

By 2009, 41 percent of respondents noted that the environment should be a top priority for the Congress. Forty-two percent said it was important but a lower priority, and 12 percent said the environment was not too important. Interestingly, 3 percent reported that the Congress should not focus on the environment at all.[270]

In 2009, results of a CNN/Opinion Research Corporation poll indicated that 28 percent of respondents reported that it was "extremely important" for Congress to deal with the environment in the upcoming year. In that poll, 34 percent said it was very important, 25 percent agreed that it was moderately important, and 11 percent reported it as "not that important."[271]

Of course, most people have an opinion as to who could perform the job of protecting the environment better: the membership of the Republican Party or that of the Democratic Party. Even though members within both political parties have recognized and responded to rising public concern about environmental degradation,[272] the Democratic Party is more widely seen as the party that is more likely to act to protect the environment. In 1979, a poll by the Republican National Committee showed that only 14 percent of respondents believed that the Republicans would do a better job of handling the problem of protecting the environment. It also showed that 32 percent believed that the Democrats would do better.[273] Through the 1980s, this trend remained the same, with most respondents to similar polls indicating that the Democrats would do a better job of protecting the environment than the Republicans. As Table 1.1 indicates, anywhere from 33 to 62 percent of respondents believed the Democrats would do a better job, but only 14 to 37 percent thought the Republicans would be better.

However, a poll taken in 1995 showed that 59 percent of respondents approved of the way the Republicans were handling the issue of the environment, whereas only 38 percent disapproved.[274] Different results were shown in a 1996 poll, which indicated that 48 percent of respondents trusted Bill Clinton to protect the environment, whereas only 27 percent trusted the

TABLE 1.1. What Party Would Do a Better Job of Handling the Environment?

Year	Democratic	Republican
1981	60	25
1982	62	24
1983	60	25
1985	51	37
1986	49	32
1987	59	26
1989	36	15
1990	33	14

Source: Harris Survey, September 1981, conducted by Louis Harris and Associates, September 19–24, 1981, and based on 1,249 telephone interviews, sample: national adult, http://webapps.ropercenter.uconn.edu/CFIDE/cf/action/ipoll/questionDetail.cfm? (accessed July 25, 2010);

Harris Survey, September 1982 conducted by Louis Harris and Associates, April 16–April 22, 1982, and based on 1,258 telephone interviews, sample: national adult, http://webapps.ropercenter.uconn.edu/CFIDE/cf/action/ipoll/questionDetail.cfm? (accessed July 25, 2010);

Harris Survey, June 1983, conducted by Louis Harris and Associates, June 7–11, 1983, and based on 1,249 telephone interviews, sample: national adult, http://webapps.ropercenter.uconn.edu/CFIDE/cf/action/ipoll/questionDetail.cfm? (accessed July 25, 2010);

Harris Survey, July 1985, conducted by Louis Harris and Associates, July 25–28, 1985, and based on 1,254 telephone interviews, sample: national adult, http://webapps.ropercenter.uconn.edu/CFIDE/cf/action/ipoll/questionDetail.cfm? (accessed July 25, 2010);

Harris Survey, April 1986, conducted by Louis Harris and Associates, April 5–8, 1986, and based on 1,254 telephone interviews, sample: national adult, http://webapps.ropercenter.uconn.edu/CFIDE/cf/action/ipoll/questionDetail.cfm? (accessed July 25, 2010;

Harris Survey, February 1987, conducted by Louis Harris and Associates, February 20–24, 1987, and based on 1,250 telephone interviews, sample: national adult, http://webapps.ropercenter.uconn.edu/CFIDE/cf/action/ipoll/questionDetail.cfm? (accessed July 25, 2010);

NBC News/*Wall Street Journal* Poll, November 1989, methodology: conducted by Hart and Teeter Research Companies, November 4–7, 1989, and based on 1,512 telephone interviews, sample: national registered voters, http://webapps.ropercenter.uconn.edu/CFIDE/cf/action/ipoll/questionDetail.cfm? (accessed July 25, 2010);

Time/CNN/Yankelovich Clancy Shulman Poll, October 1990, survey by *Time*, Cable News Network, methodology: interviewing conducted by Yankelovich Clancy Shulman on October 10, 1990, and based on 500 telephone interviews, sample: national adult, http://webapps.ropercenter.uconn.edu/CFIDE/cf/action/ipoll/questionDetail.cfm? (accessed July 25, 2010).

Figures are given in percentage of respondants

Republican Congress.[275] By 2002, it was clear, from a poll by the *Los Angeles Times,* that most people supported the Democratic approach to protecting the environment. This showed that 48 percent of respondents trusted the Democrats in Congress to protect the environment, whereas only 19 percent trusted President Bush.[276] A poll taken the following year indicated that only 34 percent of respondents trusted Bush to do a better job of handling the environment, but 60 percent believed that to be true of the Democrats in Congress.[277]

In 2004, most people (49%) responding to a poll believed that the environment should be a top priority for the president and Congress, but 40 percent thought the environment was important but lower priority. In that poll, 10 percent of those responding thought the environment was not too important.[278] Two years later, only 4 percent of those responding approved strongly of the way Congress was dealing with the environment; 11 percent approved somewhat, 32 percent reported neither approval or disapproval, 22 percent disapproved somewhat, and 31 percent disapproved strongly.[279] Finally, a poll taken in 2007 asked those responding if they believed the Republican Party or the Democratic Party was more likely to protect the environment. To that question, 14 percent answered the Republican Party, 57 percent the Democratic Party, and 14 percent neither.[280]

Conclusion

Environmental policy has evolved throughout the past century as our knowledge of the effects of pollution has grown. During that time, new environmental problems have arisen, such as hazardous waste, and old problems, including the quality of the air and water, have remained. Every modern president has suggested ways to improve the environment, and Congress has made many proposals of its own. Much of the action has been in response to public concern about pollution. The following chapters describe the action in more detail.

Chapter 2

The Politics of Air Pollution

Air pollution has become one of the most pervasive and persistent environmental problems in the United States and worldwide. Trepidation about the quality of air has become widespread among many, especially those living in or near urbanized and industrial areas surrounding major cities.

There are many sources of air pollution, and they have different impacts. The sources are both natural and man-made.[1] In nature, volcanoes can produce toxic materials, hot springs create gases that contain sulfur, and forest fires produce carbon dioxide and small amounts of carbon monoxide.[2] Humans have also been the source of many types of air pollution. Gases called sulfur dioxide and nitrogen dioxide are the two major man-made air pollutants.[3] Sulfur oxides are emitted from burning fossil fuels (coal, oil, and natural gas) that are used to produce electricity. These chemicals can travel hundreds of miles in the atmosphere. They react with water, oxygen, and oxidants and form acidic compounds, which then fall to the earth as acid rain. Acid rain affects other natural systems such as aquatic ecosystems, forests, crops, and even buildings, and it can affect the health of humans and the marine life in lakes and rivers.[4] The carbon dioxide released into the air by burning coal and by automobile and truck exhaust has also been linked to global warming, a controversial environmental issue of modern times.

In the 1970s, environmentalists began to warn people about the possible dangers of the thinning ozone layer, caused by the release of chlorofluorocarbons (CFCs) into the air. CFCs, invented in the 1940s, are useful as refrigerants, propellants in spray cans, and solvents to clean machine parts and have been increasingly used in modern society. When released into the air, CFCs effectively eat away at the ozone layer that lies about 20 to 30 miles above the Earth and blocks out most of the sun's harmful ultraviolet light. Environmental groups put pressure on the federal government to eliminate CFCs after scientists in the 1980s discovered a hole in the ozone layer over Antarctica that was the size of the United States. In 1988, a panel of scientists convened by the National Aeronautics and Space Administration (NASA) concluded that CFCs were destroying the Earth's ozone shield, causing an increase in sunburns, skin cancer, eye diseases and cataracts, and crop failure.[5] At this point, the full extent of damage from CFCs is unknown.[6]

The production and importation of CFCs was banned in the United States as part of an international treaty to eliminate their manufacture and use worldwide. In 1987, 140 nations signed the Montreal Protocol, agreeing to reduce their CFC production by 50 percent by 1998. Later, as more evidence of a deterioration of stratospheric ozone was discovered, nations agreed to eliminate all CFC production by the year 2000, a date later changed to 1996. Developing nations were given an extra 10 years to eliminate their CFC production.[7]

The long-term consequences of air pollution include effects on human health, the economy, and the health of animals and plants in our environment.[8] U.S. policies to protect air quality date back to the late 19th century. At that time, major U.S. cities including Chicago, New York, and Pittsburgh started to regulate smoke emissions. Even as far back as the 1930s and 1940s, people recognized an apparent link between asbestos, an airborne particle, and lung disease.[9] Over the past three decades, scientific understanding of the effects of air pollutants on public health has improved greatly. However, there are still debates over what to do to protect air quality and how much to spend to improve it.[10] In more recent times, Congress has passed many bills to address air quality. One of the first was the Air Pollution Control Act (PL 84-159), passed in 1955. Since then, there have been many legislative attempts to reduce pollution in the air. This chapter provides a description of the proposals made by presidents and the Congress to address concerns about air pollution. It is easy to see that the proposals have changed over time as we have learned more about the problems associated with air pollution. Support for cleaner air has come from environmentalists and the general public, although legislation has been politically controversial because of its impact on industry and economic growth.

President Truman

President Truman noted in early 1950 that air pollution had become a serious problem that affected all segments of the population and that the effects of air pollution included the destruction of crops and damage to property. He called for more research into the problem. Truman suggested that each local government should study its individual problems and draft laws for its own conditions but that the federal government could help by developing standards for states to follow.[11] In doing so, Truman was stating that pollution was a state issue rather than federal one.

President Eisenhower

In his 1956 Annual Budget Message, Eisenhower recommended increased appropriations to attack the problem of air pollution.[12] He noted that because of expanded industrial growth and urban development, the atmosphere in some places was approaching the limit of its ability to absorb air pollutants. He recommended increased funding for studies into air pollution and called for more research on finding solutions for controlling the problem.[13]

1955–56: 84th Congress

During this session of Congress, the members passed S 928, the Air Pollution Control Act (PL 84-159) in July 1955. It authorized $5 million in federal funds to assist the states to pay for research and technical assistance for pollution control as well as training of technical and managerial personnel relating to air pollution control. It did not set any regulations for air quality nor create punishments for violators of potential regulations. In the end, the new law did little to prevent air pollution, although it provided for a better understanding of the problem. However, because the law regulated air pollution from the federal level, it brought national recognition of the dangers pollution posed for public health.[14]

1959–60: 86th Congress

In 1959, both chambers of Congress agreed to extend the Air Pollution Control Act of 1955 for four years. This time, the bill authorized $5 million a year for research studies into the causes and control of air pollution.[15] The bill (HR 7464/PL 86-365) was signed by Eisenhower on September 22.[16]

President Kennedy

Kennedy argued that prevention of air and water pollution was essential to the nation's health. To do that, he urged Congress to enact three bills, the first of which was S 455, which would allow the federal government to conduct studies and hold conferences concerning air pollution issues. Second, Kennedy wanted to establish a National Environmental Health Center to act as a clearinghouse for nationwide activities related to air and water pollution. Finally, he wanted to increase appropriations for research on pollution.[17]

Research into the causes and solutions for pollution was imperative for Kennedy. He proposed legislation to support new research on the causes, effects, and control of air pollution. He sought to provide federal grants to state and local agencies to assist in the development of programs, and he gave the federal government the authority to conduct studies and hold conferences concerning air pollution.[18] After reviewing scientific reports that described overwhelming evidence linking air pollution to heart conditions and chronic respiratory diseases, Kennedy called for greater government action. He recommended legislation authorizing the Public Health Service (in the Department of Health, Education and Welfare [HEW]) to engage in more research, provide financial assistance to states, and take action to abate interstate air pollution.[19]

1961–62: 87th Congress

In this session, Congress took action on S 455 (PL 87-761), a proposal to expand federal air pollution control programs. The bill authorized the surgeon general to investigate and recommend solutions to air pollution problems that had interstate ramifications. One part of the bill directed the surgeon general to study the amounts and kinds of substances discharged from motor vehicle exhausts and to determine how much could be safely discharged into the atmosphere. The bill extended the federal Air Pollution Control Act of 1955 for two years, authorizing federal grants and assistance for air pollution control. It also retained the existing $5 million limit on annual federal appropriations.[20]

President Johnson

In December 1963, Johnson signed the Clean Air Act (HR 6518/PL 88-206) to control air pollution. With the support of the administration, the bill expanded the role of the federal government in preventing air pollution. In the final version of the bill, Congress authorized $95 million in appropriations from fiscal year 1964 to 1968 for expanded research, the establishment of more pollution control agencies and programs, and federal financial grants to states for programs to control and prevent air pollution. The law also allowed the federal government to take legal action against anyone who was found responsible for sources of pollution. The bill reinforced that the states held the primary responsibility for dealing with air pollution, and it encouraged states and municipalities to cooperate with each other in ending air pollution. The new legislation provided money in aid for states and local governments to establish air pollution control programs.[21]

By 1965, Johnson wanted Congress to amend the Clean Air Act of 1963 to allow the secretary of HEW to investigate potential air pollution problems before they happened.[22] In addition, Johnson, in his Economic Report, promised to issue an executive order covering air pollution from federal installations.[23] He stated that he also wanted to begin discussions that would lead to an effective elimination or substantial reduction of automobile exhaust pollution,[24] and he intended to institute discussions with industry officials and other groups that would lead to that goal.[25]

Clean air remained on Johnson's agenda even toward the end of his administration. In his 1967 State of the Union address, Johnson said the country should vastly expand the fight for clean air with a total attack on pollution at its source.[26] He gave another speech on pollution in which he described the problem across the country as getting worse. He said that the states had to commit themselves to solving the problems, but he also asked Congress to pass the Air Quality Act of 1967 to control emission levels for industries that contribute heavily to air pollution and allow the secretary of HEW to develop industry-wide emission levels. He also wanted to create regional air quality commissions to enforce pollution control measures.[27]

In his last year as president, Johnson continued to be concerned about air pollution. On March 8, 1968, he gave a speech to Congress on conservation, called "To Renew a Nation." In it, he discussed his plans to protect the air quality. He noted that "air pollution is a threat to health, especially of older persons. It contributes significantly to the rising rates of chronic respiratory ailments. It stains our cities and towns with ugliness, soiling and corroding whatever it touches. Its damage extends to our forests and farmlands as well." He discussed his past accomplishments to protect the air quality, including the Clean Air Act of 1963 and the Air Quality Act of 1967, but then noted that he sought an additional $128 million for fiscal year 1969 to carry out further efforts to fight air pollution.[28]

1963–64: 88th Congress

In 1963, Congress enacted HR 6518 (PL 88-206), the Clean Air Act. This new law updated and replaced the 1955 Pollution Control Act by expanding and strengthening the national program to control and prevent air pollution; it centered on the harm done to people by air pollution.[29] The law stated that "the growth in the amount and complexity of air pollution brought about by urbanization, industrial development, and the increasing use of motor vehicles, has resulted in mounting dangers to the public health and welfare, including injury to agricultural crops and livestock, damage to and the deterioration of property, and hazards to air and ground transportation."[30] The primary purpose of the law was to protect and enhance the quality of the nation's air resources and promote the public health, welfare, and productivity.[31] The secondary purpose was to "preserve, protect and enhance the air quality in national parks, national wilderness areas, national monuments, national seashores, and other areas of special national or regional natural, recreational, scenic or historic value."[32]

The authors of the law wanted to control air pollution at its source but believed that the primary responsibility for that lay with the state and local governments. The authors recognized that federal financial assistance and leadership were essential to develop cooperative programs between federal, state, regional, and local governments to prevent and control air pollution.[33] Thus, the bill authorized $95 million in federal assistance to states for the development of their air pollution control agencies, and it also expanded air pollution research programs, established more state and local pollution control agencies, and provided for legal action against any person responsible for sources of pollution. Additionally, the bill authorized the U.S. attorney general to bring suits to halt air pollution that originated in one state but affected people in other states.[34] In other words, the new legislation introduced a way for the federal government to assist states when cross-boundary air pollution problems arose.[35]

The bill set emission standards for new automobiles and called for an administrator to "prescribe... in accordance with the provisions of this section, standards applicable to the emission of any air pollutant from any class or classes of new motor vehicles or new motor vehicle engines, which in his judgment cause, or contribute to, air pollution which may reasonably be anticipate to endanger health or welfare."[36] Another section set standards for emissions from urban buses and called for the use of low-polluting fuel.[37]

Under the new Clean Air Act, the states would be allowed to enter into agreements, or compacts, to control air pollution, but they would need to be approved by Congress to become binding. There would also be cooperation between federal, state, and local governments to encourage air pollution prevention, as well as more uniform laws.

More research and program development on air pollution would occur under the new law. This would involve additional investigations, experiments, studies, and demonstration programs on pollution control by both public and private agencies. They would conduct studies of specific air pollution problems if there was an interstate effect. The secretary of HEW would be responsible for making grants available to those agencies or individuals interested in conducting the research and would oversee that research. The secretary would then collect and distribute the findings.

A program for pollution abatement was also established in the new law. A series of steps were provided that local, state, or federal governments could take to end air pollution that threatened citizens' health. If the pollution originated in one state but affected people in another state, the secretary of HEW could initiate the process. As a last step, the attorney general could obtain a court order requiring that the pollution be stopped. In cases where the pollution affected only one state, the secretary of HEW could call a conference and make recommendations to solve the problem. In some cases, technical or financial assistance could be provided.

Finally, when it came to pollution caused by automobiles, the law provided for the secretary of HEW to appoint a committee to develop fuels and other devices to limit the problem. Representatives from the car manufacturing industry would be on the committee. The committee would report to the Congress about the progress made toward eliminating exhaust pollution each year.[38]

1965–66: 89th Congress

New federal legislation to control air pollution was approved by Congress in this session. The bill, S 306 (PL 89-272), or the Motor Vehicle Air Pollution Control Act of 1965, had two parts. The first part directed the secretary of the federal Department of HEW to establish standards to control the emission of substances that could pollute the air from new motor vehicles, including carbon monoxide, hydrocarbons, and other air pollutants. The second part authorized a national research program to develop improved methods of disposing of solid wastes.[39] This legislation was the start of federal involvement in controlling air pollution from mobile sources, but it stopped short of imposing vehicle emission standards.[40]

Congress in 1966 also enacted administration-supported legislation (S 3112/PL 89-675) that expanded federal air pollution reduction activities under the Clean Air Act of 1963. The bill authorized money to maintain and improve air pollution programs,[41] including a new program of grants to states and local and regional air pollution control agencies to help them maintain pollution control programs. The maintenance grants were established to supplement existing federal grants for developing, establishing, and improving air pollution prevention and control programs.

1967–68: 90th Congress

In this session, Congress continued to study the problems of water and air pollution but enacted no significant legislation on these subjects.[42] A special task force on environmental health released a report in 1967 that recommended a federal program to deal with problems such as air and water pollution. The report called the "danger to environmental quality" one of the most important current domestic problems.[43]

Because many states had no meaningful policies to prevent pollution, Congress enacted S 780 (PL 90-148) to increase federal responsibilities for pollution control. Called the Air Quality Act of 1967, the new law amended the Clean Air Act of 1963, incorporating many of President Johnson's recommendations for pollution control programs. It expanded federal grants to states to help them plan and implement different pollution control strategies. Under the new law, states were required to establish air quality control regions that shared common air quality concerns. If states did not establish standards, the secretary of HEW would do so. It also directed the Department of HEW to investigate and publish information about the adverse health effects associated with air pollutants so that the states could then set air quality standards for them. Moreover, the Department of HEW was to devise possible pollution control techniques so that each state could begin to regulate polluters in order to attain its air quality standards.[44]

However, the new law did not authorize the federal government to set national emission standards as Johnson originally requested.[45] It allowed states to set their own. The federal government was to get involved only if the states failed to act. But the law authorized the secretary of HEW, in times of "imminent and substantial" danger to public health from air pollution, to seek a court injunction to halt further emissions into the atmosphere.[46] It also established the President's Air Quality Advisory Board.

President Nixon

During the 1970s, it was recognized that industrial pollutants from coal-fired power plants, automobile exhaust, and steel mills were causing acid rain. Acid rain, in turn, was harming the surfaces of buildings and statues and causing the deaths of fish and other marine life. The pollutants were found in the air across Europe, Canada, and the United States. Scientists also discovered serious depletion of the stratospheric ozone layer, which protects life from dangerous ultraviolet radiation. The depletion of the ozone layer was caused by CFCs that were released into the air. The CFCs rose into the stratosphere, where they were broken down by intense ultraviolet light. In the stratosphere, the atoms destroyed the ozone molecules. The CFCs were reducing the ozone layer's ability to block ultraviolet radiation from reaching the Earth's surface. Evidence that the ozone layer was disappearing had been growing for many years. It was more evident at the North and South Poles. However, some experts were critical of the evidence and maintained that the threat to the ozone layer was exaggerated.[47] CFCs were then also linked to a possible change in the world climate because they are a greenhouse gas.

Overall, Nixon was not a strong advocate for the environment. But he proposed that the federal government establish nationwide air quality standards and give states one year to prepare plans to meet those standards.[48] Nixon proclaimed that the automobile was the worst polluter of the air. In order to control that, he fought for research to improve engine design and fuel composition. He wanted to set strict standards and strengthen enforcement procedures.[49] Moreover, the secretary of HEW was asked to publish a notice of new, more stringent motor vehicle emission standards for new cars. New legislation required that samples of production vehicles be tested throughout the model year.[50]

That theme continued throughout his presidency. On February 8, 1970, Nixon gave a speech to Congress on the environment. In that speech he described his intent to strengthen pollution control programs by making changes to sulfur oxides emission standards and instituting a tax on lead in gasoline. The tax would supplement regulatory controls on air pollution.[51] In another speech to Congress in 1972, Nixon reiterated his idea of putting a tax on lead used in gasoline in order to bring about a gradual transition to the use of unleaded gasoline. He noted that this was essential if the automobile emission control standards that were scheduled to be effective for the 1975 model-year automobiles were to be met at a reasonable cost. In his speech he told the members of Congress that he was going to propose legislation for a special tax to make the price of unleaded gasoline lower than the price of leaded gasoline.[52]

1969–70: 91st Congress

During this session, Congress passed a new law (HR 17255/PL 91-604), the Clean Air Act of 1970, which became the most comprehensive air pollution control bill in U.S. history. It was intended to protect and enhance the quality of the air in the nation to protect the public's health. The law established legal deadlines for the reduction of certain hazardous automobile emissions. The law also set up new research programs and aid to states to establish abatement programs, and it established primary air quality standards to be administered nationally.[53] Both the Nixon administration and the automobile industry opposed the imposition of specific deadlines by law.[54]

There were many provisions to the law. One was that the Environmental Protection Agency (EPA) was now required to identify air pollutants that were hazardous to human life and, based on that, to set National Ambient Air Quality Standards (NAAQS) for the six most common air pollutants: sulfur dioxide, nitrogen oxides, carbon monoxide, particulates, lead, and ozone. The air quality standards would be based solely on protecting the public's health without regard to cost. The EPA was to set both national primary and secondary ambient air quality standards. The primary standards would define the levels of air quality that are necessary to protect the public's health, while the secondary standards would define levels necessary to protect the public from any known or anticipated adverse effects from a pollutant. The NAAQS reflected the EPA's opinions about the amount of pollutants that could be in the air without harming the health of the population, particularly vulnerable citizens such as the elderly or people with respiratory ailments.

States would be required to develop implementation plans, subject to EPA approval, to achieve these standards. Each state had to show that each of its air quality districts had met the standards or was taking steps to attain them.[55] It also required that every new "point source" of air pollution would need to obtain a federal permit, which would be based on technology rather than risk. The law required that the point source use the "best available" technology to control pollution.

There were some other provisions in the law. One provision required that some emission limits would be set, such as for automobile emissions. This meant that automobile manufacturers would have to achieve a 90 percent reduction in carbon monoxide and hydrocarbon emissions from any new vehicles, and an 82 percent reduction for nitrogen oxides. Deadlines were set for the production of low-pollution car engines.[56] Another provision extended for one year a section of the Air Quality Act of 1967 that authorized research on the control of air pollution resulting from fuel combustion in gasoline- and diesel-powered vehicles.[57] Finally, the new law increased the federal operating subsidies to state air pollution control programs.[58]

The Clean Air Act of 1970 was different from past laws in a few ways. First, it defined national goals in the form of ambient standards (levels of air pollution in the surrounding air) that all parts of the country had to meet. Before this, the standards were set by the states, if any were set at all. Second, the law directed the EPA to set national standards for controlling emissions of toxic air pollutants. Third, it set limits for emissions from mobile sources (cars and later trucks) as "percentage reductions" from existing levels. Fourth, it gave the EPA the authority to set national standards for emissions from new sources of pollutants. It also directed the states to prepare state implementation plans (SIPs) describing how they would reduce emissions from existing sources in areas (air quality districts that are defined in the act) where the air pollution levels exceeded the ambient air quality standards.[59]

The act included a provision that the EPA should issue quarterly reports on its investigations of firms that had been forced to close due to pollution control requirements. In the past, some companies claimed they were forced to shut down because of the new requirements. The Congress wanted to determine whether the job losses were due to the new air quality requirements. In the end, the reports found that job losses were due to other problems in the companies rather than the pollution control requirements themselves.[60]

The provisions of the new law were controversial and were eventually challenged in the courts. The new source performance standards were challenged in *Portland Cement Association v. Ruckelshaus* (486 F.2d 375, 1973). The plaintiffs argued that the standards set by the EPA and applied to the Portland Cement plant discriminated against them economically when compared with standards set forth for power plants and incinerators. They argued that the economic costs for the cement company were not adequately taken into account by the EPA as it created the standards, which economically hurt the cement plants more than the standards created for power plants. The plaintiffs also argued that the EPA had not proven that the standards were achievable. The cement company even produced experts who testified to errors in the EPA data and the possible significance of these errors. The federal district court decided that the EPA was not required to justify different standards for different industries. But the case was remanded to the EPA on the grounds that they had not demonstrated that the emission standards were achievable since they were based on erroneous data.[61]

The EPA's secondary ambient sulfur oxides standards were challenged in *Kennecott Copper v. EPA* (462 F.2d 846, 1972). Again, the defendants claimed that the EPA had made errors in the calculations on which the standards were based. The court again remanded the case to the EPA. As a result, the EPA rescinded the relevant portion of its secondary standard.[62]

The third case, *Anaconda Company v. Ruckelshaus* (482 F.2d 1301, 1973), revolved around the ability of Anaconda's copper smelter, which emitted sulfur oxides, to conform with the emission standards created by the EPA. The federal district court decided that since EPA's standards affected only Anaconda, the company should have been consulted many times during the time when the standards were being made. They also stated that there was not enough evidence to support the standards that the EPA established. However, the court of appeals overturned the district court's decision, finding the EPA had followed due process.[63]

In a smaller bill (S 3072), The Senate attempted to encourage the development of low-emission cars and trucks by requiring the federal government to purchase available low-emission vehicles, thereby creating an assured market. The House membership did not take action on it.[64]

The date by which the emission standards had to be met was challenged in *International Harvester Co. v. Ruckelshaus* (478 F 2d 615). The law required the EPA administrator to set emission standards by 1975, but Congress included a special clause for industries, giving them the power to request a one-year extension. International Harvester requested a suspension but was denied because the automaker had failed to show that effective control technology was not available. International Harvester, along with other major car companies, including General Motors, Ford, and Chrysler, initiated a legal challenge concerning the timeline for meeting the standards. A report from the National Academy of Sciences concluded that the necessary technology to meet the standards was not available to International Harvester at that time, but it noted that it was not impossible that the large car manufacturers might progress rapidly enough to produce cars that would be in compliance by the deadline. The court stated that the EPA could not reject the report unless it could demonstrate that its technological methodology was more reliable than that of the National Academy of Sciences, so the EPA could not reject the academy's conclusion that the technology was not available. In the end, the EPA decided to suspend the standards for a year and instead established interim standards.[65]

1971–72: 92nd Congress

The Senate Commerce Subcommittee on the Environment held a hearing on the role of the EPA in the inner city. The subcommittee chairman expressed his disappointment that the EPA had rejected a recommendation that all lead be removed from gasoline by 1977, which had been issued by a previous task force that had examined environmental problems in the inner city. The chairman stated that airborne lead, coming principally from motor vehicle emissions, was a significant threat and should be addressed by the EPA.[66]

1973–74: 93rd Congress

In this session of Congress, the members passed a bill (HR 5445/PL 93-15) that authorized $475 million for fiscal year 1974 to fund air pollution and automobile emission control programs that were created under the 1970 Clean Air Act. The law continued funding for the program at fiscal year 1973 authorization levels.[67]

In the June 1974 issue of *Nature* magazine, research was presented that illustrated more clearly the role that CFCs played in depleting the ozone layer. In response, legislation was proposed in Congress that would limit the use of CFCs in aerosol cans. The proposal was sent to a House committee for deliberation, but the congressional session ended before hearings could be held.[68]

The Congress passed, and Nixon signed, HR 14368 (PL 93-319) to temporarily delay some of the clean air standards established in the 1970 Clean Air Act. The new law, the Energy Supply and Environmental Coordination Act, allowed the EPA, under special circumstances, to delay a plant's compliance schedule until January 1, 1979. It also delayed final automobile emission standards. The Congress established a civil penalty of a $2,500 fine for each violation of the act and a $5,000 fine for each willful violation. It also provided for a $50,000 fine, six months in prison, or both for persons who willfully violated a coal allocation order and who had previously been subjected to the $2,500 fine for an earlier violation.[69] The final version of HR 14368 provided that major fuel-burning facilities with the ability to use coal could be barred from burning oil or natural gas. Other provisions of the law barred the EPA from using parking surcharges as air pollution control measures.[70]

One bill that did not become law this session was S 2680. Debated in the Senate Public Works Committee, the proposal would have authorized the EPA administrator to suspend certain requirements of the Clean Air Act of 1970 relating to the burning of coal. But the bill did not go anywhere.[71] In 1974, Congress considered the proposed Energy Supply and Environmental Coordination Act (HR 13360). This proposal would have amended the

Clean Air Act to allow auto manufacturers more time to meet emission deadlines. The bill did not pass.

President Ford

During the mid-1970s, a new term was coined to describe the emergence of major air pollution problems in most large cities. The new term, *smog*, described the visible combination of smoke and fog that settled over Los Angeles and other towns in the mid-1970s.

By the middle of 1975, an interagency task force was announced in Congress to examine the validity of a previous study that linked CFCs with possible health dangers. The task force confirmed the findings of the research presented in *Nature* magazine concerning CFCs. Although the task force acknowledged the possible dangers of CFCs, it also acknowledged that many questions remained. In the end, the task force members could not reach a final decision on the possible threat from CFCs to human health. Another report was released in early 1976 by the National Research Council. This report also verified the danger of CFCs, but they, too, were unable to call for the regulation of CFCs without more evidence.[72]

By 1975, photochemical smog was beginning to appear in many cities. Officials at the EPA put forth a regulation to improve the situation and announced that the newly designed catalytic converter had to be installed in most cars manufactured in 1976. The following year the requirement applied to every new automobile and truck manufactured in the United States. However, they mandated converters before they were fully perfected. Converters installed in the first few years after the mandate wore out before the cars did.

The Food and Drug Administration (FDA) wrote a regulation based on the Federal Food, Drug and Cosmetic Act of 1938 that stipulated that all containers of toxic ingredients must carry a warning. The new regulation stated clearly that CFCs were toxic and that every product containing that chemical was to be considered dangerous. Hence, the public saw warnings on hair spray, shaving cream, and all aerosols using CFCs. The EPA also established restrictions on the import and export of CFC material. Congress followed suit and proposed the Toxic Substance Control Act. Attached to it was a provision that all canisters that contained CFCs would have to be labeled. This act would, in effect, classify CFCs as toxic substances.[73]

The Ford administration favored legislation that would stabilize auto emission standards for three years at the levels specified by the EPA for model year 1977 and then impose stricter standards for two years thereafter.

He thought that standards were needed to achieve a better balance among environmental, energy, and economic needs.[74] Ford's approach was based on his belief that it was reasonable to increase automobile efficiency by 40 percent and still achieve an increase in environmental emission standards. Based on that, he recommended that the Congress change the law concerning emission standards. Simultaneously, automotive manufacturers agreed to increase automotive efficiency 40 percent in the next five years, meaning that cars would get 40 percent more miles per gallon but have a higher emission standard than cars at the time.[75]

Ford was also looking for ways to ensure the safe and environmentally sound production of coal. He sought improved technology to burn coal directly without producing environmental damage.[76]

1975–76: 94th Congress

A new version of the bill to limit CFCs was submitted in 1975, and hearings were held on it.[77] Some members questioned the validity of previous study results that linked CFCs with health concerns and requested additional information.[78] They asked the president to establish a blue-ribbon interagency task force to investigate the dangers of CFCs with respect to the depletion of the ozone layer, and Ford complied. The task force was composed of members from 12 government organizations and was headed by the representative of the National Academy of Sciences.

The Congress considered HR 3118, the Ozone Protection Act, but did not have enough votes to pass it. The proposal would have directed the administrator of the EPA to conduct a study of the cumulative effect of all substances and practices that might affect the ozone layer. During the hearings for the bill, witnesses agreed the issue needed further study because there were insufficient data to warrant an immediate ban on aerosol spray can propellants.[79]

In this session, Congress considered but did not pass legislation amending the 1970 Clean Air Act (S 3219). The bill was complex and would have lengthened the time limits for compliance with pollution standards for cars and industrial plants but not to the extent requested by the Ford administration and many of the affected industries.[80] The proposal was very controversial, with environmentalists generally supporting the proposal, although many would have preferred a stronger measure. There was also strong opposition from utilities and other business groups, who claimed it would shut off development of energy resources and other necessary growth in vast areas of the country. Partly because of this, the bill did not pass.[81]

President Carter

Carter noted that the health effects of air pollution could include asthma, chronic lung disease, respiratory illness, and cardiovascular attacks. He said that "we cannot hope to have a successful public health program in this country without a major effort to reduce pollutant levels in our air." He proposed the use of emissions technology to control automobile emissions.[82]

To address the problem of acid rain, Carter announced a 10-year research program called the National Acid Precipitation Assessment Program. The program was estimated to cost $600 million and would involve 2,000 scientists. Although Carter understood the importance of using coal in manufacturing and for energy, he noted that we need to allow for "increased reliance on coal without sacrificing the environment."[83]

Carter recognized that air pollution was not solely a state responsibility. In speaking to residents of Denver, Colorado, about their Denver Air Project, the president said, "I believe that we can deal with this problem not through heavy-handed government prohibitions, but rather through a positive demonstration of how Federal, State, and local resources and, of course, those of the private sector as well, can be brought to bear in a coordinated way."[84] In this statement, Carter was showing that air pollution is not simply a state matter but needs to be addressed at all levels of government.

1977–78: 95th Congress

Congress passed, and Carter signed, HR 6161 (PL 95-95), the Clean Air Amendments, which reauthorized and amended the Clean Air Act of 1970. The new bill relaxed the clean air standards by extending the existing allowable levels for major automobile exhaust pollutants through the 1979 models. New standards were set for "prevention of significant deterioration" in areas with good air quality. These standards would prevent dirty industries from simply moving from polluted regions to clean ones. The amendments gave the EPA the ability to award waivers covering carbon monoxide and nitrogen oxide emissions for 1980 and 1981 model cars, sometimes to the "lowest achievable emissions rate." The bill also extended the deadline for cities to meet national clean air standards from 1977 until 1982, and in some cases until 1987.[85] The new law affected virtually all industrial and transportation activity.[86] Finally, the National Commission on Air Quality was established in the new law.

1979–80: 96th Congress

In 1980 the Congress funded a 10-year study called the National Acid Precipitation Assessment Program. The intent was to collect information on the extent and effects of acid rain. The study eventually concluded that there was insubstantial evidence that acid rain caused a decline of trees other than red spruce trees at high elevations.[87] In another act (SJ Res 188/PL 96-300), the Congress agreed to extend the life of the National Commission on Air Quality in order for it to submit a report about the status of the quality of the air and how to improve it.

President Reagan

Reagan supported passage of the Clean Air Bill, which, while protecting the environment, would make it possible for industry to rebuild its productive base and create more jobs.[88] In 1985, Reagan sent to Congress the Vienna Convention for the Protection of the Ozone Layer for approval. The convention would provide for an international cooperative effort to protect the ozone layer from further damage.[89] A few years later, in 1987, Reagan also promised to continue to study the issue of stratospheric ozone depletion. He wanted to work with private industry, the scientific community, and officials in Canada to monitor and find solutions to acid rain, and he even proposed an increase of 112 percent in research funds for acid rain.[90] His administration also developed proposals that made use of market incentives to control air pollution and the causes of acid rain.[91]

1981–82: 97th Congress

The National Commission on Air Quality released the results of a $9.5 million study on the Clean Air Act. The report contained 433 findings and 109 recommendations, including a recommendation to eliminate national deadlines, change pollution regulations in clean-air areas, and eliminate requirements that new plants in dirty-air areas install the tightest possible emission controls. It found that although the nation's air was cleaner and continuing to improve, costs for air pollution control were likely to rise in the future.[92]

The commission recommended, among other things, that it was important to continue to establish national air quality standards. They believed that the EPA should review its standards to determine whether it was necessary to control fine particles in the air, whether standards for long-term exposure to ozone should be established, and whether there should be a separate carbon monoxide standard for high-altitude areas. Additionally, they recommended eliminating the requirements that mandated that new industries achieve the lowest possible emission rate. Instead, new industries would be required to install less stringent pollution control equipment referred to as the "best available control technology." The commission recommended requiring inspection and maintenance of automobile pollution control devices in certain large cities. Other recommendations included the relaxation of the carbon monoxide emission standard and a "significant reduction" of sulfur dioxide emissions over the next 10 years. Finally, the panel recommended that an interagency task force be created to study the effect of indoor pollution on health risks.[93]

Despite that report, members of Congress in 1981 fought over a rewrite of the Clean Air Act, HR 5252, even though it was one of Reagan's top five economic priorities. The most important issue was whether to change the way the EPA set national air quality standards.[94] The new bill required that the EPA set NAAQS. Any area not meeting the deadline for standards faced certain penalties or sanctions.[95]

1983–84: 98th Congress

Congress again made many attempts to rewrite the Clean Air Act, but no proposals had enough support to be passed.[96] S 768 and S 431 were two bills that were stalled by debate over the way to combat acid rain.[97] Another bill to amend the Clean Air Act was HR 150, which would have changed provisions relating to automobile emissions control devices and fuel additives. It would have allowed certain exemptions from the provisions of the law relating to emission standards for new automobiles, the removal of automobile emissions control devices, and the control of fuel additives. The bill did not pass.

Congress considered HR 151 to amend the Clean Air Act. This proposal would have repealed the requirement that states implement obligatory periodic inspections of motor vehicles, but it did not pass. And, finally, another proposal to amend the Clean Air Act was HR 305, which would have promoted the use of a blend of unleaded gasoline and alcohol as a motor vehicle fuel and as an additive to motor vehicle fuels.

1985–86: 99th Congress

In the summer of 1986, news of the ozone hole was reported in the media, causing great concern among environmentalists. Nonetheless, Congress was unable to renew the Clean Air Act during this session. Bills were proposed to reduce acid rain–causing pollution, but they failed to get through Congress (HR 4567/S 2203). The members of Congress representing the Northeast, where acid rain was the worst, favored strict controls, but members from the Midwest, the area that produced the pollution, were leery of the economic costs.[98]

1987–88: 100th Congress

During this session, Congress once again tried to overhaul the Clean Air Act, but they could not come to an agreement on the proposal, and it stalled. At issue was a deadline by which cities had to achieve air that was clean enough to meet the standards set by the EPA. The deadline had expired at the end of 1987. Those cities that failed to meet the standards faced penalties, including bans on construction of certain factories and possible loss of federal highway and sewer grants.[99] Environmentalists, utilities, automakers, and other groups opposed the final compromise efforts, so the bill did not pass.[100]

However, an extension for cities that would have to achieve clean air or face penalties was included in HJ Res 395 (PL 100-202), a fiscal year 1988 omnibus appropriations bill. Without the extension, about 100 cities would have faced the loss of federal grants or bans on new air-polluting factories. This bill was called the Clean Air Act Amendments of 1990.[101]

President G.H.W. Bush

President Bush promised to break the gridlock and support renewal of the Clean Air Act. He believed that clean air is too important to be a partisan issue. He sent to Congress a proposal that would reduce sulfur dioxide emissions by 2 million tons more than legislation then in the House. The legislation also called for tough standards to ensure that every industrial plant had the best available control technologies. According to the president, his proposal created a permit system to ensure that all sources of pollution met applicable limits, and it provided for sanctions for those cities that did not make progress in cleaning the air.[102]

Bush proposed the Clean Air Act Amendments of 1990 as a way to reduce air pollution. Upon signing the legislation, he said that it provided the world's most advanced, comprehensive, and market-oriented laws to address air pollution. To help communities and industries meet the objectives of the act, Bush announced a "cash for clunkers" program, to allow states and industries to buy old, high-polluting cars, take them off the road, and use the resulting pollutant reductions to satisfy federal clean air standards.[103] In another effort to clean the air, Bush proposed the Clean Air Interstate rule to reduce the major causes of ozone and fine particles by 70 percent. He also sought $710 million for the Clean Coal Technology Program, which would encourage the development of new methods to reduce sulfur dioxide and nitrogen dioxide emissions.[104] Additionally, President Bush met with leaders in Canada to discuss environmental problems, particularly the precursors of acid rain.[105]

1989–90: 101st Congress

Members of Congress responded to the public's growing concern about air quality, to new scientific research that demonstrated the health risks of polluted air, and to reports of worsening ozone in urban areas by proposing a bill (S1630)[106] to amend the Clean Air Act. After being passed by both the House and Senate, the proposal was signed into law by President Bush (PL 101-549).[107]

The Clean Air Act Amendments of 1990 reauthorized the 1970 Clean Air Act but also overhauled the nation's antipollution law by adding stronger provisions aimed at addressing air quality by limiting urban smog in major metropolitan areas, toxic air pollutants emanating from industrial and chemical plants that were linked to health threats, and the acid rain problem that originated primarily from coal-fired Midwestern plants.[108] The basic objectives of the law were to overhaul the nonattainment provisions, to create a technology-based control program for toxic air pollutants, to address acid precipitation and power plant emissions, to mandate the phaseout of CFCs, and to strengthen the enforcement powers of regulatory agencies. The law set new controls on smog, acid rain, and toxic air pollutants.[109]

There were many elements in the new bill. First, it revised the provisions for attaining and maintaining the NAAQS for ozone, carbon monoxide, and fine particulates. It did this by imposing strict federal standards on toxic emissions, including emissions from motor vehicles. The law set tighter limits on tailpipe emissions in order to lower the levels of hydrocarbons, carbon monoxide, nitrogen oxide, and particles in the air. At the same time, it set standards for a newer generation of clean fuels.

The law also imposed new standards for urban smog, specifically for toxic air pollutants and stratospheric ozone depletion. This was important because many areas across the country had not attained the ozone emission standards established previously. The nonattainment areas were placed in five categories, according to the severity of the level of ozone in a city's air. The categories were "marginal," "moderate," "serious," "severe," and "extreme." Those in the "severe" category were given more time to attain the NAAQS but had to meet more stringent control requirements. Plans for the dirtier areas had to include tighter controls, cover more sources, use economic incentive programs, and require more controls on mobile sources and on transportation.[110] The new law also included new provisions for air toxins. It directed the EPA to set maximum available control technology (MACT) standards for 189 toxic air pollutants for specific categories of sources (such as petroleum refineries) emitting those pollutants.

Another major aspect of the Clean Air Act Amendments dealt with acid rain control. Large industrial sources would often release sulfur oxide and nitrogen oxide into the air. These pollutants were the cause of acid rain that fell over New England and Canada. The goal of the act was to reduce the annual emissions of sulfur oxide to 10 million tons and nitrogen oxide to 2 million tons below their 1980 levels. It also required use of the best technology for reduction of hazardous air pollutants. The amendments added automatic penalties for nonattainment of goals and the enforcement necessary to make the law effective.[111] Most of these reductions would be achieved through an innovative allowance-trading system known as a national cap and tradable allowances.[112] Those who reduced their emissions were rewarded with emission credits, which they could sell to others who might want to buy them.

One of the unique parts of the 1990 Clean Air Act was a permit program for larger sources that released pollutants into the air. Under the law, air pollution would be managed by a new national permit system. According to the system, permits would be issued by states. The permits included information on which pollutants would be released, how much could be released, and what steps the source's owner or operator was taking to reduce pollution, such as plans to monitor the pollution levels. The permit system simplified and clarified the obligations of the business in limiting air pollution.

Another unique part of the act was the creation of a "bubble policy" that created a bubble over several sources of pollution. Within the bubble, emissions of industries or other pollution sources were combined and treated as if they came from one emission point.[113] Utilities companies were nonetheless required to reduce emissions. The law provided for interstate commissions on air pollution control, which were to develop regional plans for cleaning up air pollution.

The 1990 amendments represented another major effort by the Congress to address many of the complex and controversial issues related to clean air. The new law was expected to have far-reaching effects on federal facilities and industry. It was estimated that the costs to meet the new standards would be between $20 and $25 billion per year. Since the law covered the entire United States, it ensured that all Americans had the same basic health and environmental protections. Individual states could have stronger pollution controls, but they could not have weaker pollution controls than those set for the country. The new law also addressed the problem of international air pollution. It covered pollution that originated in Mexico and Canada and drifted into the United States, as well as pollution from the United States that reached Canada and Mexico.

1991–92: 102nd Congress

In 1991, Canada and the United States signed the United States-Canada Air Quality Accord that would limit the destructive impact of acid rain in Canada, about half of which was caused by smokestack emissions in the United States.[114] Also in 1991, the Senate passed a bill (S 455) that sought to reduce the exposure of Americans to indoor air pollutants. But a companion measure in the House (HR 1066) never made it to the full chamber. The Senate bill would have authorized $242.5 million for programs related to indoor air pollution. The Bush administration opposed the bill.[115] In 1992, Congress amended the Clean Air Act again through SJ Res 187. This time, they directed the EPA to set specific standards for particulate pollution as limits on the number of particles in a given size range per cubic yard of air.[116] The president signed the act (PL 102–187), which he supported.[117]

President Clinton

President Clinton was very concerned about protecting the quality of the air across the United States and seeing that Americans "live in a world with clean air."[118] He believed that people shouldn't have to worry about the quality of the air they breathed.[119] He also stated that clean air is needed to develop a strong economy.[120]

One way Clinton sought to clean the air was to encourage car companies to develop automobiles that had lower emissions.[121] He developed a partnership with GM to produce a "clean car,"[122] and suggested that people purchase vehicles that ran on alternative fuel, especially in cities where air pollution is severe.[123] In 1999, Clinton announced a proposal to require all passenger vehicles to meet tough pollution standards. The plan demanded that all vehicles produced be 75 to 95 percent cleaner than those then being manufactured.[124] This was later expanded to trucks and busses.[125] He worked with other countries to help solve the problems relating to air pollution, including China.[126]

In declaring National Cancer Control Month, Clinton recognized some health problems, including some forms of cancer may be caused by, or aggravated by, air pollution.[127]

Clinton sometimes blamed Congress for passing legislation harmful to the air. He called an appropriations bill being considered in the House Appropriations committee unacceptable because it cut funding from the EPA, which would diminish their ability to protect the public from air pollution.[128] Clinton also blamed Congress for passing legislation that cut enforcement of environmental laws. In the end, Clinton said that such a choice would result in more smog in the air.[129]

Despite this, during the final years of his administration, Clinton took credit for improving the quality of the air across the country.[130]

1993–94: 103rd Congress

In this session, Congress proposed HR 881, a proposal to ban smoking in federal buildings. Specifically, the bill would prohibit smoking in any building owned or leased for use by a federal agency except in specified areas. This did not pass. In a similar vein, another bill (HR 710), the Preventing Our Kids from Inhaling Deadly Smoke (PRO-KIDS) Act of 1993, would direct the administrator of the EPA to issue guidelines for enforcing a nonsmoking policy prohibiting smoking at every federal agency, except in a part of the building that was ventilated separately. This did not pass. Another bill that did not pass was HR 1930, called the Indoor Air Quality Act of 1993. This was a proposal to direct the EPA administrator to establish a national research, development, and demonstration program to ensure the quality of indoor air.

1995–96: 104th Congress

During this session, the EPA announced a total ban on CFC production. Officials in other industrialized nations followed suit.[131] The EPA also lowered the acceptable rate of emissions for acid-rain causing airborne pollution by public utility companies and at the same time awarded the utilities certificates that permitted them to exceed their emissions allowances.

Utility companies that met or exceeded their emissions goals were encouraged to sell their pollution credits to companies that were unable to meet the stricter standards.

In the House, members considered HR 46, a proposal to delay the implementation date for enhanced vehicle inspection and maintenance programs under the Clean Air Act for two years. It would also extend the date by which the EPA administrator was to reissue regulations relating to such programs. This was not passed.

Another bill that did not see support was HR 307, the Clean Air Flexibility Act of 1995. This proposal would direct the EPA administrator to revise, update, and republish state guidelines for the motor vehicle inspection and maintenance programs required by the Clean Air Act. Additionally, the proposal would have required states to submit revisions to their programs to meet the new standards as written by the EPA within two years.

1997–98: 105th Congress

In 1997, delegates to a conference in Kyoto, Japan, wrote the Kyoto Protocol, which, for the first time, established binding national limits on greenhouse gas emissions. The document was signed in 1998 by over 160 nations, including the U.S. delegate. However, before it could become an international treaty, the national assembly of each nation had to ratify the document. In the United States, the treaty faced opposition on Capitol Hill. Opponents pointed out that, under the treaty, the United States would have to reduce its greenhouse gas emissions to 7 percent below 1990 levels by 2012, whereas developing nations would not be required to adhere to any firm limits.[132]

The Congress considered amendments to the Clean Air Act (HR 130) but was not able to gather enough support to pass them. The original law mandated that the EPA administrator review the NAAQS for each air pollutant under question; once a standard had been set, the administrator would have to review that standard, and the criteria on which it was based, every five years. On July 18, 1997, the EPA administrator revised the standards for particulate matter and ozone. The EPA created a standard for tiny airborne particles of soot produced by sources such as coal-fired power plants and diesel engines. The EPA argued that since ozone is a "nonthreshold pollutant" (any amount in the air harms public health), the standards should be very stringent, even though they might cost hundreds of millions of dollars to implement nationwide. Proponents of the bill, including environmental and health groups, said the new rules were based on strong scientific evidence and would protect the health of the public. Critics argued that the Clean Air Act was working well as it was, that the new standards were based on flimsy science, and that they would impose millions of dollars in compliance costs with no appreciable benefit. Because of the controversy, nothing was passed in Congress.[133]

1999–2000: 106th Congress

A Senate bill (S 880/PL 106-40) to amend the Clean Air Act was passed in this session. Called the Chemical Safety Information, Site Security and Fuels Regulatory Relief Act, the amendment removed flammable fuels from the list of substances for which reporting and other activities were required under the risk management plan program. The bill prohibited the EPA administrator from listing flammable substances that might cause death or have serious health or environmental consequences in the case of accidental release, if they were used as fuels or held for sale as a fuel at a retail facility.

President G. W. Bush

Bush planned to reduce air pollution with a program he termed the "Clear Skies." This was a proposal that set new standards for reducing air pollution. In introducing the idea, Bush acknowledged that the Clean Air Act had helped reduce the problem but said that it was necessary to do more at the federal level. Bush proposed to place limits on sulfur dioxide and nitrogen oxides, which contribute to acid rain, as well as mercury, which is toxic to humans. He took a market-based approach that rewarded innovation. He claimed that the plan would cut emissions of pollutants by 70 percent over 15 years, reduce smog and mercury emissions, and stop acid rain, all while reducing costs.[134] The new proposal would make the first major changes to the Clean Air Act since 1990.

President Bush supported the International Convention for the Prevention of Pollution from Ships. Part of the convention included steps to control and prevent emissions of harmful air pollutants from ships, including discharge of nitrogen oxides from large marine diesel engines.[135] Another method used by Bush to clean the air was to support research into clean coal technology. He provided $2 billion to discover methods to remove virtually all pollutants from the emissions of coal-fired power plants.[136]

2001–02: 107th Congress

In 2001 at the Kyoto Protocol meetings in Bonn, Germany, 178 countries (but not the United States) approved more precise rules that brought the treaty closer to adoption. The Bonn agreement required the industrial countries to reduce the six greenhouse gases to 5.2 percent below 1990 levels for the years 2008–12. They also agreed that that carbon-absorbing "sinks" such as forests and farmland could be counted as reductions and that greenhouse gas trading could be used to meet most of the reductions. If the rules were approved by the legislatures of at least 55 countries, they would then go into effect. It was thought that this could happen by the end of 2002.[137] President G. W. Bush announced that he would not submit the Kyoto Protocol to the Senate for ratification. He and other officials argued that the protocol was unfair because it imposed carbon dioxide emission limits on highly industrialized nations, while developing nations would be allowed to delay such actions.

Senators John McCain (R-AZ) and Joe Lieberman (D-Conn) co-sponsored the Climate Stewardship Act to cap greenhouse gas emissions by power plants, refineries, and other industries, using tradable allowances. It was blocked from becoming a law.

Supreme Court Cases

The higher standards for air quality passed in the 105th Congress via the Clean Air Acts were reviewed by the court this year. Many groups, including the American Trucking Association, business groups, and even some states (Michigan, Ohio, and West Virginia) challenged the new standards in the U.S. Court of Appeals for the District of Columbia Circuit, and the case then went to the U.S. Supreme Court in *Whitman v. the American Trucking Association* (531 U.S. 457, 2001). The groups argued that the law that gave the head of the EPA the authority to set standards was unconstitutionally vague. They also argued that the EPA should be required to perform a cost-benefit analysis when setting national air quality standards as a way to keep costs down. The justices on the Supreme Court upheld the EPA's authority to set standards as well as the regulations. In the decision, the justices explained that the statute was not overly vague. Further, according to the justices, no cost-benefit analysis was needed.[138]

2003–04: 108th Congress

Congress passed a new law (S 551/PL 108-336) that provided for the implementation of air quality programs developed through an intergovernmental agreement between the Southern Ute Indian Tribe and the state of Colorado revolving around air quality on the Southern Ute Indian

Reservation. The bill was titled the Southern Ute and Colorado Intergovernmental Agreement Implementation Act of 2004.

During this session, the EPA published new regulations on particulate emissions, referred to as the Clean Air Rules of 2004. These new rules updated the 1955 Air Pollution Control Act and were aimed at improving the air quality over the following 15 years. The rules addressed five areas: interstate air, mercury, nonroad diesel emissions, ground-level ozone, and fine particulates. Three of the new rules specifically dealt with the transportation of pollution across state borders. The first, the Interstate Clean Air Rule, provided states with a solution to problems associated with power plant pollution that drifted from one state to another. The new regulation used a cap-and-trade system to reduce pollutants by 70 percent. Second, the Mercury Clean Air Rule regulated mercury from power plants—the largest domestic source of mercury emissions. This was the first time that mercury emissions from power plants were regulated. The third new rule was the Nonroad Clean Air Rule that altered the way diesel engines functioned in order to reduce emissions and changed the way diesel fuel was refined to remove sulfur.

The EPA's new rules also included new regulations on ozone that designated those areas where the air did not meet the health-based standards for ground-level ozone. The ozone rules classified the seriousness of the problem and required states to submit plans for reducing the levels of ozone in areas where the ozone standards had not been met. The last new rule was the Fine Particle Rule, which designated those areas where the air did not meet the health-based standards for fine particulate pollution. This required states to submit plans for reducing the levels of particulate pollution in areas where the fine particle standards had not been met.

2005–06: 109th Congress

In 2005, the Senate Environment and Public Works Committee deadlocked in a 9–9 vote on President Bush's Clear Skies initiative, a top environmental priority of the White House. The vote meant that the measure would not advance to the full Senate for further action. Another House bill (HR 6/PL 109-58), the Energy Policy Act of 2005, became law. It included provisions to develop clean-coal technologies to reduce pollutants. It also included grants to encourage cleaner fuels and clean school buses, as well as money to develop efficient engine technology for aircraft. The money could also be spent on programs to promote bicycling.[139]

A bill proposed in the Senate (S 2047) was called the Healthy Communities Act of 2005. The proposal would promote healthy communities by requiring the secretary of health and human services to review environmental health data to assess the impact of federal policies on health. Those communities identified as at risk would be awarded grant money to allow them to improve their environments. The bill was referred to committee but went no further.

Another proposal that did not receive enough support for passage was S 1265, the Diesel Emissions Reduction Act of 2005. This bill would have given the EPA administrator the authority to grant loans to local agencies to implement programs to reduce emissions.

2007–08: 110th Congress

During this session, the Congress passed a bill (S 2146/PL 110-255) to authorize the administrator of the EPA to accept certain diesel emissions reduction projects. If there was a settlement of an alleged violation of environmental law, the EPA administrator could accept diesel emissions reduction projects if the projects were shown to protect human health or the environment, were related to the violations, and did not provide funds for the EPA staff to carry out the organization's operations.

The Senate Environment and Public Works Committee approved S 2191, the Lieberman-Warner Climate Security Act of 2007 (S 2191). The bill proposed a cap on nationwide emissions of greenhouse gases from power plants, manufacturers, petroleum refiners, and other sectors of the economy. It set as a goal the reduction of emissions by 18 to 25 percent below 2005 levels by 2020, and by 62 to 66 percent by 2050. Under the proposal, the EPA administrator would establish a federal greenhouse gas registry that would be managed by an advisory body made up of members from private enterprise, agriculture, environmental groups, and state, tribal, and local governments. The EPA would publish the information from the registry on the Internet for the public to track, except if the information should remain private because of national security concerns. The proposal did not pass through the Senate.

The Congress again proposed the Healthy Communities Act as they did in the previous session. The bill was largely the same proposal as before. As in the previous session, the proposal went to committee but no further.

President Obama

Shortly after becoming president, Obama stopped the last-minute "midnight regulations" of the Bush administration. These involved the EPA's use of a "reference man," or an average-sized white male, in determining safe radiation levels. Bush's changes revolved around allowing a higher level of radiation than the National Academy of Sciences recommended for taking protective actions after a nuclear accident or attack, such as from a radiological "dirty bomb." The changes would allow radiation levels in drinking water that were higher than the EPA's standards. A year after Obama's actions to stop the new regulations, environmental activists were concerned that the EPA still planned to change its guidelines to increase the levels of radiation it considered safe for humans.[140]

As his term continued, President Obama disappointed liberals by choosing not to endorse a plan to cap carbon emissions.[141] Obama's energy policy was centered around increasing wind and solar power and promoting low-carbon energy sources.[142] At the same time, Obama indicated that he wanted to take a more activist role in curbing greenhouse gases than had the previous Clinton and Bush administrations.[143] He issued an EPA order that would trigger the regulation of greenhouse gases under clean air laws if Congress failed to act.[144]

President Obama made stemming global warming a major policy goal.[145] His position was strengthened at the Copenhagen climate change summit, where developing nations such as China asked richer, more developed countries to promise up to $200 billion a year to help poorer nations deal with the effects of global warming, including drought and rising sea levels. At first, the United States insisted on international verification of carbon emission reductions. When it appeared the delegates would fail to agree, Secretary of State Hillary Clinton announced that the United States would lead an effort to raise $100 billion in annual aid by 2020 for those developing nations but only on the condition that they make their carbon-accounting books available for inspection. After that announcement, China and other countries agreed to the deal. Members of the U.S. Congress expressed their concerns about the possible sources of the funds.

Under the cap-and-trade system proposed in Congress, certain polluters such as power plants and oil refineries would be able to meet the emission standards by purchasing permits to allow for emissions from the federal government. Another way to meet standards was to pay for carbon-offset projects, which are activities that seek to extract carbon from the atmosphere or even prevent carbon from being produced at all. These can include planting trees, building wind and solar electric plants, and improving the energy efficiency of a building or factory.[146]

2009–10: 111th Congress

One proposal that Congress considered to aid the ailing automobile industry and improve their emissions was a bill called the Consumer Assistance to Recycle and Save Program, otherwise known as "Cash for Clunkers." HR 3435 (PL 111-47), the bill would provide incentives for consumers to trade in old vehicles for newer, fuel-efficient cars. The bill was proposed after it was reported that the original $1 billion appropriation was exhausted after only one week. If the bill passed, an additional $2 billion would be made available for the program. The new bill would approve the transfer of money from the economic stimulus law (PL 111-5) enacted earlier to the auto trade-in program. The program offered up to $4,500 in cash rebates to people who traded in old, fuel-inefficient cars or SUVs for newer, more fuel-efficient vehicles.[147] Those opposed to the bill argued that the action could have impacts on other businesses, including automobile repair businesses who might see their business drop off.[148]

As part of a comprehensive energy bill, the House passed a proposal that would limit or cap emissions of greenhouse gases that cause global warming (HR 2454). Called the American Clean Energy and Security Act, it would create a cap-and-trade system for emissions and require electric utilities to produce a percentage of their power from renewable alternative sources of energy such as wind, solar, and geothermal. The proposal would cap greenhouse gas emissions at 17 percent below current levels by 2020, 42 percent below current levels by 2030, and 83 percent below current levels by 2050. Utilities, refineries, and factories would have to hold government-issued emissions allowances, but the credits could be traded as commodities in the marketplace. Utilities would be required to generate a growing share of electricity from renewable sources while improving efficiency. In addition, the bill would allow states to use 1 percent of the expected government revenue from the sale of emissions allowances for "green transportation" projects such as mass transit systems, bicycle paths, or high-occupancy vehicle lanes. The money would amount to about $530 million a year at the start but was projected to increase to about $1 billion annually as the allowances become more valuable in the market.[149]

The Senate version of the bill (S 1462: The American Clean Energy Leadership Act of 2009) was designed to cut carbon pollution by 17 percent by 2020 and by 80 percent by 2050. It relied on a sector-by-sector approach to limiting emissions, addressing emissions from the utility, manufacturing, and transportation sectors in different ways. Electric utilities would be subject to a carbon cap and be required to purchase pollution allowances from the government. Their cap would begin in 2013. The allowances could also be bought and sold. Manufacturers would be subject to a separate but similarly structured cap, to be phased in four years after the power plant cap (by 2016). Transportation fuels would not be subject to a cap, but oil companies or consumers would pay some form of fee based on the fuel's carbon content.

The Senate bill included new incentives to expand nuclear power, offshore drilling, and investments in technology to capture the carbon emitted by coal-burning power plants and safely store it underground. The legislation would create $2 billion in program offsets each year, which are projects that remove carbon from the atmosphere or prevent it from being released. Under the plan, companies could purchase offsets in place of carbon credits. Additionally, the bill would provide $54 billion in loan guarantees for new nuclear power generators, streamline the permitting process, and give tax breaks to new nuclear plants. It would also allow some new offshore drilling off the Atlantic coast. States could choose to prohibit drilling within 75 miles of their shores and ban it entirely if they could show that their coastlines could be at risk from an offshore drilling accident. Florida would be permitted to keep an existing ban on drilling within 125 miles off its Gulf coast. States that allowed drilling would receive 37.5 percent of the royalties paid by oil companies drilling off their coasts.[150] Neither the House nor the Senate version of this bill became law.

Another bill, proposed by Senators John Kerry and Barbara Boxer (S 1733: The Clean Energy Jobs and American Power Act), referred to as the Kerry-Boxer Bill, would require a 20 percent reduction in emissions from 2005 levels by 2020, compared with a 17 percent cut as required in the House bill, which sought an 83 percent drop from 2005 levels by 2050. The Kerry-Boxer bill included a cap-and-trade model, requiring polluters to purchase allowances for each ton of greenhouse gas they emitted and creating a market in which allowances could be bought and sold. The idea was to set a price on emissions to encourage more manufacturers to shift to cleaner energy sources. The allowances would be priced through a structure that the lawmakers termed a *soft collar:* setting a price floor of $10 per ton and requiring the EPA to flood the market with additional allowances once the cost reached $28 per ton after 2012.[151] The proposal was not signed by the president.

A Senate panel approved proposed legislation (S 3373: the Air and Health Quality Empowerment Zone Designation Act of 2010) that would authorize $20 million annually from fiscal year 2011 through 2015 for the EPA to provide grants to areas with extreme air pollution to assist them in replacing polluting vehicles or engines and engaging in other activities that would improve the quality of the air. Interestingly to some, only the San Joaquin Valley (the home of chairwoman Barbara Boxer) would have qualified for the grants.[152] The proposal did not receive the support of the entire Senate.

Other bills concerning air quality passed the Senate but went no further. One was the Sustainable Schools Pollution Reduction Act of 2010 (S 3362), which would have created a program to provide competitive grants to public school districts to allow for the implementation of technologies to reduce air pollutants. Another proposal, SJ Res 26, would have overturned the EPA's finding that greenhouse gases are hazardous to human health.[153]

Finally, the Congress passed S 1660 (PL 111-1-969: the Formaldehyde Standards for Composite Wood Products Act) to create a national emission standard for formaldehyde in domestic and imported composite wood products, the most common source of the chemical in homes, according to the EPA.[154]

Conclusion

Most would agree that the quality of the air across the nation has improved significantly since Congress approved the Clean Air Act in 1970. Scientific data taken from thousands of monitoring stations across the country indicate that the quality of the air has improved over the past decade.[155] Sulfur oxides that cause acid rain seem to have declined as well, partly because of agreements between the government and managers of electric power plants to limit emissions. Now, if a plant exceeds its limit, the managers face fines. If it produces less emissions, the manager can sell the pollution permits to another plant.[156] Nitrogen oxides are also on the decline. This is probably because new exhaust gas processing has been adopted, and because catalytic converters are used.[157] Lead in the air has also decreased because of the use of unleaded gasoline.[158] There has also been a decline in CFCs.[159] Many manufacturers stopped producing CFCs before the EPA formally banned their production and use.

However, less industrialized countries continue to produce them. As a result, CFC smuggling has become a serious problem.[160] CFCs are being smuggled into the United States from countries that continue to manufacture them, allowing for substantial profits. CFCs can be produced in developing countries for only $1 a pound and then resold in the United States, disguised as recycled material, for $20 a pound. There is a thriving black market for the product.[161] Production of CFCs is due to stop in all countries in 2010. However, every air conditioner and refrigerator made before 2000 has CFCs in its cooling system. The CFCs in such equipment can be captured and resold, and there is no legal penalty for doing so.[162]

As we learn more about how pollutants in the air affect the quality of human health, Congress may need to establish more restrictions on the amount of emissions that can be allowed into the air. There is no doubt that Congress and the president will continue to be concerned with the nation's air quality and make laws to ensure that harmful emissions into the atmosphere are restricted.

Chapter 3

The Politics of Water Pollution

The Florida Everglades are a lush, tropical wilderness area full of marshes, mangrove forests, lakes, and tree islands. In the early 20th century, it was thought that that the wetlands were only good for draining and as farmland. Developers began building canals and dikes to siphon off the region's water to create an agricultural and real estate boom. But in the 1950s it was recognized that these areas were valuable habitats for many species of plants and animals. Today, half of the Everglades' wetlands have been lost, the water is polluted by runoff from farms, and there is a high risk of losing wildlife. Calls for protecting these valuable resources have become prevalent.

There were more calls for action to protect water resources after an oil tanker called the *Exxon Valdez* ran aground in Alaska and over 11 million gallons of crude poured into the water of Prince William Sound in March 1989. A week after the spill, the slick covered an area the size of Rhode Island. In some places, the oil was three feet thick, causing significant environmental damage. Shorelines and coastlines were covered in oil. When sea otters and waterfowl swam in the oil, their feathers became coated and could no longer protect the animals' skin from cold temperatures. This, in turn, caused hypothermia. Many of the wild animals died from exposure to the cold water. Others died from ingested oil that caused harm to internal organs. By the end of September, 33,000 seabirds, 1,000 sea otters, and 193 bald eagles had died, but thousands of others may have died without being recorded.[1] The captain may have been under the influence of alcohol, too tired from 24 hours of loading the tanker, or just trying to avoid colliding with drifting icebergs.[2]

Exxon was subject to civil suits from local fishermen who claimed damages from the spill. However, much of the money that Exxon was forced to pay did not go to the victims. Exxon paid $125 million in fines to the federal government and the state of Alaska. In addition, Exxon was forced to pay $900 million into a fund to be doled out by government officials for environmental projects, habitat protection, and scientific research, among other things. In May 1994, $38.7 million was used to create a new state park.

Another accident in January 1990 prompted congressional action. A pipeline connecting the Exxon Bayway refinery at Linden, New Jersey, to the Bayonne, New Jersey, plant burst, and 567,000 gallons of oil leaked into the Arthur Kill. Even though a red warning light went off in the Exxon facility, it was ignored because it had misfired several other times. Storm winds carried the oil down the kill into the tidal creeks and onto the nearby marshes. At first, Exxon denied responsibility but eventually accepted it.[3]

The availability of clean water is vital for the health and safety of the world's inhabitants, not only in the Everglades, but everywhere. Clean water is essential for a strong economy, as such resources are necessary to support agriculture, industry, electric power, recreation, navigation, and fisheries.[4]

Although most people would agree about the need for clean water, debate rages over how to best protect and maintain clean water sources. The debate sometimes surrounds what constitutes "clean" water, while at other times it revolves around how to maintain those standards. The debate over clean water began many years ago, and, since then, many laws have been passed to address specific water problems. The first federal water pollution control bill, the Refuse Act, was enacted in 1899. It was passed to protect navigation and barred the discharge or deposit of refuse matter in navigable waters without a permit. Then, in 1912, the Public Health Service and Oil Pollution Act was passed that contained provisions authorizing investigations of water pollution related to disease and controlling harmful oil discharges in coastal water.

In 1920, Congress passed the Water Powers Act (S 936/PL 226). This law established federal regulation of hydroelectric power and created the Federal Power Commission. A few years later, in 1924, Congress passed the Oil Pollution Control Act. This law regulated oil pollution from seagoing vessels only and not from stationary sources. That year Congress also passed the Alaskan Fisheries Act (HR 8143/PL 204), which established federal control over management of ocean fisheries. Congress passed the first inland water pollution control bill in 1938, which would have created a division of water pollution in the Public Health Service and provided grants-in-aid to states and municipalities for sewage treatment facilities. Although the president at the time, Franklin Roosevelt, approved of a division of water pollution control, he vetoed the bill on the grounds that appropriations for such projects should be based on estimates submitted in the budget.[5]

Since these early laws were passed to protect the quality of water, Congress has taken many other actions to ensure the quality of the nation's water. This chapter reviews these proposals made by presidents and members of Congress to protect water supplies. It shows how the proposals have changed over time to address new, emerging water issues and in response to new information about the health effects of polluted water on humans, wildlife, and the economy.

President Truman

In a special message to Congress about a health plan for the nation, President Truman noted that the national health must be improved through the provision of safe water systems, sewage disposal plants, and sanitary facilities. He said that the country's streams and rivers must be safeguarded against pollution.[6] As a way to reduce pollution in waterways, Truman called on Congress in 1946 to provide grants and loans to the states to help them develop water pollution control methods.[7]

When asked in a news conference if he supported a bill to control pollution in harbors and streams, Truman noted that he voted for a similar bill when he was in Congress.[8] However, he chose not to sign a bill (HR 597) that would have prohibited dumping garbage from outside the United States into waterways. He explained that even though dumping would be prohibited unless the garbage had been reduced to fluid form, he believed that this was inadequate to prevent garbage that was infected with human or animal diseases from being dumped into waters that were sometimes used for human consumption.[9]

Truman also called for reorganizing some agencies of the executive branch to more effectively deal with pollution issues. Under Plan 16, certain water pollution control functions would be assigned to the Federal Security Agency, which was already responsible for the preparation of water pollution plans for interstate streams, for the conduct of surveys and research, for the maintenance of relationships with state water pollution control agencies, and for the approval of sewage treatment projects.[10]

1945–46: 79th Congress

In 1945, many bills aimed at limiting water pollution went unreported by different committees, including the Commerce Committee, the Finance Committee, and the Rivers and Harbors Committee. These included S 330, S 535, S 1037, S 1462, S 1536, HR 519, and HR 587. None of these passed into law.

During this session, Congress considered Truman's options for reorganizing agencies that had some responsibility for protecting water quality within the executive branch. They approved Plan 16, which became effective on May 24, 1950.[11]

1947–48: 80th Congress

The first major water pollution bill passed by Congress was the Water Pollution Control Act (PL 80-845). The objective of the bill was to create a national program to restore and maintain the chemical, physical, and biological integrity of the nation's waters. It was the first act that provided a comprehensive federal, state, and local approach to protecting waters. It focused on limiting pollution in interstate waters by encouraging states to enter into interstate compacts that would become binding upon approval by Congress. The bill also created a Water Pollution Control Advisory Board. Additionally, Congress allocated funds for treatment programs and construction projects related to water pollution.[12] Provisions also included federal enforcement procedures for pollution abatement by allowing the Department of Justice to file suits against polluters.

The Water Pollution Control Act was the first federal legislation to deal explicitly with conventional forms of water pollution. It gave the federal government some role in abating interstate water pollution, although that role was secondary to that of the states.[13] The federal government could engage in research, investigation, and surveys concerning the problem of water pollution, and the law authorized the federal government to make loans to local governments to help them construct municipal sewage treatment facilities. But there was no federal authority to establish water quality standards, limit discharges, or engage in any form of enforcement.

Some argued that the bill actually protected those who polluted the water because federal inspectors were not given the right to enter the property of accused polluters but had to get permission to do that. This meant that polluters could prevent inspectors from entering their property until any incriminating evidence was removed. The polluters were not taken to court for a period of between two and six months, at which point the government had to prove that the polluters' actions were causing harm outside of the state in which they were located. This was an extremely tough standard to meet. Additionally, the act did not allow for specific penalties on those who violated the law.[14] The provisions for federal enforcement in the law were extremely cumbersome and largely unenforceable.[15]

President Eisenhower

Water pollution was a concern to President Eisenhower, because, as he noted, as the nation's population grew, more cities used more chemicals that were dumped into the water supplies. He proposed greater assistance to states for water pollution programs, more research into the problems of water pollution, and a strengthening of the Water Pollution Control Act, which was set to expire.[16] He signed a bill in July 1956 that authorized measures for controlling pollution and provided federal grants to local governments to pay part of the cost of constructing municipal sewage treatment plants.[17]

In 1958, the president proposed discontinuing grants for construction of waste treatment facilities under the Water Pollution Control Act Amendments of 1956.[18] But his budget for fiscal year 1961 included substantial increases in allocations for activities to fight water pollution. He promised to make legislative recommendations to strengthen the enforcement provisions of the Water Pollution Control Act and to authorize greater federal leadership in combating air pollution.[19]

However, Eisenhower vetoed HR 3610, a bill to amend the federal Water Pollution Control Act. The bill would have provided for an increase in federal grants to municipalities to help them fund construction of sewage treatment works. According to Eisenhower, because water pollution is a local concern, the primary responsibility for solving the problem should not lie with the federal government but rather must be assumed by state and local governments. Eisenhower argued that the bill would tempt local governments to delay essential water pollution abatement efforts in order to wait for federal funds. Eisenhower explained that the federal government's role should be to stimulate state and local action rather than provide excuses for inaction.[20] States should play a major role in water pollution control, according to the president, because water pollution "is strictly local in its character, and I think it belongs to local government."[21]

1955–56: 84th Congress

In this session, Congress passed HR 8750 (PL 84-1018), which was then signed into law by the president on August 7, 1956. The bill broadened the Watershed Protection and Flood Prevention Act of 1954. Among other things, it allowed for federal aid for pollution abatement programs.

Another bill passed by the Congress and signed by President Eisenhower was S 890/HR 9540 (PL 84-660), the Water Pollution Control Act Amendments of 1956. The bill amended the Water Pollution Control Act of 1948 to increase federal grants available to municipalities for the development of water pollution control programs. Under the program, local governments would share in the costs of constructing sewage treatment facilities. The money would be allocated to states based on their population as well as the extent of pollution. It also required the surgeon general to help states prepare comprehensive water pollution control programs and conduct research projects. In the new law, a Water Pollution Control Advisory Board was established. The nine members would be appointed by the president and would be responsible for approving the water quality standards established by each state.[22]

The amendments included provisions for water pollution that occurred in one state and affected another state. The new law provided for an "enforcement conference" between those involved in a dispute over water pollution as a way to force dischargers to clean up any pollution they caused. If there was no action within six months after a conference, or if a serious water pollution problem was recognized in interstate waters, then a public hearing could be called. That meeting could include Public Health Service officials, state and local officials, the alleged polluters, and any other interested parties. The participants would make recommendations as to who would be responsible for cleaning up the pollution, and to what extent. However, because the conference relied on voluntary participation by those involved and relied on those members coming to a consensus, the conferences did not often result in the establishment or enforcement of pollution control requirements. So the cases would be forced into the courts, as required by the law.[23]

In 1956, Congress included in an omnibus water pollution control bill a provision authorizing up to $50 million a year for 10 years (fiscal years 1957–66) in federal matching grants to state and local governments for the construction of community sewage treatment plants. Although the president did not support the bill, he signed it nonetheless. Later, in his budget message to Congress in January 1958, Eisenhower asked Congress to abolish the grant program, stating that that states should assume full responsibility for the construction.[24]

1959–60: 86th Congress

In 1960, the Congress passed HR 3610, which was intended to amend the 1956/1948 Water Pollution Control acts. The bill increased federal funds to states for the construction of sewage plants and other water projects. It increased federal grants to $90 million a year beginning with fiscal year 1961 and extended the authorization for grants through fiscal year 1970. The president vetoed the bill because he believed that water pollution was a local concern for which state and local governments must assume responsibility; because the House did not override the veto, the bill did not become law.[25]

President Kennedy

President Kennedy agreed that water resource programs and water pollution control programs required priority attention. But, unlike his predecessors, Kennedy said that sound and uniform standards needed to be developed for sharing costs between federal, state, and local governments.[26]

In 1961, President Kennedy gave a speech in which he described his approach to natural resources policy and the need to preserve natural resources. He said that because of the country's rapid growth and change, the pollution of our country's rivers and streams was alarming. Water pollution had become a serious problem for which corrective efforts to date were not adequate. He promised to establish a Presidential Advisory Committee on Natural Resources to seek the advice of experts outside the government. He also proposed increasing federal financial assistance for water pollution control agencies, for construction of waste treatment facilities, for strengthening abatement programs for seriously polluted areas, for more intensive research efforts to determine sources of the pollution and their adverse effects on people, and for a new special agency within Health, Education, and Welfare (HEW) to research and control pollution.[27]

In that same speech, Kennedy proposed new legislation that would provide federal financial assistance for interstate water pollution control agencies and that would strengthen enforcement procedures to abate serious pollution situations of national significance and broaden research efforts.[28] In July 1961, Kennedy signed the Water Pollution Control Act Amendments, which addressed some of those problems.[29]

Kennedy then gave a speech on conservation on March 1, 1962, in which he focused partly on the country's water resources. He said, "Our goal, therefore, is to have sufficient water sufficiently clean in the right place at the right time to serve the range of human and industrial needs." He urged Congress to enact the Water Resources Planning Act, which would authorize federal grants to assist the states in water resource planning. He authorized the establishment of river basin commissions to prepare and keep current basin plans, and he established a Water Resources Council that included his top cabinet officers to coordinate federal river basin planning and development policies.[30]

1961–62: 87th Congress

A proposal for new water pollution legislation, HR 6441/S 120, was passed by Congress and became law (PL 87-88). The Water Pollution Control Act Amendments increased federal grants to help communities develop pollution abatement plans and to help them construct sewage treatment plants. The law increased the existing $50 million annual authorization for federal grants to $80 million for fiscal year 1962, $90 million for fiscal year 1963, and $100 million for each of the fiscal years from 1964 to 1967. The law stipulated that 50 percent of the funds had to be used for grants to communities with populations of 125,000 or fewer.[31]

In addition, the new law established a Water Pollution Control Advisory Board and authorized the secretary of HEW, in cases of interstate pollution, to conduct investigations and make recommendations on abatement. Before moving forward with a case, the secretary would be required to get the consent of the governor of the state involved in cases before any action could be taken. The secretary was authorized to bring a suit on behalf of the United States to secure pollution abatement, if needed. The law also authorized the establishment of water pollution demonstration and research facilities, and it required an annual report on the operations of the program.[32]

President Johnson

Upon becoming president, Johnson supported environmental policy. He said, "The water we drink, the food we eat, the very air that we breathe, are threatened with pollution."[33] However, he believed that the solutions to the problems of pollution did not rest solely with the federal government, nor could they rely solely on the strained resources of local authority. Instead, they required cooperation between the Capitol and the leaders of local communities, something he called "creative federalism."[34] He said, "Now, it is a time for action. It is a time for Federal action, but that never substitutes for State or local action."[35]

Johnson supported a bill (S 649) that provided new federal authority and money to control water pollution. The legislation established a Federal Water Pollution Control Administration to identify the nation's long-term needs and recommend policies to meet those objectives. The bill provided grants for research and increased controls over pollution of interstate waters. Another bill backed by the president, S 1111, was designed to assure the establishment of plans for development, conservation, and use of water and related land resources.[36] Johnson later transferred the Water Pollution Control Administration to the Department of the Interior.[37]

Johnson gave a speech to Congress that focused on many types of pollution, including water. He asked Congress to set new standards in water quality and effective enforcement procedures to create a national program to prevent pollution at its source rather than attempting to clean up pollution after it occurred.[38] Johnson wanted to increase grants for states to help pay for water pollution control programs and provide research programs.[39] The president wanted to provide additional incentives for multi-municipal projects to control water pollution and for new research programs into water pollution. Johnson also said he would work with Canada to develop a pollution control program for the Great Lakes and other border waters.[40]

Another issue that concerned Johnson was pollution of rivers. In both his 1966 State of the Union address and his Economic Report, President Johnson made his position clear. He said that the continued poisoning of our rivers was shameful. He wanted to undertake a cooperative effort to end pollution in several entire river basins by making additional federal funds available to help create plans and construct the plants needed to make the entire river systems clean. The program was called the Clean Rivers Demonstration Program. The government would also adopt water quality standards for the rivers.[41]

On March 8, 1968, Johnson gave a message to Congress on conservation, called "To Renew a Nation." In it, he talked about water pollution. He said, "Our major rivers are defiled by noxious debris. Pollutants from cities and industries kill the fish in our streams. Many waterways are covered with oil slicks and contain growths of algae that destroy productive life and make the water unfit for recreation. 'Polluted Water—Not Safe for Swimming' has become a familiar sign on too many beaches and rivers. A lake that has served many generations of men now can be destroyed by man in less than one generation."[42] To solve the problem, he suggested a multipronged approach. First, he wanted water quality standards set for entire bodies of water that would vary from place to place depending on the water's use. Second, Johnson argued that the standards must be enforceable and must apply to both municipalities and industries. His third suggestion had to do with waste treatment plants. He said that they must be constructed and that those and other methods should be developed to prevent pollutants from reaching the water. There had to be new methods of cooperation and enforcement established at all levels of government, because Americans are "determined to resist the flow of poison into their rivers and streams."[43] To help build new and improve old water treatment plants,

Johnson proposed an appropriation of $225 million for grants under the Clean Water Restoration Act.

In another speech, Johnson proposed the Safe Drinking Water Act of 1968 to "develop, adopt and enforce improved standards relating to chemical contaminants in drinking water; conduct a comprehensive study of the safety of public drinking water supplies in the United States, and determine whether any additional steps are necessary in this area."[44]

Johnson identified the need for a comprehensive system to control oil pollution and provide for prompt cleanup by building on the Oil Pollution and Hazardous Substances Control Act of 1968. He sought to make the discharge of oil illegal if it occurred from a shore facility or a ship operating within 12 miles of shore. He proposed a system to impose on the oil polluter the responsibility for cleaning the beaches and waters that were harmed by oil. He also suggested giving the federal government the power to clean up oil spills whenever the owner or operator failed to act. In those cases, the polluter would be forced to reimburse the government for the cleanup costs. Last, the proposal would authorize the government to establish regulations for shipboard and related marine operations to reduce the possibility of oil leakage at the source.[45]

1963–64: 88th Congress

Congress debated, but did not pass, S 649, the Federal Water Pollution Act, designed to strengthen the 1948 Water Pollution Control Act. The new bill would have increased federal assistance to states to help them deal with their water pollution problems. It would have created a new Federal Water Pollution Control Administration within the Department of HEW; the administration would be responsible for the prevention and control of interstate pollution and navigable waters.[46] Federal financial assistance and regulatory action would have been expanded with the allocation of $20 million in annual grants to explore solutions to the problem of pollution caused by runoff from storm water sewers.

By passing the law, Congress wanted to initiate steps to combat pollution. One way to do this was to focus on the problem of non-decomposing detergents. The bill provided for establishment and enforcement of federal standards for water quality in interstate waters.[47] The proposal was passed by the Senate only and did not become law.[48] Congress passed, and the president signed, S 2 (PL 88-379), the Water Resources Research Act, which was designed to foster water research in the hope of avoiding severe water shortages and authorized grant programs for water research.[49]

1965–66: 89th Congress

A new law passed by Congress during this session was the Water Quality Act of 1965 (S 4/PL 89-234). Requested by President Johnson, the new law strengthened controls over water pollution. For the first time, the federal government had supervisory control of the water pollution control activities of state and local governments.[50] It was the first federal law to mandate state actions with respect to water pollution control policy. New ambient water quality standards were created for interstate streams. Once the standards were established, authorities could then use court action to prevent individuals or industries from discharging matter into streams that would lower the water quality below the established standard. In the bill, states were required to establish water quality standards, which had to be approved by the federal government. The standards were to specify permitted levels of pollution in all interstate waters.[51] The HEW secretary could impose standards on those states with outdated standards. States were also required to establish implementation plans that would place limits on discharges from individual sources. Thus, there would be state responsibility in the setting of the standards, the development of implementation plans, and enforcement. It was hoped that by setting water purity standards, pollution would be prevented before it occurred.[52]

The bill also created a new federal water pollution control agency within HEW to administer the program.[53] Additionally, the 1965 act provided federal funding for state planning efforts that promoted regionwide water quality management. The new act expanded the construction grants program to $150 million per year and increased the maximum federal subsidy per project from $600,000 to $1.6 million.

In the second year of the congressional session, members passed a bill (S 2947/PL 89-753) that increased the federal government's water pollution control activities. The original Water Quality Act of 1965 had passed water purity standards for interstate waters. The new bill, the Clean Water Restoration Act of 1966, provided money to help communities pay for the costs of abiding by those standards. It increased the amount of funds available for grants to communities for construction of sewage treatment plants.[54] For the fiscal years 1967–71, $3.9 billion was authorized for federal water pollution control activities. As stipulated, the federal government would pay up to 50 percent of construction costs if states contributed funds and set water quality standards. Most of the money was set aside to help communities pay for sewage treatment plants.[55] S 1766 (PL 89-240) was another bill backed by the Johnson administration. It authorized money for a new program to help states develop water supply and waste disposal system in rural areas. The grants were to go to towns or areas with small populations.[56]

1967–68: 90th Congress

In 1967, a special task force on environmental health released a report that recommended a federal program to deal with problems such as air and water pollution. The report called the "danger to environmental quality" one of the most important domestic problems at the time.[57] In 1968, Congress continued to study the problems of water and air pollution but enacted no significant legislation on these subjects.[58]

In the Senate, a new bill was proposed (S 2760) to combat lake, mine, and oil pollution, but no action was taken on it. Called the Federal Water Pollution Control Act Amendments of 1967, the bill would have authorized pilot programs to prevent and control lake pollution and acid or other mine water pollution.[59] S 3206, which had to do with financing of water pollution treatment facilities and would have amended the Federal Water Pollution Control Act of 1948, also failed to pass.[60]

In 1968, Congress established a nonfederal National Water Commission to review water resource problems and to identify possible solutions.[61] In S 20 (PL 90-515), a seven-member committee was established to make a comprehensive study of the nation's water resources problems.[62]

Through the Wild and Scenic Rivers Act of 1968 (S 119/PL 90-542), some of the nation's largest rivers were protected from development. The law essentially protected rivers designated for their wild and scenic value from activities that might adversely impact that value and provided for a mechanism to determine whether a river met certain eligibility requirements for protection as a wild and/or scenic river. The 1968 Estuary Preservation Act (HR 25/PL 90-454) authorized the Department of the Interior to make agreements with states for preservation of estuarine areas (marshlands, bays, and sounds) and directed the department to make a study of methods to preserve estuaries.[63]

President Nixon

In January 1969, a leaking offshore oil well off the coast of Santa Barbara, California, spread oil over 800 square miles of the Pacific Ocean. The oil coated the beaches, killed marine life, and damaged property. Then on September 15, 1969, a tugboat pulled out of Tiverton, Rhode Island, on its way to a power plant on the Cape Cod Canal. The towline broke, and the barge hit submerged boulders near the mouth of the West Falmouth Harbor. Before the oil could be removed from the barge, 175,000 gallons of light refined oil spilled onto the shores of Buzzards Bay.

In 1969, the Cuyahoga River in Ohio became so polluted that the river actually caught fire and burned. An ember from a railcar landed on an oil slick in the heavily polluted river, sparking a blaze that burned for hours. These events focused the public's attention on the need for more regulation of water pollution.

President Nixon addressed the problem of water pollution frequently throughout his presidency. He understood that the federal government played a role in the control of water pollution. In one speech,[64] Nixon addressed the relationship of federal and state governments in pollution control, saying that "whatever Federal regulations are adopted, there must be a joint State and Federal supervision of those regulations to be sure that they are adhered to."[65] He also said, "The major responsibility to reduce pollution rests appropriately with State and local governments and the private sector. However, the Federal Government must exert leadership and provide assistance to help meet our national goals."[66]

At the same time, Nixon recognized the need to cooperate with other countries. He said, "We are all grappling with problems of a modern environment, which are the byproducts of our advanced technologies—problems such as the pollution of air and water, and the congestion in our cities. Together, we can dramatically advance our mastery of these problems. By what means can we best cooperate to bring this about?"[67] He proposed that all nations adopt a treaty under which they would renounce all national claims over the natural resources of the seabed beyond the point where the high seas reach a depth of 2,000 meters and would agree to regard these resources as the common heritage of humankind.[68]

On February 8, 1970, Nixon gave a message to Congress on the environment. In that speech he noted that he wanted to strengthen water pollution control programs with a $12 billion national program and strengthen standard-setting and enforcement authorities.[69] He explained that the country needed "adequate treatment of the large volume of commercial, industrial and domestic wastes that are discharged through municipal systems," which "requires a great expenditure of funds for construction of necessary facilities." He stated that "the urgency of this need and the severe financial problems that face many communities, require that construction of waste treatment facilities be jointly funded by Federal, State and local governments. We must also assure that adequate Federal funds are available to reimburse States that advanced the Federal share of project costs.

I propose that $6 billion in Federal funds be authorized and appropriated over the next three years to provide the full Federal share of a $12 billion program of waste treatment facilities.... I again propose the creation of an Environmental Financing Authority so that every municipality has an opportunity to sell its waste treatment plant construction bonds."[70]

When it came to water quality standards, Nixon said,

> Water quality standards now are often imprecise and unrelated to specific water quality needs. Even more important, they provide a poor basis for enforcement: without a precise effluent standard, it is often difficult to prove violations in court. Also, Federal-State water quality standards presently do not apply to many important waters. I again propose that the Federal-State water quality program be extended to cover all navigable waters and their tributaries, ground waters and waters of the contiguous zone. I again propose that Federal-State water quality standards be revised to impose precise effluent limitations on both industrial and municipal sources. I also propose Federal standards to regulate the discharge of hazardous substances similar to those which I proposed and the Congress adopted in the Clean Air Amendments of 1970. I propose that standards require that the best practicable technology be used in new industrial facilities to ensure that water quality is preserved or enhanced. I propose that the Administrator of the EPA be empowered to require prompt revision of standards when necessary. We should strengthen and streamline Federal enforcement authority to permit swift action against municipal as well as industrial and other violators of water quality standards.[71]

In addition to these proposals, Nixon asked that the administrator of the Environmental Protection Agency (EPA) be authorized to issue abatement orders and impose fines in those situations in which water pollution could threaten the water quality. He asked for swift hearings in place of time-consuming enforcement hearings and for a tripling of federal grants to state pollution control enforcement agencies.[72]

As a response to the oil spills, Nixon gave a message to Congress on marine and beach pollution from oil spills. He said, "The threat of oil pollution from ships—both at sea and in our harbors—represents a growing danger to our marine environment."[73] He asked the Senate to act on a treaty with the Intergovernmental Maritime Consultative Organization, an arm of the United Nations that would empower the United States, by an international agreement, to take preventative action against any vessels on the high seas that threatened imminent pollution danger to our coasts. A second treaty Nixon discussed in that speech was the amendments to the

1954 Convention for the Prevention of Pollution of the Sea by Oil. That treaty focused mainly on the international discharge of oil or oily wastes on the high seas.

In addition, Nixon wanted to see action on the International Standards for Ship Construction and Operation and on the Ports and Waterways Safety Act. This legislation would protect against oil spills by allowing the Coast Guard to control vessel traffic in the inland waters and territorial seas of the United States. Nixon asked for increased surveillance in offshore areas with the highest spill potential, along with a new Harbor Advisory Radar System that would improve traffic control in waterways.

Further action was needed, according to Nixon. This included increased research and development of emergency oil transfer and storage systems and cooperation between private industry and port authorities in reducing oil spills. He wanted to install radiotelephones in all vessels to increase communication with other vessels, the licensing of towboat operators, and new methods to finance the cleanup of oil pollution.[74]

That same year (1970), Nixon gave a message on the environment in which he outlined a 37-point program that included 23 major legislative proposals and 14 new measures being taken by administrative action or executive order in five categories. One of those categories was water pollution control. Nixon proposed ideas such as a new Clean Waters Act, the creation of a new Environmental Financing Authority, the creation of new water quality standards, and court action for violation of established requirements.[75]

Nixon mentioned water pollution in his State of the Union address in 1970 and promised to propose to Congress a $10 billion nationwide clean waters program to put modern municipal waste treatment plants in every place in America where they were needed.[76] Nixon wanted new quality standards to be set and, with that, new penalties for violators, including fines of up to $10,000 per day. According to Nixon's plan, violations would be considered sufficient cause for court action.[77] In 1970, Nixon proposed a Clean Water Act that authorized $4 billion to cover the full federal share of the cost.[78] Nixon attempted to help clean up the Great Lakes when he promised to transmit legislation to the Congress that would stop the dumping of polluted dredged spoil into those lakes.[79]

In another speech on the environment, on February 10, 1970, Nixon spoke about water pollution. He said that there were three sources of water pollution: municipal, industrial, and agricultural. According to the president, the most troublesome source of pollution is the agricultural industry, including animal wastes, eroded soil, fertilizers, and pesticides. But, to Nixon, the most acutely damaging sources were municipal and industrial waste. To deal with municipal waste, he proposed increased federal assistance and the creation of an Environmental Financing Authority that would ensure that every municipality had an opportunity to sell its bonds for construction of a waste treatment plant. To deal with industrial pollution, Nixon suggested a seven-point program to enforce control of water pollution resulting from industrial waste. That included new federal–state water quality standards that would impose precise effluent requirements on all industrial and municipal sources, with violations being sufficient cause for court action. Under the program, failure to meet established water quality standards or implementation schedules could be result in fines of up to $10,000 per day. The secretary of the interior would be authorized to seek immediate injunctive relief in emergency situations in which severe water pollution constituted an imminent danger to health. Finally, all federal pollution control programs would be extended to include all navigable waters.[80]

1969–70: 91st Congress

Partly because of the Santa Barbara oil spill and other incidents, the Interior tightened the restrictions for granting leases for offshore drilling on federally owned property during this congressional session.[81] The 91st Congress considered action on a major bill to control water pollution. Called the Water Quality Improvement Act of 1969 (HR 4148), the bill would make changes to the Water Pollution Control Act.[82] It mandated that anyone who owned or operated ships or offshore facilities, such as petroleum companies, would be held liable for the costs of cleaning up oil spills, for up to $14 million for a vessel and $8 million for an offshore facility, unless the owners could prove it was not their fault.[83] It also outlawed flushing raw sewage from boat toilets. The law created an Office of Environmental Quality that would act as a staff for the President's Council on Environmental Quality. Nuclear power plants would be forced to comply with state water quality standards through the strengthening of restrictions on thermal pollution from these plants. In another provision of the bill, criteria covering the effects of pesticides in streams, rivers, and other waterways had to be developed. The proposal passed in 1970 and became law (PL 91-224).[84]

Specifically, the new law declared that no oil should be discharged into navigable waters in the United States. Penalties would be assessed if an oil spill occurred. The president would be responsible for the cleanup of the spill if it was not properly done by the person responsible for the spill. Liability amounts were established, including up to the lower of either $100 per gross ton of oil or $14 million for a vessel. However, if a spill resulted from willful negligence or misconduct, there would be unlimited liability.

Under the new law, the federal government could enter into agreements with states to carry out research to eliminate or control acid or other mine water pollution resulting from active or abandoned mines. The federal government could also work with states to carry out research on new ways to eliminate or control pollution within the watersheds of the Great Lakes.

1971–72: 92nd Congress

The Senate passed S 3994 to help in the provision of safe drinking water to all citizens. The legislation would have established a program within the EPA to regulate the safety of drinking water by setting minimum standards with maximum limits of contaminants in water. The House took no action, and it did not become law.[85]

The Senate also passed S 2770 (PL 92-500), the Clean Water Act Amendments, to amend the Federal Water Pollution Control Act. The bill was aimed at making the nation's waterways virtually pollution-free by 1985. The earlier water laws that had been enacted in 1948, 1956, and 1965 had limited the role of the federal government to guiding, supporting research and investigation, and providing grants to states. The 1972 law changed this by creating a centralized program with federal oversight of the states. The law became the most comprehensive and expensive environmental legislation in history.[86]

The authors of the bill wrote that there was a need for an interim goal of water quality to provide for the protection and propagation of fish, shellfish, and wildlife and for recreation in and on the water.[87] They also wrote that "due regards shall be given to the improvements which are necessary to conserve such waters for the protection and propagation of fish and aquatic life and wildlife, recreational purposes, and the withdrawal of such waters for public water supply, agricultural, industrial, and other purposes."[88] Therefore, Congress mandated that all waters be "fishable and swimmable" by 1983, with zero pollution discharge by 1985.

To do this, Congress set forth certain mandates. One had to do with setting standards for discharges into waters. For the first time, the EPA was directed to set uniform, national standards and limits on effluents for all major sources of water pollution. The standards were called "effluent guidelines" and were designed to make industries create and use the latest available technology to reduce pollution. All direct dischargers into navigable waterways had to satisfy two different standards, one relating to water quality, the other to effluent limits. The water quality standards, established by the states according to guidelines issued by the EPA, were to identify the use for a body of water into which a polluter was discharging (such as recreation, fishing, boating, waste disposal, irrigation, and so forth) and to establish limits on discharges in order to ensure that use. Effluent standards, established by the EPA, were to identify what technologies any discharger had to use to control its effluents. In meeting these dual requirements, the polluter was required to achieve whichever standard was stricter. A different set of standards was established for municipal wastewater treatment facilities.

The authors of the bill set limits on the amount of contaminated wastewater that industrial plants could release into waterways. All point-source emitters of pollution would be required to obtain a National Pollution Discharge Elimination (NPDE) permit issued by the EPA or an EPA-approved state agency. With the exception of city waste treatment plants, all existing dischargers were required to have technological controls, prescribed by the EPA, which were to be the "best practicable control technology currently available" by July 1, 1979, and the "best available technology" by July 1983. Specifically, the law mandated that all treatment plants in existence on July 1, 1977, were required to have "secondary-treatment" levels but that all facilities, regardless of age, were required to have "the best practicable treatment technology" by July 1, 1983. All new sources of discharge, except municipal treatment plants, were required to use control technologies based on "the best available demonstrated control technology, operating methods or other alternatives."[89] The only qualification was that the limits and technologies they were based on must be "economically achievable."[90]

Indirect dischargers were also included in the legislation. Many pollutants, including toxic chemicals, are released into municipal wastewater systems by industrial and commercial sources and later enter waterways through city sewage treatment plants unable to eliminate them. The law required the EPA to establish pretreatment standards, which were to prevent the discharge of any pollutant through a public sewer that "interferes with, passes through or otherwise is incompatible such works." The purpose of this provision was to compel such indirect dischargers to treat their effluent before it reached the city system.[91]

Another goal of the new laws was "that Federal financial assistance be provided to construct publicly owned waste treatment works."[92] Under the law, all municipalities were required to have secondary sewage treatment systems by 1977. To help municipalities pay for the systems, the bill authorized $16.8 billion for construction of sewage treatment plants. Congress authorized the federal government to pay 75 percent of the cost of local plants.[93] In doing so, the act expanded the federal role in financing construction of municipal waste treatment facilities.

Both civil and criminal penalties were established for those industries that violated the antipollution standards. But there would be both federal and state enforcement of the provisions in the law. The EPA was authorized to delegate responsibility for enforcing most regulatory provisions to qualified states, which would issue permits to all polluters specifying the conditions for their effluent discharges.[94] The dumping of agents of chemical and biological warfare into the waterways as well as the dumping of radioactive materials or wastes would be prohibited, and ocean dumping of wastes would be subject to federal control.[95] Additionally, dumping of toxic pollutants was prohibited.[96]

Nixon criticized the bill because it gave too much power to the federal government and not enough to the states.[97] The bill passed Congress and was vetoed by Nixon, who said it had an unreasonable price tag.[98] The Congress overrode his veto, making it law (PL 92-500).

The new law, which was intended to restore and maintain the integrity of the nation's waters, sought to eliminate pollutant discharges into U.S. waters and make them safe for fish, shellfish, wildlife, and recreation by 1993. The final bill had many provisions. When it came to research, the EPA administrator would make grants available to states to allow them to develop comprehensive water quality control plans for river basins and lakes. The research would involve investigations, advisory committees, pilot programs, and the like to study the problems of water pollution. Interstate cooperation was encouraged.

With regards to standards, the law made the discharge of any pollutant by any person unlawful. Limits for pollution discharges would be based on the "best available technology" as determined by the EPA administrator. The discharge of radiological, chemical, or biological warfare agents or high-level radioactive waste would be prohibited. If this was done, criminal penalties of between $2,500 and $25,000 per day, one year in prison, or both, could be established. For a second offense, the penalties increased to $50,000 per day, two years in prison, or both. Civil penalties of up to $10,000 per day could also be imposed.

President Ford

In his budget, Ford proposed spending $6.5 million to help local communities build modern wastewater treatment plants. That was a 65 percent increase over the amount for the current fiscal year and a 90 percent increase over that in the previous fiscal year.[99] He told the public that his administration was pursuing a massive program to clean up the nation's waterways, and, because of that, the rivers, lakes, and coastal waters were being cleaned up. He supported legislation that would make loans available to municipal governments so that they could construct municipal wastewater treatment plants.[100]

President Ford supported the Safe Drinking Water Act to increase the protection of the public's health. In signing the bill (S 433), he noted that the new law would give the states the primary responsibility to enforce the standards and ensure the quality of the drinking water. The federal government could step in when the states failed, but Ford said that this would seldom be necessary. Thus, according to Ford, the water quality was a state rather than a federal responsibility.[101]

1973–74: 93rd Congress

The Congress passed the Safe Drinking Water Act (S 433/PL 93-523) in 1974. The act would enhance the safety of public drinking water by establishing new national standards. It directed the EPA to set maximum contaminant levels (MCLs) for the public drinking water supply based on health-based standards. Treatment techniques would be established to meet the standards. Local water utilities would be required to both monitor for contaminants and install EPA-approved technology to control them.[102] Any underground injection of wastewater would have to be authorized by a permit. Such a permit could not be issued until the applicant could prove that such disposal would not affect drinking water sources. The act also authorized additional funds for research and grants to help implement new standards to be set by the EPA.[103] Ford signed the bill.[104]

Congress sent the president a bill (S 2812/PL 93-243) to provide a formula to allocate funds to the states for the construction of sewage treatment facilities. With the formula, every state would receive a certain minimum amount. The remaining amount would be determined based on the state's needs.[105]

Finally, another bill passed the Congress but did not become law. Nixon vetoed the Rural Water and Sewer Grant Program Bill (HR 3298), which was passed to restore the rural water and sewer grant program. Ford argued that the bill was a disservice to the taxpayers of this country and that it enlarged the federal responsibility in an ineffective way.[106] A motion to override the president's veto failed in the House by a vote of 225–189.

1975–76: 94th Congress

The National Commission on Water Quality sent Congress a report that recommended some changes to improve the 1972 water pollution control law. Among other things, the committee recommended delaying the discharge requirements for certain types of polluters and putting off the requirements for up to 10 years so another commission could study the need for imposing them. They also reported that Congress should authorize the EPA to certify states to take over the administration of water quality planning, as well as discharge permit and construction grant programs, provided that the states met certain standards.[107]

President Carter

President Carter believed that the country needed to focus on three areas to reach the goal of having fishable and swimmable waters. He proposed additional funding for construction of municipal sewage systems, more funds for water quality management programs in state and local governments, and new legislation to impose penalties on firms that had failed to abate their pollution. However, Carter noted that the job of preventing water pollution and abating pollution must be done at the state and local levels rather than the federal level.[108]

President Carter promised to continue to work with the Congress to pass legislation needed to improve federal water resources programs and to support the states in their primary responsibilities for water allocation and management.[109] On January 21, 1980, Carter gave a speech about his agenda in which he mentioned the problem of water pollution. He said that "sound water management is vital to the economic and environmental health of our nation.... It should be clear that my Administration supports sound water resources development, and has taken several steps to improve the quality of projects sent to Congress for authorization and funding."[110]

1977–78: 95th Congress

Congress cleared legislation (HR 3199/PL 95-217), the Federal Water Pollution Control Act Amendments, to amend and extend the Clean Water Act Amendments of 1972. The purpose of the original legislation was to achieve fishable and swimmable waters by 1983.[111] The bill authorized money for non-research programs and grants to states for enforcement and aid for planning waste treatment programs and for cleaning up polluted lakes. It also authorized funding for alternative treatment facilities and extended controls on oil pollution from 12 miles from shore to 200 miles from shore.[112] The amendments delayed some deadlines for industries and cities to meet treatment standards, and they gave more weight to the control of toxic pollutants.[113] The new law also set national standards for industrial pretreatment of waste, increased funding for sewage treatment construction grants, and gave states flexibility in determining priorities. The control of toxic pollutants was expanded.

The Clean Water Act Amendments established three categories of water pollution: toxic, nonconventional (pesticides, metal compounds, etc., whose toxicity had not yet been established), and conventional (sediment, oxygen-demanding wastes, and plant nutrients). The act retained the controversial 1985 zero-discharge goal but applied it only to toxic pollutants. The deadline for industrial installation of the best available technology for conventional pollutants was extended to 1984, and the best-available-technology deadline for nonconventional pollutants was extended to mid-1987, with waivers allowed when costs exceeded benefits. The deadline for toxic pollutant control was set for 1984.[114]

Carter signed a new law (S 1528/HR 6827/PL 95-190) to amend the Safe Drinking Water Amendments. It authorized $152 million in fiscal years 1978 and 1979. The largest share of the money went for grants to the states for public water supervision programs. Other amendments to the act provided for EPA training grants to state and local personnel. The bill also extended the expiration date for certificates of need for water treatment chemicals, by which states could receive chemicals that were not readily available. The law also applied the act and state and local regulations to federally owned or maintained public water systems.[115]

Two other bills passed during this session. The first (S 2701/HR 11655/PL 95-404) authorized funds for fiscal year 1979 for the Water Resources Planning Council. The bill was called the Water Resources Planning Act. The second bill to pass during this session was S 2704/HR 11126 (PL 95-467), the Water Research and Development Act, promoted a more adequate and responsive national program of water research and development.

1979–80: 96th Congress

In this session, Congress considered a revision of the federal criminal code. As part of that revision, members raised the violations of federal clean water and waste disposal laws from misdemeanors to felonies punishable by a two-year maximum sentence, a fine, or both.[116] A bill that was considered but did not pass was S 2710, a proposal to amend the Hazardous Materials Transportation Act. The bill would have authorized funds for programs under the 1972 Federal Water Pollution Control Act. Although members from the House and Senate met several times to discuss provisions of the bill, they were unable to come up with a compromise version.[117]

Two other bills concerning funding authorizations did pass. One of those, S 1146 (PL 96-63), was a reauthorization of the Safe Drinking Water program. It included $24 million for the cleanup of hazardous waste spills in public drinking water supplies.[118] Another was S 1640 (PL 96-457), which authorized funds for fiscal years 1981 and 1982 for water resources research and development and saline water conversion research and developmental programs.

Another Senate bill that became law was S 901 (HR 4023/PL 96-148), which extended until June 30, 1980, the existing moratorium on industrial cost recovery (ICR) charges that were levied on users of treatment facilities built with federal funds. In most situations, Congress gave itself six months to decide whether to charge industrial users of municipal wastewater treatment plants a share of the cost of the plant's construction. The moratorium on ICR charges was due to expire on June 30, 1979, but when Congress was writing the Water Pollution Control Act in 1972, it added the ICR provision so that industries would have to pay their share of the construction costs, as well as the operating expenses, of the waste treatment plants they used.[119]

Another extension was granted in HR 8117 (PL 96-502), which gave cities three extra years to meet the federal standards for cleaning up cancer-causing substances in public drinking water supplies. The standards had been set in the 1974 Safe Drinking Water Act. In 1974, approximately 13,600 water systems were in violation of one or more of the standards and could not meet the original deadlines. Cities planning to regionalize their water treatment systems were also given an additional three years to comply with the act. The measure also exempted natural gas storage wells from the underground injection control program mandated by the act. Another provision gave the EPA the discretion to approve state underground oil and gas injection programs, without subjecting them to federal regulations prepared under the act, if those plans adequately protected underground water supplies.[120]

President Reagan

Reagan chose not to sign the Water Resources Research Bill (S 684), which would have authorized an ongoing program of research on water resources.

In doing so, Reagan explained that the bill would have authorized appropriations for a variety of research activities on water resources, including a new, separate authorization of grants for the development of water technology, which was not an appropriate federal activity.[121] It became public law without his approval.

The Reagan administration unveiled its proposals for rewriting the Clean Water Act. These included many new provisions. One of them extended the deadline for those industries that dump wastes directly into rivers and streams to install the best available technology for treating toxic wastes and the "best conventional technology" for treating nontoxic pollutants. Another provision specified that "new source performance standards" (NSPS) would apply to a plant only if construction began after the NSPS rule was originally issued. The EPA administrator was given the authority to assess civil penalties of up to $10,000 a day for clear violations of the act. Criminal penalties were also set. Those included up to $50,000 per day and two years' imprisonment for knowing violations of certain parts of the act.[122]

1981–82: 97th Congress

Congress did not pass a proposal this session to create a new Water Policy Board.[123] They also did not take action on proposals to rewrite the Clean Water Act (HR 6344/HR 6670) that would have extended the deadline for industries that dump wastes directly into rivers and streams. The proposal would have expanded the scope of the president's authority to exempt federal facilities from pollution control rules when he found it in the "paramount interest" of the United States for military operations.[124]

1981–82: Courts

During this time, many cases regarding different provisions of the Water Pollution Control Act Amendments were heard in the Supreme Court. One of those was *Environmental Protection Agency v. National Crushed Stone Association* (449 U.S. 64; 101 S.Ct. 295; 66 L.Ed.2d 268, 1980). This case revolved around the discharge limitations for the coal mining industry and other industries that the EPA had created in response to the new law. Those challenging the law sought to have the courts review the regulations, challenging the substantive standards and the economic capability of industries to meet the new standards. The Supreme Court decided that Congress, in adopting the standards for clean water, did not intend for the

EPA to concern itself with the economic impact of compliance with those standards.[125]

1983–84: 98th Congress

The Clean Water Act had last been overhauled in 1977, but the authorization for the money to implement parts of the act lapsed in 1982. Congress voted to fund the law through annual appropriations bills while it considered how to revise the law.[126] In 1984, the House passed a rewrite of the Clean Water Act (the Water Quality Renewal Act: HR 3282), but it did not pass through the Senate. At the same time, the Senate did not act on its own bill (S 431). The members of the Senate Environment and Public Works Committee passed their version, but the entire Senate could not agree, and the bill died.[127]

The House passed HR 5959 to amend the 1974 Safe Drinking Water Act. The bill would have set deadlines for the EPA to set standards. Failure to meet the deadlines would expose the agency to lawsuits. It also would bar many underground deposits of toxic wastes near underground sources of drinking water. It provided matching funds to protect these water sources. But the Senate passed another version of the bill, and when compromise efforts failed, the bill died.[128]

1985–86: 99th Congress

The House and Senate both considered bills to beef up federal protections for drinking water. HR 1650 and S 124 were both passed by the full chambers, and conferees were able to reconcile the different bills for the president to sign (PL 99-339).[129] The Safe Drinking Water Act Amendments of 1986 forced the EPA to set national standards (MCLs) limiting 83 different contaminants in drinking water supplies in the next three years and to see that states used the best available technology to meet the standards. The maximum civil penalties that could be imposed for violations of the new regulations were increased from $5,000 per day to $25,000 per day. The Congress authorized $10 million annually during fiscal years 1987–91 for technical assistance to small public water systems so they could comply with the new EPA regulations.[130]

Congress debated two bills (HR 8/S 1128) that would have reauthorized and strengthened the Clean Water Act,[131] but Reagan vetoed the version sent to him arguing the bill was too expensive. The controversy surrounded $18 billion in federal money to help local governments build sewage treatment plants.[132] Reagan claimed the bill exceeded acceptable levels of intended budgetary commitments.[133]

1987–88: 100th Congress

The Water Quality Act of 1987 (HR 1/PL 100-4) reauthorized the Clean Water Act Amendments, the country's main law to control water pollution. It reauthorized a 10-year program of federal aid to state and local governments for construction of sewage plants. The new law tightened the maximum penalties for violations of the Clean Water Act and expanded the EPA's authority to enforce it. For the first violation, criminal penalties ranged up to a fine of $25,000 per day and a year in jail. For subsequent violations, the fine was set at up to $50,000 per day and two years in prison. If the violation was deemed to be "knowing," the penalties jumped as high as $100,000 per day and six years in jail. The civil penalties for violations of certain provisions were raised from $10,000 per day to $25,000 per day. The bill also postponed deadlines for technology-based effluent standards. Reagan chose to veto the bill because it contained many projects that were "pork barrel" projects (those that were not related to water quality) and because the bill was too expensive. He said the bill was "well-intentioned" but included "short-term programs (that) balloon into open-ended, long-term commitments costing billions of dollars more than anticipated or needed." But he also said that his veto would have no impact whatsoever on the immediate status of any water quality programs.[134] Despite his concerns, Congress overrode his veto, and the bill became law.[135]

The amendments required each state to have a plan approved by the EPA for controlling pollution from nonpoint sources. Such plans must include "best management practices," but state officials were permitted to decide whether to require owners and managers to use such practices or whether to make them voluntary.[136] Further, the amendment delayed some of the deadlines for nonpoint sources of pollution such as discharges from urban storm sewers. The new law also expanded water quality programs for lakes and estuaries, including special programs for the Great Lakes and other areas.[137]

The House easily passed a bill (HR 791) that would create an interagency research committee to help states fight groundwater contamination, the source of almost half of the nation's drinking water. The bill would create an Interagency Groundwater Research Committee to coordinate the work of several agencies of the federal government. It would assess groundwater contamination and coordinate cooperation between state and local governments and federal agencies.[138] In the end, this bill did not become law.

Another bill passed by Congress in this session (HR 4939/PL 100-572) was the Lead Contamination Control Act of 1988. This focused on water coolers and school drinking fountains that added toxic lead to water. At the time, the EPA estimated that 42 million Americans drank water with unsafe amounts of lead. This was happening because the lead often entered the water after it left the treatment plant, so normal treatment regulations did not address the problem. The bill attempted to control several sources of lead in drinking water. These were water coolers and school drinking fountains that were made with lead solder or lead-lined tanks. The new legislation required the Consumer Product Safety Commission to recall any drinking water coolers that had lead-lined tanks. To support states and school districts in finding and fixing such problems, the bill authorized grants of $30 million annually for fiscal years 1989–91.[139]

President G.H.W. Bush

Bush supported the Clean Water Act and promised to work with the Congress to get it passed. He sought out more opportunities to incorporate innovative, market-oriented provisions to solve the problem of polluted waters.[140] He also chose to sign the Great Lakes Critical Programs Act, which provided essential tools to clean up the Great Lakes. This was important to Bush because of the significance of these areas for people for both recreational and historical reasons.[141]

1989–90: 101st Congress

In response to the *Exxon Valdez* oil spill in Alaska in 1989, Congress debated but did not pass the Environmental Crimes Act of 1989 (HR 3641), a new law that would have stiffened criminal penalties for individuals and groups responsible for environmental disasters that caused death, injury, or severe damage. The new law would have made it a felony to knowingly create a risk of death by polluting. Prior to this, negligent endangerment, or "gross deviation from the standard of care that a reasonable person would exercise," would have been a misdemeanor. Under the proposed law, penalties could have reached fines of $2 million for corporations or $500,000 for individuals. Furthermore, repeat offenders could have faced jail terms of up to 30 years. The administration opposed the bill, stating that tougher penalties could backfire by making juries less likely to convict defendants in environmental cases.

A similar bill did pass Congress (HR 1465/PL 101-380): The Oil Pollution Act was created to prevent oil spills and punish the spillers by increasing the spillers' liability and imposing stiffer civil and criminal penalties. It required spillers to pay for cleaning up oil spills and to compensate parties economically harmed by them. The law authorized using money from a federal fund, subject to annual appropriations, to pay for cleanup and compensation costs not covered by spillers.[142]

The Great Lakes Critical Programs Act of 1990 (HR 4223/PL 101-596) amended the Water Pollution Control Act relating to water quality in the Great Lakes. The new law implemented parts of the Great Lakes Water Quality Agreement of 1978, signed by the United States and Canada, in which the two nations agreed to reduce certain toxic pollutants in the Great Lakes. That law required the EPA to establish water quality criteria for the Great Lakes addressing 29 toxic pollutants with maximum levels that are considered to be safe for humans, wildlife, and aquatic life. It also required the EPA to help the states implement the criteria on a specific schedule. Another law passed during the session was HR 1101 (PL 101-397) to extend the authorization of the Water Resources Research Act of 1984 through fiscal year 1993.

President Clinton

President Clinton sought to protect the water quality for all Americans, so that the water people used did not make them sick. He recognized that polluted water can affect an individual's health, and that polluted water can hamper the growth of a strong economy.[143]

The Clinton administration supported the Clean Water Legislation that would require the use of the best available technology to keep our nation's water clean.[144] He called for changes to the Safe Drinking Water Act aimed at giving states money for improvements to water systems, which he signed into law. In a radio address in 1999, Clinton announced action to ensure that "every river, lake, and bay in America is clean and safe."[145] He promised to work in conjunction with states to achieve that goal.

Clinton showed his disapproval for a bill that passed the House Appropriations Committee because it cut funds for the EPA, and he claimed that Congress cut provisions for enforcing environmental laws, both of which resulted in more water pollution.[146] He recognized the need to conserve water, and promised to landscape federal buildings in such a way that water use would be at a minimum.[147] He also sought to conserve water through supporting recycling across the country.[148]

1991–92: 102nd Congress

The Congress worked on the Clean Water Act during this session, but nothing was done.[149] HR 2221 was a proposal to amend the Clean Water Act and require the director of the EPA, at the request of the governor of Washington, to establish a grant to implement and update the Puget Sound Water Quality Management Plan. This would allow the state to implement a plan to monitor the water quality of Puget Sound and support research to gain a better understanding of the impact of pollution on the resources there.

1993–94: 103rd Congress

In 1993, an estimated 300,000 people became ill in Milwaukee from bacteria in the city's drinking water supply. But federal efforts to help Milwaukee and other cities update their outdated drinking water facilities were quickly bogged down in a turf battle between the Energy and Commerce Committee and the Public Works and Transportation Committees. The Energy and Commerce Committee passed a bill (HR 1701: the Drinking Water and Public Health Enhancement Amendments of 1993) that called for a new loan fund to assist states in helping cities and local governments to prevent and detect drinking water contamination. The Public Works Committee approved another bill (HR 1865) that included a revolving fund to be used only for construction, rehabilitation, and improvement of city water supply systems. In the end, no bills advanced in this area.[150]

1995–96: 104th Congress

In this session, members of the House proposed HR 961, the Clean Water Reauthorization Act. It was an overhaul of the federal Clean Water Act, but it not only was blocked in the Senate but also drew a veto threat from President Clinton. The bill would have allowed federal, state, and local officials to waive some regulatory requirements for facilities that discharged wastes into waterways, provided that the actions did not harm the environment. The bill would also have repealed a requirement that facilities and municipalities obtain federal permits to discharge polluted storm water into waterways. Risk assessments and cost-benefit analyses would have been required for many pollution regulations.[151]

The Congress considered S 1316, a major revision of the Safe Drinking Water Act. The bill addressed many long-standing problems with the drinking water program and dealt more realistically with regulating contaminants based on their risk to the public's health. It repealed an existing requirement that the EPA set standards for 25 new contaminants every three years. Instead, the EPA would have the option of issuing regulations for at least five contaminants every five years. The EPA would be required to set drinking water standards for arsenic, radon, and sulfates, and, if water systems did not comply with the standards, they would be required to notify customers within 24 hours. Any pipe or fixture that was not "lead-free" would be banned. Further, $7 billion was made available for state-administered loans. Water systems were required to provide their customers with annual reports on the safety of local water supplies and to notify their customers of any possible contamination problems.[152] The bill passed both houses, and the president signed it (PL 104-182).

Under a bill considered in the Senate (HR 2747: the Water Supply Infrastructure Assistance Act of 1996), the director of the EPA would establish grants to states to allow them to develop water supply infrastructure. This would include the construction, rehabilitation, and improvement of water supply systems or methods to address pollutants in navigable waters to make such waters usable by water supply systems. This bill did not have enough support to pass. HR 2024 was a bill to phase out the use of mercury in batteries and establish new labeling requirements for rechargeable batteries. This was needed because, when disposed of in a landfill, the mercury broke down and polluted the groundwater. The bill was entitled the Mercury-Containing and Rechargeable Battery Management Act, and became a law after the president signed it (PL 104-142).[153]

Finally, in this session, Clinton signed the Coast Guard Authorization Act of 1996 (S 1004/ PL 104-324)to protect the environment from plastic pollution and oil spills. The act preserved the federal government's right to recover the costs of oil spills from responsible parties. The act also contained new requirements for inspections of vessel waste reception facilities and required additional safety equipment on certain barges.[154]

1999–2000: 106th Congress

The Congress passed HR 999 (PL 106-284) to amend the Water Pollution Control Act to improve the quality of coastal recreation waters. The Beaches Environmental Assessment and Coastal Health Act of 2000 required states that had coastal recreation waters to adopt new water criteria and standards for pollutants. The EPA director would also be required to undertake research to assess the potential health risks from exposure to pathogens in beach waters.

President G. W. Bush

President Bush vetoed a bill (HR 1495) called the Water Resources Development Act. He said that the even though he supported federal funding for water projects, this bill lacked the fiscal discipline needed to do that. He also claimed that the bill did not have priorities set and would actually hinder the nation from meeting its water resource needs.[155] The proposal authorized funding for 900 Army Corps of Engineers' navigation, flood control, and environmental restoration projects. Additionally, the law provided money for the hurricane-damaged coastal areas in Louisiana. Since the Congress voted to override Bush's veto, the bill became law.

2005–06: 109th Congress

During this session, the Senate considered S 1995, a proposal to amend the Water Pollution Control Act and enhance the security of wastewater treatment facilities. If it had passed, the bill would have allowed the EPA administrator to provide grant money to local wastewater treatment facilities to assess their plants and prepare and implement security plans. It would also allow for emergency response plants and deal with any possible security threats against the facilities. The bill did not have enough support to become law.

Another bill the Senate considered was S 728, the Water Resources and Development Act of 2005. Under the proposal, the secretary of the army would be asked to construct projects that would allow for improvements to rivers and harbors in the nation. If the bill passed, it would provide for ecosystem restoration and flood or storm damage reduction in states such as Alaska, Florida, Louisiana, Texas, and others. However, the bill did not pass during this session and did not become law.

2007–08: 110th Congress

In 2008, Congress voted to amend the Federal Water Pollution Control Act (S 2766/HR 5949/PL 110-288) in the Clean Boating Act. The new law addressed certain discharges that were incidental to the normal operation of a recreational vessel. It specified that no permit should be required by the

EPA under the national pollutant discharge elimination system for the discharge from a recreational vessel of cooling water, weather deck runoff, or effluent from properly functioning marine engines, which is incidental to the normal operation of the vessel.

Congressional members also considered S 2994, the Great Lakes Legacy Act of 2008. This bill would amend the Water Pollution Control Act and provide for the cleanup and monitoring of contaminated sediment in the Great Lakes and other regions. A similar bill, HR 6460, the Great Lakes Legacy Reauthorization Act, was introduced into the House. The House version made it through both chambers and became law (PL 110-365).

President Obama

The Obama administration considered limiting the amount of deicing fluid used on planes before they pull away from gates in winter because the EPA feared the chemicals may contaminate water supplies. The proposal would permit airlines to use no more than 25 gallons on a plane before it pulls away from the gate. The idea has been resisted by airlines.[156]

2009–10: 111th Congress

During this session the House passed the Water Quality Investment Act of 2009 (HR 1262) to authorize $13.8 billion in federal funds over five years for wastewater treatment grants and loans as part of a broader $19.6 billion package of water quality measures. If passed, the proposal would reauthorize the Clean Water State Revolving Fund, which provides low-interest loans and grants to communities for building and maintaining wastewater treatment facilities for the first time in 15 years. The bill would provide $300 million each year through fiscal year 2014 in state management assistance and $100 million in annual grants to nonprofit organizations to provide technical and management assistance to help improve wastewater treatment systems in rural areas, small municipalities, and tribal communities. As amended, the five-year authorization also included $500 million a year in grants for projects to prevent sewer overflows, $150 million a year to address containment of sediment in the Great Lakes watershed, and $50 million annually to provide grants for pilot programs testing alternative methods for enhancing water supplies, such as wastewater reclamation and reuse.[157] The bill did not have enough support to become law.

A new drilling technology called hydraulic fracturing—also known as *hydro-fracking*—was discussed in Congress. The new technology has allowed reserves of once-inaccessible natural gas to be accessed, but it faces growing backlash from environmentalists, local activists, and important members of Congress, who fear that widespread use of the technique could pollute drinking water and cause other environmental damage. Some of the chemicals used in the process are known carcinogens that could pollute the water supply. Those who use hydro-fracking say that the chemicals are not harmful to the environment. In the beginning, hydro-fracking was exempt from many of the federal regulations that regulate other forms of energy extraction, including a requirement that companies disclose what chemicals they inject into the ground. In the House, a bill was proposed (HR 2766) that would end hydro-fracking's exemption and require companies to disclose the chemicals they use in extraction. The Senate also debated their version of the bill (S 1215). They heard testimony that hydro-fracking can have a toxic effect on the water supplies.[158] Despite all the work on the proposals, they did not pass.

A bill introduced into the House would establish a strategy against harmful algal blooms found in freshwater such as the Great Lakes. The blooms release toxins that can be deadly to wildlife and humans who swim in or consume contaminated water or eat shellfish from it. The blooms can grow quickly as nutrients in water increase as a result of changes in water quality, temperature, and sunlight. HR 3650 would authorize $205 million over five years and also address hypoxia, or severe oxygen depletion, in seawater and freshwater. In the Senate, S 952 was introduced to address these issues.[159] Neither bill became law.

The Senate Environment and Public Works Committee passed a proposal (S 787) that would restore broad federal protections over the nation's waters. Some of the protections had been limited by previous Supreme Court decisions that precluded the EPA from using the provisions of the Clean Water Act to regulate pollution in some wetlands because they were not considered to be navigable bodies of water. The cases were *Solid Waste Agency of Northern Cook County v. US. Army Corps of Engineers* (531 U.S. 159, 2001) and *Rapanos v. United States* (547 U.S. 715, 2006). The Supreme Court decided that the phrase "navigable waters of the United States" applied to continuously flowing bodies of water and did not apply to channels where water flows intermittently or that periodically provide drainage for rainfall. The new law would have changed the definition of water in the Clean Water Act (PL 80-845) to clarify the federal government's jurisdiction over all water sources. Past references to "navigable waters of the United States" would now be replaced with "waters of the United States."[160] The bill did not become law.

A bill was proposed in the House (HR 2868) to protect chemical plants and water systems from acts of terrorism. The bill would extend and modify the authority of the Department of Homeland Security and the EPA to enhance security at chemical facilities and wastewater treatment and drinking water plants.[161] The proposal was not sent to the president for his approval.

Conclusion

Since Congress began passing policies to reduce water pollution and increase the quality of water across the nation, there has been a major reduction in the raw pollution in surface waters.[162] This is important not only for the health of citizens but also for a strong economy. Safe drinking water and unpolluted waterways are necessary for a healthy lifestyle and a strong economy. Members of Congress have debated the best way to ensure clean water, and they will probably continue to do so in the future. Nonetheless, they will continue to pass legislation that is geared toward protecting water supplies in all states.

Chapter 4

The Politics of Ocean Pollution

In 1988, many beaches along the Atlantic coast, especially in New York and New Jersey, were forced to close after several kinds of medical waste washed up on beaches, including blood bags, syringes, hypodermic needles, and surgical waste. Many resort areas suffered financial losses as tourists cancelled their vacation plans. Pleas for government intervention came from beach owners, local businesses, environmentalists, and residents, asking for help to clean up the water and prevent waste from washing up onto the beaches in the future.

The protection of sensitive coastal environments from dredging, draining, and filling is today one of the top national concerns. Unfortunately, these delicate areas were not always recognized as being important for wildlife, fisheries, and the overall integrity of land and water resources. In recent years legislation to safeguard such areas from pollution has become essential.

Oceanography and the questions surrounding how to protect ocean waters received little attention until the late 1950s. At that time, it was becoming clear that knowledge about the ocean environment (including tides, minerals, and food resources) was vital. Because of that, the federal government began to spend more on understanding the ocean and marine life. This chapter provides a review of congressional action that has been taken to protect the oceans and marine animals from pollution.

1959–60: 86th Congress

In 1960, a few members of the Senate began to work on legislation to create an organization that would coordinate federal oceanographic activities. A bill to do this (S 2692) was passed in the Senate in 1960.[1] The bill would set up a Division of Marine Sciences in the National Science Foundation. The agency would be provided with $534 million over 10 years to coordinate a research and survey program of the oceans. The House took no action on it, so it was not enacted.[2]

President Kennedy

President Kennedy asked Congress for over $106 million in appropriations for programs in oceanography in fiscal year 1962. He requested the additional funds in a speech given on March 29, 1961, in which he said that "we have thus far neglected oceanography" although "our very survival" may hinge upon charting the oceans for military and other purposes.[3]

In 1963, the Kennedy administration issued a report through the President's Office of Science and Technology that developed a plan for the 20 federal agencies involved in oceanographic research. The report created a new coordination of federal oceanographic efforts to ensure a more efficient and complete analysis of the ocean's resources.[4]

1961–62: 87th Congress

During the 1961 congressional session, the House passed HR 8181, a proposal to authorize a National Fisheries Center and Aquarium either in the District of Columbia or somewhere close to it. The aquarium would be overseen by a nine-member, nonpartisan board and would help to educate people about aquatic life and contribute to oceanographic research. The Senate did not take any action on the proposal in 1961,[5] but work continued on the bill in 1962. It was eventually signed into law on October 9, 1962 (PL 87-758).[6]

In 1961, the Senate passed a bill that would strengthen the federal oceanography program by establishing a 10-year research program for the ocean and the Great Lakes. S 901 would create a Division of Marine Sciences in the National Science Foundation to coordinate the program. As passed by the Senate, the bill would, among other things, create a long-range national oceanography program for the "benefit of mankind, national defense, efficient operation of the Navy, rehabilitation of U.S. commercial fisheries and increased utilization of aquatic resources." The bill would also support studies of the ocean and its resources. But President Kennedy pocket vetoed the bill, and the White House did not release his reason for doing so. Later, it said that the veto had to do with problems of antisubmarine warfare and other national security considerations.[7]

The Congress sent the president a bill (HR 6845) that would extend the Coast Guard's authority to conduct oceanographic research and to collect and analyze data. The bill was signed into law (PL 87-396).[8]

1963–64: 88th Congress

On August 5, 1963, the House passed a bill (HR 6997) similar to S 901, which had been pocket vetoed by the president in 1962. This bill was the result of the administration's report on oceanography, noted earlier, and was thus given the support of the administration. The bill required the president to issue a statement of national goals on oceanography and to develop a program of oceanographic activities. It also authorized the president to appoint a seven-member Advisory Committee for Oceanography that would review the existing national oceanography program and make recommendations about how to improve it. Each year, the president would be required to report to Congress on the status of oceanography, including the status of research and a financial analysis of the program. The Senate did not act on the bill in 1963.[9]

President Johnson

On March 8, 1968, Johnson gave a message to Congress on conservation, called "To Renew a Nation." In it, he noted that the oceans have potential wealth in food, minerals, and resources. He asked the secretary of state to consult with other nations on the steps that could be taken to launch a historic and unprecedented adventure—an International Decade of Ocean Exploration for the 1970s. He also asked Congress to approve a request for $5 million in the Coast Guard budget for fiscal year 1969 to improve their ocean buoys program to improve information on weather and climate. An additional $6 million was requested by the president to fund the National Sea Grant College and Program Act.[10]

1965–66: 89th Congress

Despite opposition from the administration Congress passed a bill originating in the Senate (S 944/PL 89-454) that would upgrade the existing federal oceanographic efforts and coordinate federal oceanographic programs. The Senate and House (H Rep 1025) passed different versions of the bill to provide for expanded research in the oceans and Great Lakes and to establish a National Oceanographic Council. The Senate version included a cabinet-level National Council on Marine Resources and Engineering Development in the executive office of the president. The council would be headed by the vice president and would coordinate, stimulate, and give direction to the activities of different federal oceanographic agencies.[11]

The administration supported HR 16559 (PL 89-688), the National Sea Grant College and Program Act of 1966. This also authorized a program of federal assistance to promote the development of marine resources. Under the bill, the National Science Foundation would disburse the federal monies as either grants or contracts. They would be given to institutions of higher education that had already implemented major oceanographic programs. These institutions would be considered "sea grant colleges." Other labs and organizations would also be eligible for money under certain conditions. Those institutions that received the money would need to increase their oceanographic education and research programs, train marine technologists, and establish programs to improve U.S. technology related to oceans. This law amended PL 89-454, the Marine Resources and Engineering Development Act of 1966, which was intended to upgrade and accelerate the nation's oceanographic programs.[12]

President Nixon

President Nixon outlined actions that he believed the Congress should take to reduce the risks of oil pollution. These included international conventions, international standards for ship construction and operation, the Ports and Waterways Safety Act, increased surveillance, and research into emergency oil transfer and storage systems.[13]

On February 8, 1970, Nixon gave a message to Congress on the environment. In it, he said that in the previous year,

> the Council on Environmental Quality extensively examined the problem of ocean dumping. Its study indicated that ocean dumping is not a critical problem now, but it predicted that as municipalities and industries increasingly turned to the oceans as a convenient dumping ground, a vast new influx of wastes would occur. Once this happened, it would be difficult and costly to shift to land-based disposal. Wastes dumped in the oceans have a number of harmful effects. Many are toxic to marine life, reduce populations of fish and other economic resources, jeopardize marine ecosystems, and impair aesthetic values. In most cases, feasible, economic, and more beneficial methods of disposal are available. Our national policy should be to ban unregulated ocean dumping of all wastes and to place strict limits on ocean disposal of

> harmful materials. Legislation is needed to assure that our oceans do not suffer the fate of so many of our inland waters, and to provide the authority needed to protect our coastal waters, beaches, and estuaries.

He then recommended a national policy that would stop unregulated ocean dumping of all materials. He further recommended that Congress pass legislation to require a permit from the administrator of the Environmental Protection Agency (EPA) for the dumping of any materials into the oceans, estuaries, or the Great Lakes. The EPA administrator would be authorized to ban the dumping of wastes that are dangerous to the marine ecosystem.[14]

Not long after that, Nixon issued an executive order to create the National Oceanic and Atmospheric Administration (NOAA), which he argued would help to improve our understanding of the resources of the sea.[15] In a speech on July 9, 1970, Nixon told the American public that the agency would be housed within the Department of Commerce. It required "pulling together into one agency a variety of research, monitoring, standard setting and enforcement activities now scattered through several departments and agencies." The new agency would incorporate the responsibilities of other bureaus, including the Environmental Sciences Services Administration, some portions of the Bureau of Commercial Fisheries, the Marine Minerals Technology Center, the Office of Sea Grant Programs, the National Oceanographic Data Center, and some other agencies.[16]

Later that year, Nixon again discussed the report from the Council on Environmental Quality. In a speech on ocean pollution on October 7, 1970, Nixon reported that the study concluded that the then-current level of ocean dumping was causing serious environmental damage in some areas and that the volume of wastes dumped in the ocean was increasingly rapidly. Further, a vast new influx of wastes was likely to occur as municipalities and industries turned to the oceans as a convenient sink for their garbage. The trends indicated that ocean disposal could become a major nationwide environmental problem, and unless the country began to develop alternative methods of disposing of these wastes, institutional and economic obstacles would make it extremely difficult to control ocean dumping in the future.[17] He told Congress that he would seek legislation to regulate all dumping of wastes into the oceans, the Great Lakes, and estuaries.

President Nixon took an international approach to controlling pollution in the oceans. For example, he recommended legislation that would allow the United States to regulate foreign fishing off American coasts to the fullest extent authorized by international agreements. It would also permit federal regulation of domestic fisheries in the U.S. fisheries zone and in the high seas beyond that.[18] Nixon also transmitted the Convention on the Prevention of Marine Pollution by Dumping Wastes and Other Matter to the Senate for ratification. This treaty was designed to establish a national system for regulating the disposal of wastes in the ocean.[19] He also sent to the Senate the Convention on International Trade in Endangered Species of Wild Fauna and Flora, which was written to establish a system by which states could control the international trade in specimens of species in danger of becoming extinct and monitor the trade in specimens of species that might be expected to become endangered.[20]

Before leaving office, Nixon signed HR 5451 into law (PL 93-119). This law created amendments to the 1954 International Convention for the Prevention of Pollution of the Sea by Oil. The bill, called the Oil Pollution Act Amendments of 1973, amended the Oil Pollution Act of 1961 and implemented the 1969 and 1971 amendments to the 1954 international convention. The convention was the first international agreement dealing with problems of oil pollution of the ocean caused by discharges from vessels. Discharge restrictions were extended by amendments to the convention in 1969 and 1971.

1969–70: 91st Congress

The House and Senate both passed HR 8794 (PL 91-15), which extended the National Council on Marine Resources and Engineering Development for one year. The council had been established by the Marine Resources and Engineering Development Act of 1966 to coordinate all governmental marine science activities.[21]

The Senate held hearings on S Res 33, "Principles Governing the Use of Ocean Space." One member of the Subcommittee on Ocean Space, the subcommittee that held mark-up on the bill, Claiborne Pell (D-RI), had a long-standing interest in obtaining an international treaty governing the use of oceans. Recent scientific activities and discoveries, he noted, raised questions as to whether exploration carried a preferential right to exploitation.[22]

A federal reorganization plan proposed by President Nixon was accepted by Congress in 1970. This plan, Reorganization Plan No. 3, established the EPA. In doing so, Nixon sought to consolidate all major programs to combat pollution into a single agency that would be independent of all existing departments. Under the plan, the EPA would be comprised of the Federal Water Quality Administration, certain pesticide research programs that were housed within the Department of the Interior, the National Air Pollution Control Administration, parts of the Environmental Control Administration (specifically the Bureau of Solid Waste Management, the Bureau of Water Hygiene, and part of the Bureau of Radiological Health) and the Food and Drug Administration (the sections dealing with pesticide research and standards, then in the Department of Health, Education, and Welfare), the pesticide registration authority of the Department of Agriculture, the standard-setting functions of the Atomic Energy Commission pertaining to radiation protection, and some research authority of the Council on Environmental Quality. Also included would be all of the functions of the Federal Radiation Council, which advised the president on radiation matters affecting health, assisted federal agencies to formulate radiation standards, and helped states establish programs concerning radiation.

Nixon explained that the reorganization was needed because there were overlapping or closely related responsibilities between many agencies. He explained that the EPA would have an estimated budget of $1.4 billion in fiscal year 1971 and approximately 6,000 employees. As he envisioned the agency, it would have responsibility for research and standard setting for pollution emissions, formulation of a coordinated policy for pollution control across levels of government, recognition of new environmental problems as they emerged and development of new programs to meet them, integration of pollution control and enforcement, coordination with other governments, and clarification of the responsibilities of private industry to prevent pollution.

Another federal reorganization plan proposed by President Nixon was accepted by Congress in 1970. Reorganization Plan No. 4 established the National Oceanic and Atmospheric Agency (NOAA), to be incorporated into the Department of Commerce. The new agency would take over the responsibilities of the Environmental Science Services Administration (in the Department of Commerce); the Bureau of Mines's Marine Minerals Technology Program, most of the programs of the Bureau of Commercial Fisheries, and the marine sports fishing program; parts of the Army's U.S. Lake Survey; the Navy's National Oceanographic Data and Instrumentation Center; and the national data buoy program of the Department of Transportation.[23]

1971–72: 92nd Congress

The House members were the first to debate a bill to create a federal permit program for the killing of marine mammals.[24] The bill was passed during the second year of the session (HR 10420/PL 92-522). The final version set a permanent moratorium on most killing of ocean mammals and on the importation of their products.[25] The bill generated controversy among environmentalists, as some endorsed complete protection for sea mammals and others preferred a more scientific program of wildlife management. The final bill was a compromise that emphasized protection of animals but also granted several exceptions. For example, under the new law, permits could be granted for the killing or capture of ocean mammals for scientific research or for public display in zoos or commercial marine exhibits. The most controversial exemption was granted to the commercial fishing industry, primarily tuna fishing, in which thousands of porpoises were inadvertently killed each year when they became tangled in tuna nets. The industry was exempt from the moratorium for two years while a plan was developed that would reduce porpoise deaths.[26]

A bill to control the dumping of waste materials in the oceans and coastal waters became law in 1972 (HR 9727/PL 92-532). Called the Marine Protection, Research and Sanctuaries Act of 1972, the new law called for a continuing program of research into the effects of pollution on ocean ecosystems. The final bill expressed the congressional finding that unregulated dumping of wastes in the ocean endangered human health and welfare, the marine environment, and the oceans' economic potential. It also declared it national policy to regulate ocean dumping to prevent adverse effects on humans or the marine ecosystem by controlling the transportation and dumping of materials in the territorial waters, or other waters within U.S. control under international law.[27]

In a similar vein, HR 8140 (PL 92-340) authorized standards and regulations to prevent oil spills and other accidents in the nation's ports and waterways.[28] The new legislation gave the secretary of transportation the ability to establish regulations for the design, construction, maintenance, and operation of vessels in order to protect the marine environment. Congress set January 1, 1976, as the latest date for which the initial design and construction standards would take effect for vessels in foreign trade or foreign registry. The secretary of transportation was required to establish and operate marine traffic systems for waterways that were congested and to prescribe safety equipment and procedures for docks and other structures. Last, the bill provided a civil penalty of up to $10,000, as well as criminal penalties of between $5,000 and $50,000, five years imprisonment, or both, for violations of the act.[29]

In 1972 the House passed HR 15627 to amend the Oil Pollution Act of 1961 (PL 87-167) and to implement the 1969 amendments to the 1954

International Convention for the Prevention of the Pollution of the Sea by Oil. The purpose of the bill, known as the Oil Pollution Act Amendments of 1972, was to expand the penalties of the 1961 act. The bill would extend the idea of a total prohibition of oil pollution (subject to certain exemptions) throughout the oceans and eliminate any "free zones." However, an exception would be granted to oil tankers that limited oil-containing discharges to areas more than 50 miles from the nearest land, with a discharge rate of not more than 60 liters per nautical mile while en route, and a maximum discharge of 1/15,000 of the total cargo capacity. For any vessels other than tankers, the bill limited discharges to areas as far as practicable from land, to not more than 60 liters per nautical mile en route, and to discharges containing not more than 100 parts per million of oil. It would abolish the exemption for discharge of residue arising from purification or clarification of fuel or lubricating oil. It would provide for a simplified oil record book for tankers and other vessels, establish criminal penalties for willful oil discharges of up to $10,000 per violation or one year in prison or both, and establish civil penalties of up to $10,000 for discharges and up to $5,000 for other violations.[30] The Senate did not act on the bill, and it went no further in this session.

The Senate Commerce Committee reported a bill (S 582) that would create a grant program for states to develop and operate coastal zone management programs. Under the proposed program, states along the Atlantic and Pacific coasts, the Gulf of Mexico, Long Island Sound, and the Great Lakes would be permitted to establish coastal zone management programs and to acquire and operate estuarine sanctuaries programs.[31] This did not pass.

Two international treaties dealing with oil pollution were ratified by the Senate during this session. One was the Convention Relating to Intervention on the High Seas in Cases of Oil Pollution Casualties, and the other was the amendments to the 1954 International Convention for the Prevention of Pollution of the Sea by Oil. The first established the right of a coastal nation to take action against the threat of oil pollution after a maritime accident such as a collision at sea. The second, the amendments to the 1954 convention, altered existing international law governing the intentional discharge of oil and specified a rate-of-discharge formula to limit the amount of oil a ship was legally permitted to discharge into the sea.[32]

1973–74: 93rd Congress

As signed by the president, HR 5451 amended the Oil Pollution Act in three ways. First, it extended the principle of "prohibited zones" where oil discharge was restricted. After the bill was passed, the prohibited zones were expanded to cover entire oceans instead of areas within 50 miles of the nearest land. Oil discharge would be permitted subject to specified rates and amounts of discharge. Second, the new bill provided special protection for the Great Barrier Reef off the northeastern coast of Australia. The Great Barrier Reef would be regarded as land in order to prohibit tanker discharges of oil within 50 miles of it. Third, the bill established construction standards for new large tankers to restrict the placement of storage tanks; the purpose was to limit the maximum amount of oil spill that could result from damage sustained by a tanker in a collision or stranding.[33]

A related bill was HR 738, the Oil Pollution Act Amendments. This bill clarified the definition of actions covered under the Oil Pollution Act of 1961, and the proposed law provided construction requirements for ships built in the United States. A penalty of a $10,000 fine and/or up to one year imprisonment per violation was also specified for those who discharged oil or oily mixtures from a ship. In the end, the bill did not have enough support to pass. Another proposal to increase the penalties for discharging oil and hazardous substances was found in HR 15475, which also did not pass.

The president signed S 1070 (PL 93-248) to implement the International Convention Relating to Intervention on the High Seas in Cases of Oil Pollution Casualties of 1969. This legislation, formally titled the Intervention on the High Seas Act, authorized the secretary of the treasury, where the Coast Guard was housed, to take measures to prevent, mitigate, or eliminate any imminent dangers to the U.S. coastline from oil pollution or the threat of oil pollution caused by ocean shipping. In an emergency situation, the secretary would be authorized to destroy any ship and cargo that severely endangered or was damaging the coastline.[34]

President Ford

On July 9, 1975, Ford gave a message to Congress on oil pollution liability and compensation. In it, he proposed new legislation entitled the Comprehensive Oil Pollution Liability and Compensation Act of 1975, which would establish a uniform system for fixing liability and settling claims for oil pollution damages in U.S. waters and along the coastlines. The proposal would also implement two international conventions dealing with oil pollution caused by tankers on the high seas. He told Congress that the legislation would help protect the environment by establishing strict liability for all oil pollution damages from identifiable sources and providing strong economic incentives for operators to prevent spills. Equally important, according to President Ford, the bill would provide relief for many oil-related environmental damages that in the past went uncompensated. He explained that his proposed legislation would replace the current patchwork of overlapping and sometimes conflicting federal and state laws. In addition to defining liability for oil spills, the proposed bill would establish a uniform system for settling claims and assure that none would go uncompensated.[35]

President Ford signed HR 15540 (PL 93-472), which extended a law banning the dumping of wastes into the ocean and coastal waters for one year. The original law, called the Marine Protection, Research, and Sanctuaries Act of 1972, prohibited the dumping of hazardous wastes within a 12-mile limit anywhere along the coast of the United States. The new law also authorized a program of research on the effects of ocean dumping on marine life.[36]

1975–76: 94th Congress

President Ford signed a bill (HR 5710/PL 94-62) that amended the 1972 Marine Protection, Research, and Sanctuaries Act. The new law, formally called the Port and Tanker Safety Act, appropriated money for the ocean dumping permit program, for research concerning the impact of ocean dumping on the marine environment, and for the designation and establishment of certain ocean and coastal waters as marine sanctuaries.[37]

President Carter

In 1977, Carter gave a speech to Congress concerning oil pollution. In it, he said, "Pollution of the oceans by oil is a global problem requiring global solutions. I intend to communicate directly with the leaders of a number of major maritime nations to solicit their support for international action. Oil pollution is also a serious domestic problem requiring prompt and effective action by the federal government to reduce the danger to American lives, the American economy, and American beaches and shorelines." He proposed legislation in which he suggested corrective action to protect against oil spills from foreign vessels. That included requiring insurance coverage for vessels that come in and out of U.S. waters, working with other countries to require new tankers to have double bottoms and other construction features, and requiring the Coast Guard to inspect every single oil tanker that comes into U.S. ports to determine if it complies with standards.[38] The legislation would also require improved crew standards and training, and it would improve the federal ability to respond to oil pollution emergencies.[39]

Carter recommended ratification of the International Convention for the Prevention of Pollution from Ships. The treaty addressed problems related to protecting the marine environment by reforming ship construction and equipment standards, improving crew standards and training, developing a tanker boarding program and U.S. Marine Safety Information System, approving comprehensive oil pollution liability and compensation legislation, and improving the federal ability to respond to oil pollution emergencies.[40] Carter then sent Congress legislation to establish a single, national standard of strict liability for oil spills that was designed to replace the then-current fragmented, overlapping systems of federal and state liability laws and compensation funds. It would also create a fund totaling $200 million that would help pay for cleaning up oil spills and would compensate victims of oil pollution damages.[41]

1977–78: 95th Congress

Despite a congressional outcry at the beginning of the year, caused by several oil tanker accidents, Congress did not complete action on a bill to protect against the oil spill pollution problem. HR 6803 would have increased the liability of those responsible for oil spills, but it died during the final days of the session because of a disagreement over whether hazardous chemical spills should be covered by the bill. The oil industry opposed the bill.[42]

Conversely, S 682, also dealing with oil spills, was passed into law (PL 95-474). The bill set tougher safety standards for both domestic and foreign tankers operating in U.S. waters. It was aimed at reducing the chances of oil spills by mandating better control of ship traffic, and it required tankers to install electronic gear to prevent accidents. The bill also sought to decrease the discharge of oil during routine tanker operations. The bill authorized the secretary of transportation to require a federally licensed pilot on each vessel operating in U.S. waters when state law required one. The new law expanded the federal government's authority to investigate accidents at sea. It also prohibited any vessel carrying oil or other hazardous material from operating in U.S. waters or transferring cargo in any U.S. port if the vessel had a history of accidents, failed to comply with U.S. laws and regulations, did not meet U.S. manning requirements, or did not have at least one licensed deck officer who clearly understood English.

Another new law (S 1522/HR 10730/PL 95-136) extended the 1972 Marine Protection Act.[43] It authorized money to carry out the act during fiscal years 1979–81. The funding would be used to protect marine mammals and to fund new research projects. It also set a moratorium on most killing of ocean mammals and on the importation of marine mammal products. The new ban applied to seals, sea lions, whales, porpoises, dolphins, sea otters, polar bears, manatees, and walruses.[44]

Changes to the 1972 Marine Protection Act also came as part of the fiscal year 1978 authorization bill (HR 4297/PL 95-153). In this new bill, Congress prohibited ocean dumping of most leftovers from municipal sewage treatment plants. The dumping was to stop by December 31, 1981. The dumping of sludge loaded ocean waters with potentially lethal metals like cadmium and mercury and other harmful materials.[45] Before the session ended, Congress cleared legislation establishing a federal ocean pollution research program after citing the need to study the long-range impact of chemical pollutants on marine life. The bill (S 1617/PL 95-273), called the Ocean Pollution Research Program Act, established a five-year ocean pollution research plan administered by NOAA.[46]

In HR 3350, the House debated the idea of governing industrial mining on the ocean floor. The bill would require that those mining the seabed have government licenses or permits and that they abide by government regulations that were designed to protect the environment. U.S. mining companies would have to pay taxes on mining profits even though the minerals would come from international waters. The companies would have to turn some of their profits over to an international fund that eventually would be shared with other countries for mining operations.[47] The Senate did not act on the bill.

1979–80: 96th Congress

Three bills related to pollution of the oceans passed this session. The first, S 1148 (PL 96-572), reauthorized Title 1 of the Marine Protection, Research, and Sanctuaries Act (federal ocean dumping regulations) and prohibited the dumping of industrial wastes into the ocean after 1981.[48] The second bill was S 1123 (PL 96-381), which provided money for research on the effects of pollution on marine life.[49] The third, S 1140 (PL 96-332), provided money to designate marine sanctuaries to protect valuable offshore resources such as fishing grounds.[50]

The House passed HR 5338 to compensate U.S. victims of the Campeche Bay oil blowout off Mexico on June 3, 1979. In the accident, an estimated

140 million barrels of oil spewed from a Mexican oil rig in the Gulf of Mexico before it was brought under control. Under the legislation, up to $80 million would have been made available to victims. Fishermen, resort owners, and charter boat operators who suffered economic losses from the accident could apply to the federal government for compensation. The bill included a provision to allow the government to sue Mexico for private losses that resulted from the spill. Since the Mexican government owned the rig, the United States contended that Mexico was responsible for damages. No action was taken on the proposal in the Senate.[51]

President Reagan

Upon signing the Medical Waste Tracking Act in 1988, President Reagan promised to ensure that those who generated, handled, or disposed of medical waste were held accountable. While he did so, he noted his objections to the bill. For example, he noted that the law permitted state courts to exercise jurisdiction over federal agencies, a power he hoped would be used only on rare occasions.[52] That same year, the president also chose to sign a bill that would terminate the dumping of sewage, sludge, and industrial waste into the oceans. Together, these bills would lead to a cleaner and safer environment, according to Reagan.[53]

1981–82: 97th Congress

President Reagan signed S 1213 (PL 97-16), which authorized money for fiscal year 1982 for the Marine Protection, Research, and Sanctuaries Act. The law doubled the funds available for identifying possible sites for the ocean dumping of dredged material, sewage sludge, and industrial wastes.[54] Ocean dumping was also the topic of a House bill (HR 6113). This would reauthorize and strengthen the law regulating the ocean dumping of dredged materials.[55] It would have imposed a two-year moratorium on any dumping of low-level radioactive wastes into the ocean, but it was not approved by Senate members.[56]

In the House, HR 5906 was a proposal to clarify the limits on liability for oil spills from offshore drilling. The bill amended Title III of the Outer Continental Shelf Lands Act Amendments of 1978 (PL 95-372), which set up an Offshore Oil Pollution Compensation Fund financed by a tax on offshore oil. Persons economically damaged by offshore spills could make claims against the fund, which in turn could sue those responsible for the spill. The 1976 law set a limit of $35 million on damages that could be obtained from operators of offshore oil rigs, but it left them with unlimited liability for the costs of cleaning up the spills. The House bill changed that to a total liability limit of $75 million for both damages and cleanup costs. The Senate did not act on the proposal.[57]

Congress cleared S 1003 (PL 97-109) to continue a federal program for designating environmentally sensitive offshore areas as marine sanctuaries, where oil and resource development was strictly controlled for two years.[58] HR 6324 was a bill to reauthorize the two-year atmospheric, climatic, and ocean pollution control activities of the NOAA. It didn't clear Congress.[59]

1983–84: 98th Congress

The Congress took no final action on HR 1761, passed by the House, to tighten rules about ocean dumping of wastes. In the Marine Protection, Research, and Sanctuaries Act of 1972, Title 1 (the Ocean Dumping Act) authorized the EPA to issue permits for the dumping of waste materials into the ocean. The bill was designed to speed up the study and designation of dump sites by the EPA.[60] No action was taken in the Senate, and it did not become law.[61]

Another new law, HR 4997 (PL 98-364), called the Commercial Fishing Industry Vessel Act, reauthorized the Marine Mammal Protection Act. In doing so, Congress authorized funding to keep populations of marine mammals from being depleted and froze the animal limit on porpoises that could be accidentally taken by U.S. fishermen. The law banned the importation of fish products from nations whose standards were less protective and renewed fishing agreements with the Soviet Union and Poland.[62]

The House passed HR 2062 authorizing funding for the national marine sanctuaries program for three years. The marine sanctuaries program, created under Title III of the 1972 Marine Protection, Research, and Sanctuaries Act (PL 92-532), permitted the secretary of commerce to designate marine sanctuaries in coastal waters and control activities in them to protect the environment. Neither this bill nor its companion Senate bill, S 1102, made it to the full Senate.[63]

President Reagan pocket vetoed a bill authorizing programs for the NOAA for fiscal years 1984–86. The bill would have authorized $852 million in fiscal year 1984, $929 million in fiscal year 1985, and $258 million in fiscal year 1986. In vetoing the bill, Reagan argued that several provisions of the NOAA authorization would have hampered the agency's management ability.[64]

1985–86: 99th Congress

The House and Senate both passed oil spill liability bills in 1986 (S 2799/HR 2005), but conferees were unable to resolve differences, and the legislation died. The two chambers agreed to limit the liability of oil companies and shippers and to impose a tax on petroleum to create a $350 million fund to cover cleanup costs and damages that exceeded those liability limits, but they were unable to agree on whether federal law should preempt stricter state laws.[65]

HR 1957 passed the House but not the Senate. It would have reauthorized and amended Title I of the Marine Protection, Research, and Sanctuaries Act. Title I was known as the Ocean Dumping Act, and it gave the EPA authority to regulate ocean dumping of wastes and prohibited the dumping of any wastes the EPA considered harmful.[66]

1987–88: 100th Congress

The Medical Waste Tracking Act of 1988 (HR 3515/PL 100-582) was passed after used syringes and other medical waste washed up on beaches in New York and New Jersey.[67] In the new law, Congress attempted to control dumping of medical wastes. They set up a two-year demonstration program for tracking and handling medical wastes in New York, New Jersey, Connecticut, and the Great Lakes states and forced the EPA to issue rules requiring medical waste to be separated from conventional garbage and to be properly packaged and labeled. It also increased the EPA's enforcement powers over violators.[68] Top civil penalties were $25,000 per day of violation, and normal criminal penalties were $50,000 per day and two years in jail. In cases where the dumper knowingly endangered the health of another person, those penalties escalated to a $250,000 fine and 15 years in jail, or a $1 million fine if the dumper was an organization.[69] Reagan signed the bill because it would help to ensure that those who generated, handled, or disposed of medical waste were held accountable for properly disposing of their potentially dangerous waste.[70]

Members of Congress also passed legislation (HR 3674/PL 100-220) that banned the dumping of plastics at sea by any vessel within 200 miles of the U.S. coast and by U.S.-flag vessels anywhere in the world. The bill, called the United States-Japan Fishery Agreement Approval Act also banned the dumping of certain other garbage within 12 miles of the U.S. coast.[71]

A conference version of another bill (S 2030/PL 100-688), the Ocean Dumping Ban Act, was intended to end ocean dumping of sewage and industrial waste was eventually signed by Reagan and became law. The bill was aimed at ending all dumping of U.S. sewage sludge into the oceans after 1991. Such dumping was originally supposed to stop at the end of 1981 under a law passed in 1977. But New York City had contended that it was unable to find any place on land to put its sludge, so it continued to dump sludge from barges. New Jersey complained. Under the conference agreement, fees on the ocean disposal of sewage sludge started at $100 per dry ton in 1989 and escalated to $200 per ton by 1991. Most fees could be waived for a municipality that entered into an enforceable legal agreement with a schedule for phasing out the dumping. Most of the fees and penalties from each municipality went into a special trust fund, which could be spent on developing a land-based disposal facility for that particular municipality.

The Plastic Pollution Control Act (S 1986/PL 100-556) was a proposal to allow for the study, control, and reduction of the pollution of aquatic environments from plastic materials. The bill passed in the House in lieu of HR 5117. Title I of the act revolved around biodegradable plastic ring carriers and directed the administrator of the EPA to require that plastic ring carriers (such as those on canned beverages) be made of naturally degradable material, because fish and wildlife often became entangled in them. Title II of the law amended the federal law that established the San Francisco Bay National Wildlife Refuge. It would allow for expansion of the boundaries of the refuge to include additional lands.[72]

President G.H.W. Bush

Bush proposed legislation to increase penalties for those who dump waste illegally in our oceans. The legislation called for tough criminal felony sanctions against illegal dumpers.[73] He also supported and signed the Oil Pollution Act of 1990, a bill to prevent, respond to, and pay for oil spills. However, in signing the new law, Bush noted that he was unhappy with a provision that placed a moratorium on exploration for oil and natural gas off the coast of North Carolina.[74] He sent the Senate an international convention on oil pollution that would establish a global framework for cooperation among nations to prepare for combating future spills.[75]

1989–90: 101st Congress

In 1989, the Congress passed the Oil Pollution Prevention, Response, Liability, and Compensation Act (HR 1465/S 686/PL 101-380), largely in response to the *Exxon Valdez* accident. The bill set up a federal system of liability caps on tankers, barges, offshore platforms, deepwater ports, and other oil facilities. Cleanup and compensation costs beyond those paid by the spillers would come out of a fund created by taxing domestic and imported oil. The bills would also set various standards and requirements to try and prevent spills. The legislation would also provide for the compensation of damages to natural resources by requiring payment for restoring or replacing damaged resources, plus the cost of lost use of those resources before they were restored or replaced.

Any victims who suffered damages as a result of a spill would be allowed to seek compensation directly from the fund if those responsible for the spill had already reached their liability limit or refused to settle a claim for damages within 90 days. Under the law, the president would be required to ensure that spills were cleaned up. He would be permitted to demand that either federal agencies or private companies implement cleanup activities. The president could require that agencies create local oil spill contingency plans if they were located in an area where a spill was likely to occur. Tankers and oil facilities would also be required to have contingency plans regardless of their location. Any equipment designed for oil spill cleanup would be inspected periodically, and spill-response practice drills would be held in ports and other areas that could be affected by a spill. There would be at least seven regional federal oil spill strike teams in oil spill zones, probably including Alaska, Hawaii, the Pacific coast, the Gulf of Mexico, the South Atlantic coast, the North Atlantic coast, and the Great Lakes.[76]

In HR 1668 (PL 101-224) Congress reauthorized NOAA's Ocean and Coastal Programs Act in 1989, providing funding increases for most of NOAA's activities.[77] The House also considered a bill (HR 4333) aimed at providing the public with better information about beach water pollution; it passed the House. Under the proposal, the EPA would be required to develop criteria for beach water quality and to require states to test beach water and post signs if the water violated acceptable standards. The Senate did not pass the bill.[78] A third bill introduced in the House was a bill to declare the Florida Keys a national marine sanctuary (HR 5909/PL 101-605). Called the Florida Keys National Marine Sanctuary and Protection Act, the law protected the Keys from oil drilling and from certain ships that could disrupt marine life and fragile coral reefs that are located in the Keys. The bill also required that the Department of Commerce develop a comprehensive management plan and a water quality program intended for the Keys.[79]

1991–92: 102nd Congress

In 1991, the House gave voice-vote approval to a bill (The National Oceanographic and Atmospheric Administration Authorization Act, or HR 2130) that renewed weather and coastal research programs at NOAA for two years. The companion bill in the Senate did not make it to the floor. The bill would have authorized ongoing activities for two years that included establishing a national coastal pollution monitoring program. The Senate passed the bill in 1992, and it was signed by President Bush (PL 102-567).[80]

HR 12 was a measure aimed at developing and implementing quality criteria for coastal recreation waters. The proposal required the EPA to set criteria for beach water quality and mandated that states test their coastal waters. States not meeting new minimum standards were to post signs along their beaches. It was passed by the Merchant Marine and Fisheries Committee but did not proceed to the floor.[81]

President Clinton

President Clinton often spoke about maintaining the nation's oceans and stopping the destruction of wetlands.[82] He supported and signed the Ocean Pollution Reduction Act (HR 5176/PL 103-431) and proposed increasing federal efforts to enhance the health of the oceans with increased funding. Realizing the harm that offshore drilling and oil spills can do to oceans and marine sanctuaries, he promised to sign a directive to extend the nation's moratorium on offshore leasing.[83] Clinton also worked with other countries, particularly Japan and Russia, to halt ocean dumping and urged the Senate to ratify the Convention on the Law of the Sea.[84]

1993–94: 103rd Congress

Congress was able to pass S 1636 (PL 103-238), the Marine Mammal Protection Act Reauthorization. The bill was created to reduce accidental killing of marine mammals without harming the economic well-being of commercial fishermen. The final bill was hailed as a compromise between environmentalists and the fishing industry.[85] The House passed a bill (HR 4008) to reauthorize the ocean, coastal, and fisheries programs of the NOAA for two years. The proposal was for $444 million in fiscal year 1995 and $463 million in fiscal year 1996. The Senate took no action on it, and it died at the end of the session.

The House members also considered the Ocean Radioactive Dumping Ban Act (HR 3982), aimed at coordinating U.S. laws with a previous international agreement according to which the dumping of all radioactive waste in the ocean would be stopped. The Senate did not consider the measure, and it went no further. The bill was an amendment to the 1972 Marine Protection, Research, and Sanctuaries Act (PL 92-532), which was enacted to bring U.S. law into conformity with the London Convention, an international agreement that banned ocean dumping of high-level radioactive waste. The 1972 law allowed the U.S. government to dump low-level radioactive waste into the ocean with specific congressional approval of each instance. However, the United States had largely stopped dumping any radioactive waste into the oceans by 1970 and had never used the authority granted to it under the 1972 ocean dumping law. In 1993, the London Convention was expanded to include low-level waste, following the revelation that Russia had dumped waste into the waters and had been storing large amounts of high- and low-level radioactive waste in the ocean. The new treaty took effect on February 20, 1994. At first, the U.S. government resisted signing the expanded London Convention accord, but the Clinton administration reversed its stance in 1993, under pressure from environmental groups, and supported the ban's expansion.[86]

One bill that was passed into law during this session was HR 5176, the Ocean Pollution Reduction Act. This new law (PL 103-431) amended the Water Pollution Control Act to allow the city of San Diego, California, to apply for a modification of the secondary treatment requirements. However, under the law, officials in that city would need to implement a wastewater reclamation program instead.

1995–96: 104th Congress

The House passed an omnibus science authorization bill that was designed to refocus federal scientific research. The bill (HR 2405) would have brought various science programs under a single authorization bill, but it did not pass. The Coastal Zone Protection Act (HR 1965/PL 104-150) was a new law in which Congress agreed to extend through fiscal year 2000 a program of grants to states that had federally approved management plans for their coastal regions. The bill authorized $289 million for activities under the 1972 Coastal Zone Management Act.[87]

He supported and signed the Ocean Pollution Reduction Act (HR 5176/PL 103-431, and proposed to increase efforts to enhance the health of the oceans by providing $224 million. Realizing the harm that offshore drilling and oil spills can do to oceans and marine sanctuaries, he promised to sign a directive to extend the nation's moratorium on offshore leasing.[88]

1997–98: 105th Congress

The Science and Commerce committees could not agree on a bill (HR 1278) to reauthorize activities of the NOAA. Since they could not come to any agreement, no bill was sent to the president for his approval.

1999–2000: 106th Congress

In this session, members of Congress introduced HR 3310, the United States-Mexico Border Sewage Cleanup Act of 1999. The bill, which was not passed into law, proposed a procedure for the secondary treatment of waste from a treatment facility in Mexico. The law was proposed because the waste from the treatment plant was flowing into U.S. surface waters and impacting the health and safety of people in that area.

President G. W. Bush

During his term as president, Bush sent to the Senate a protocol that contained a global agreement to control pollution from ships. Among other things, this agreement included provisions to regulate the transport of oil and harmful substances.[89] Two weeks before Bush left office in January 2009, he sought to bolster his environmental record by establishing three marine national monuments in the Pacific Ocean. Taken together, they are intended to protect 195,555 square miles of sea—an area the size of Spain—from commercial fishing, mining, and other development.[90]

2001–02: 107th Congress

The North American Wetlands Conservation Reauthorization Act Amendments (HR 3908/PL 107-308) revised and updated the original law. It revised the definition of a "wetlands conservation project" and authorized additional appropriations to carry out the act in fiscal years 2003 through 2007.

2005–06: 109th Congress

Four months after pipeline corrosion caused the shutdown of the nation's largest oil field, Congress passed a bill to help prevent similar occurrences in the future. Called the Pipeline Inspection, Protection, Enforcement, and Safety Act, the bill became public law when President Bush signed the measure (HR 5782/PL 109-468). Among other things, the new law provided that a person who is engaging in demolition, excavation, tunneling, or construction in an area where there is a natural gas or hazardous liquid pipeline is prohibited from engaging in that activity without first using the state's notification system to establish the exact location of the pipeline. That person cannot engage in that activity in disregard of location information or markings placed by a pipeline operator. If damage does occur, the operator must report that damage immediately. Violators could be subject to a civil penalty of up to $1 million.[91]

Four other bills were passed in the second year of the 109th Congress. The first of those was HR 3552 (PL 109-226), the Coastal Barrier Resources Reauthorization Act. This reauthorized the Coastal Barrier Resources Act for fiscal years 2006–07. The second bill was S 260, the Partners for Fish and Wildlife Act (PL 109-294), which authorized the secretary of the interior to provide technical and financial assistance to private landowners to restore, enhance, and manage private land to improve fish and wildlife habitats through the Partners for Fish and Wildlife program. Third, S 2430 was signed into law (PL 109-326) to amend the Great Lakes Fish and Wildlife Restoration Act of 1990 to implement recommendations of the U.S. Fish and Wildlife Service contained in the Great Lakes Fishery Resources Restoration Study. The last law was S 3692 (PL 109-449), to establish a program within NOAA and the U.S. Coast Guard to help identify, determine sources of, assess, reduce, and prevent marine debris and its adverse impacts on the marine environment and navigation safety, in coordination with nonfederal entities. The formal name of this bill was the Marine Debris Research, Prevention and Reduction Act.

2007–08: 110th Congress

HR 2854, the New Jersey/New York Clean Ocean Zone Act of 2007, was a proposal to designate an area in those two states as the Clean Ocean Zone and then prohibit the EPA administrator or others from issuing permits to allow ocean dumping in that area. The bill would also prohibit underwater research or exploration in the zone unless it met certain requirements. This did not pass.

Another bill in the House was HR 21, known as the Oceans Conservation, Education, and National Strategy for the 21st Century Act. The proposal would have established a national policy to protect, maintain, and restore the health of marine ecosystems. It would also provide for more stringent federal and state requirements concerning marine ecosystems. The proposal would reestablish NOAA and create an undersecretary of commerce for oceans and atmosphere, who would be the administrator of NOAA. Along with that, other positions were established in the proposal, including a national oceans advisor, a Committee on Ocean Policy, and the Council of Advisors on Oceans Policy. An Ocean and Great Lakes Conservation Trust Fund would be established within the Treasury that would provide payments to coastal states for the development and implementation of regional ocean strategic plans. Finally, the U.S. Postal Service would be required to provide a special postage stamp to afford the public a convenient way to support efforts to protect marine ecosystems. This bill was not acted on by the full House. A similar proposal was HR 250, the National Oceanic and Atmospheric Administration Act. In it, the NOAA would be reestablished within the Department of Commerce and be headed by the undersecretary of commerce for oceans and atmosphere.

The House, in HR Res 186, made a statement to support the goals and ideals of National Clean Beaches Week and recognize the value of American beaches and their role in American culture. In doing so, the House members formally recognized the value of beaches to the American culture and the important contributions of beaches to the economy, recreation, and natural environment of the United States. The House statement encouraged Americans to keep beaches safe and clean for the continued enjoyment of the public.

President Obama

President Barack Obama understood the importance of the oceans, saying that they are critical to supporting life. In 2009, he stressed his policy toward the oceans and described his administration's ocean policy as incorporating "ecosystem-based science and management and emphasiz[ing] our public stewardship responsibilities."[92] He said that he was committed to protecting ocean resources and ensuring accountability for actions that affect them. The president proclaimed June 2009 and June 2010 as National Oceans Months to help Americans learn more about the oceans and conservation of them.[93]

In 2009, Obama formed the Interagency Ocean Policy Task Force that was to be led by the chair of the Council on Environmental Quality. The task force would develop recommendations for a national policy to protect and maintain the health of oceans and other bodies of water such as the Great Lakes.[94] The task force made its recommendations in 2010, and Obama relied on those recommendations to issue an executive order to establish a National Ocean Council that will ensure that the nation's policies seek to protect, maintain, and restore the oceans, coastal regions, and Great Lakes ecosystems.[95]

2009–10: 111th Congress

Congress continued working to keep oceans safe and clean throughout the 111th Congress. The House membership considered a bill (HR 14) known as the Federal Ocean Acidification Research and Monitoring Act of 2009. If it passed, this bill would have required the Joint Subcommittee on Ocean Science and Technology of the National Science and Technology Council to develop a plan for federal ocean acidification research and monitoring and to assess the impact of ocean acidification on marine organisms and ecosystems. The director of the National Science Foundation would continue to carry out acidification research. This did not become law.

In HR 300, the National Oceanic and Atmospheric Administration Act, the House members worked on a proposal to reestablish NOAA in the Department of Commerce. It would be headed by an undersecretary of commerce for oceans and atmosphere. This did not pass before the end of the session.

A program to explore the ocean was proposed in HR 366, the Ocean Research and Exploration Enhancement Act of 2009. The bill would require the administrator of NOAA to establish a national ocean exploration program that would promote collaboration with other federal ocean and undersea research and exploration programs. The administrator would also convene a task force on ocean exploration and undersea research technology and infrastructure, as well as appoint an ocean exploration advisory board. This did not pass and was not made into law.

Finally, HR 843 was a proposal to amend the Marine Mammal Protection Act of 1972 to repeal the goal of reducing the incidental mortality and serious injury of marine mammals in commercial fishing operations to zero and to replace it with a goal of reducing such incidental mortality and serious injury instead. This was another proposal that did not pass.

Conclusion

It is obvious that Congress has considered and passed many laws intended to keep the oceans safe for both marine life and human use. A clean ocean is essential to maintain marine animals and permit recreational use of beach areas by humans. Congress has debated many proposals to limit ocean dumping, protect marine habitats, and even create funds to help pay for cleaning up spills or other disasters. Protecting the oceans will continue to be high on the congressional agenda in the years to come as elected representatives listen to the public's concern about clean oceans.

Chapter 5

The Politics of Pesticides

In the 1970s and 1980s, many events pointed to the possible dangers of pesticides to the environment and to human health. In 1975, a spill of the pesticide Kepone into the James River in Virginia caused thousands of fish to die. High levels of polychlorinated biphenyls (PCBs) were discovered in other rivers, including the Hudson River. In 1989, the pesticide Alar, commonly used on apples, was suspected of causing cancer, especially for young children. After a serious of congressional hearings, even the Environmental Protection Agency (EPA) stressed the potential risks to children. The sales of apples and apple juice plummeted, and the manufacturer was forced to take Alar off the market.[1] These events triggered pressures on legislators to take action to protect the environment and people from harm resulting from pesticide use and overuse. Congressional members reacted to these concerns and passed legislation to reduce the dangers of pesticides. That action is reviewed in this chapter.

The first federal action to regulate pesticides was the Federal Insecticide Act of 1910. This law was passed after the Department of Agriculture raised concerns about the sale of fraudulent products. This law helped to ensure the quality of pesticides by setting standards for the manufacture of chemicals, but it also required inspections, seizure of substandard products, and penalties for violators.

1947–48: 80th Congress

In the years after World War II, synthetic pesticides were developed and widely used, leading to concerns about their potential harm to humans. In 1948, the Congress passed the Federal Insecticide, Fungicide, and Rodenticide Act (FIFRA), which gave the Department of Agriculture the responsibility for regulating pesticides. The law required that all pesticides used in interstate commerce be registered with the Department of Agriculture before they could be sold or used.[2] It also required that labels should include specific information, including directions for proper use and warning statements for users.

1957–58: 85th Congress

In 1958, when the Pesticide Research Act (PL 85-582) authorized the Department of the Interior to conduct research on the effects of pesticides on fish and wildlife, the U.S. Fish and Wildlife Service of the department began programs to study the transport of pesticides, their distribution geographically, their decomposition, and their transfer from one organism to another via the food chain.

1959–60: 86th Congress

FIFRA was amended in 1959. This altered the definition of a pesticide or "economic poison" to include nematicides, plant regulators, defoliants, and desiccants.

1963–64: 88th Congress

Partly because of Rachel Carson's book *Silent Spring*, concern over the possible long-range effects of chemicals and pesticides on humans and animals helped to increase public pressure on Congress for new pesticide controls. Congress held extensive hearings, increased appropriations for different programs, and passed new legislation. One new law (S 1605/PL 88-305) amended the 1947 FIFRA to require the labeling of federally approved pesticides with registration numbers and to repeal provisions permitting "protest registration" and marketing of pesticides declared unsafe by the Department of Agriculture. The law mandated that each pesticide product have a federal registration number and that labels contain signal words relating to the toxicity of the products, including *warning*, *caution*, or *danger*. Additionally, the secretary of agriculture was authorized to suspend the registrations of any pesticides believed to pose an imminent hazard to the public. The secretary could refuse registration to pesticides that were unsafe or ineffective and remove them from the market.

Another proposal (HR 4487/S 1251) expanded federal funds for research. Although this did not pass, the 1965 agriculture appropriation bill (PL 88-537) included $22.5 million that was requested by the administration for an intensified program to reduce the need for pesticides.[3]

1967–68: 90th Congress

In June 1968 the research program of the Department of the Interior was extended for three years with increased funding for FIFRA.

President Nixon

On February 8, 1970, Nixon gave a message to Congress on the environment in which he spoke about his desire to strengthen the controls on pesticides through comprehensive improvement in pesticide control authority.[4] He said that

> the use and misuse of pesticides has become one of the major concerns of all who are interested in a better environment. The decline in numbers of several of our bird species is a signal of the potential hazards of pesticides to the environment. We are continuing a major research effort to develop nonchemical methods of pest control, but we must continue to rely on pesticides for the foreseeable future. The challenge is to institute the necessary mechanisms to prevent pesticides from harming human health and the environment.... A comprehensive strengthening of our pesticide control laws is needed.[5]

He then went on to say, "I propose that the use of pesticides be subject to control in appropriate circumstances, through a registration procedure which provides for designation of a pesticide for 'general use,' 'restricted use,' or 'use by permit only.' Pesticides designated for restricted use would be applied only by an approved pest control applicator. Pesticides designated for 'use by permit only' would be made available only with the approval of an approved pest control consultant. This will help to ensure that pesticides which are safe when properly used will not be misused or applied in excessive quantities."[6]

Nixon proposed that the administrator of the EPA be authorized to allow the experimental use of pesticides under strict controls, in order to gather additional information concerning a pesticide before deciding whether it should be cleared for use. Additionally, the president sought easier and quicker procedures for canceling the registration of a pesticide. He proposed that the EPA administrator be authorized to stop the sale or use of harmful pesticides being distributed.[7]

1969–70: 91st Congress

High DDT residues found in Lake Michigan Coho salmon focused attention on the harmful effects of pesticides on commercial and sport fishing. The secretary of health, education, and welfare announced a ban on moving fish with DDT residues above a certain level in interstate commerce. To investigate the problem further, the Senate Commerce, Subcommittee on Energy, Natural Resources, and the Environment held hearings on the effects of pesticides on sport and commercial fishing. In 1969, Congress extended HR 15979, a program of research on the effects of pesticides and other poisonous chemicals on fish and wildlife.[8]

1971–72: 92nd Congress

The Federal Environmental Pesticide Control Act of 1972 (HR 10729/PL 92-516) was passed in this session to regulate the manufacture, distribution, and use of pesticides. Even though numerous environmental groups criticized the bill as being weaker than existing laws, it received support from the Congress and the president.[9] The law made substantial amendments to the 1947 FIFRA. First, it required that all pesticides in U.S. commerce be registered with the EPA rather than the Department of Agriculture. With few exceptions, the sale, distribution, and professional use of unregistered pesticides were now prohibited.[10] If needed, the EPA administrator could cancel or suspend the registration of a pesticide. This could happen if the administrator found that a pesticide was proven to have "unreasonable adverse effects" on both human health and the environment. Under the law, the burden was on the EPA to show unreasonable risk before it could cancel or suspend a pesticide's registration.[11] If needed, manufacturers would have to provide information to demonstrate to the EPA how the chemicals performed their intended functions without having "unreasonable adverse effects on the environment." Second, pesticides could be used only for specifically approved uses. Third, the bill divided pesticides into two types: "general use" and "restricted use." General-use pesticides were those that the EPA determined would not cause substantial adverse effects on the environment. On the other hand, restricted-use pesticides were those that could cause adverse effects and could be applied only by holders of EPA permits. The potential penalties for misuse of pesticides could be both civil and criminal. [12] Fourth, higher standards were set for newly developed chemicals. When new products were developed, the manufacturer would have to prove that they were safe. Overall, the bill was intended to protect people and the environment from potential harm from the overuse of pesticides.

President Ford

President Ford signed a bill (S 3149/ PL 94-469), the Toxic Substances Control Act that would regulate chemicals on the market. It would require testing of the chemicals for any long-term effects on the health of people. In this way, we would learn more about these chemicals.[13]

1975–76: 94th Congress

Because funding authority to carry out FIFRA was set to expire, Congress cleared legislation (HR 8841/PL 94-140) to authorize funds to extend the program through March 1977. The final bill set up a formal procedure for the Department of Agriculture to comment on EPA pesticide actions and expanded the timetable for the agency's new program of registration and classification of pesticides.[14] The EPA would notify the secretary of agriculture prior to any regulatory decisions concerning pesticides, consider the impact of possible cancellation of pesticides on agriculture, and establish a scientific advisory panel to review proposed regulations. The new law also extended the deadline for EPA approval of state plans for certification of pesticide users.[15]

President Carter

Carter noted that the dangers of pesticides were becoming clearer and that they posed an unacceptable risk to human health and the environment. He said that many pests had developed resistance to chemical pesticides, decreasing our ability to control some pests, which in turn would reduce agricultural production. In order to improve the safety of pesticides, Carter recommended two new laws. The first was a short-term solution and involved asking the EPA and the Congress to pass an amendment to FIFRA to allow the EPA to regulate 1,400 different chemical ingredients. The second was a longer-term solution; he recommended that the Council on Environmental Quality should make recommendations to encourage the development of pesticides that use natural biological control methods instead of chemical ones.[16] In signing the Federal Pesticide Act of 1978, Carter noted that the bill would allow for a streamlined and efficient process for registering pesticides and for public scrutiny of related health information on those pesticides.[17] In addition to new laws, Carter also proposed the integrated pest management system to reduce pest damage. This system relied on natural predators and parasites instead of chemicals.[18]

1977–78: 95th Congress

Both the House and the Senate passed bills (HR 7073 and S 1678) making changes in the way the EPA registered pesticides. Both laws were designed to make it easier for the EPA to evaluate the safety of pesticides on the market.[19] S 1678 made it to the president's desk, and he signed it (PL 95-396). Under the new law, called the Federal Pesticide Act, the EPA could, in some cases, grant conditional registrations of pesticides prior to the registrant's submission of all supporting data. The revisions were directed at simplifying the registration of pesticides, since the EPA claimed to have a backlog of registration applicants. The EPA also felt the law was necessary because they were becoming overwhelmed in data about the 35,000 pesticides then on the market. Under the new law, the EPA would get help from state governments, which were given new authority to enforce controls on pesticides once they met certain standards.[20]

1979–80: 96th Congress

Two bills to provide funds to control pesticide programs under FIFRA (S 717, HR 3546) were not passed and allowed to die.[21] However, money to implement FIFRA was found in HR 7018 (PL 96-539). This authorized $77.5 million in funds for the fiscal year 1981.[22]

President Reagan

Like other presidents, President Reagan wanted to limit the use of dangerous pesticides and the potential harm to citizens. Reagan claimed that his administration continued to review new chemicals as a way to control pesticides, and prevented harmful pesticides from being sold.[23] Through an Executive Order, Reagan ordered that any job where a worker is frequently exposed to highly toxic pesticides is considered to be hazardous duty.[24] He supported federal grants to states to help establish programs to clean up from the over-use of pesticides.[25] He also sought to restrict the use of some types of pesticides that resulted in the destruction of the bald eagle's habitat.[26]

1981–82: 97th Congress

Nothing was done on HR 5203, a two-year authorization of FIFRA. The proposal would require registration of chemicals used to kill insects, rodents, fungi, and plants with the federal government. There were conflicts among environmentalists, pesticide makers, and farm groups over the bill, especially whether stricter state registration laws could preempt federal statutes. It was also unclear whether individuals would be able to sue to stop violations of pesticide laws. The confidentiality of industry information that was filed with state and federal governments was also unclear.[27]

1983–84: 98th Congress

Congress passed and the president signed HR 2785 (PL 98-201), a one-year authorization of FIFRA. The bill provided $64.2 million in fiscal year 1984 for the pesticide regulation program, about $8 million more than the president had requested.[28]

1985–86: 99th Congress

A House bill that did not pass was HR 2482, which would have reauthorized and strengthened FIFRA through fiscal year 1992.[29] In addition to the provision funding the program, other provisions of the bill would have speeded up the registration process for pesticides already in use even though their health effects were still unknown, increased public access to health and safety information, given the EPA new authority to protect groundwater from pesticide contamination, regulated for the first time some hazardous pesticide ingredients previously classified as inert, and strengthened existing requirements for certifying pesticide applicators and for protecting farmworkers from exposure.[30] The bill was stalled by disagreements between environmentalists and chemical companies.

1987–88: 100th Congress

Congress reauthorized FIFRA in S 659 (PL 100-532). This time, Congress set a nine-year mandatory schedule for chemical companies to determine whether their products caused cancer, birth defects, nerve damage, or other chronic health effects.[31] The revisions directed the EPA to re-register 600 active ingredients in pesticides within that time period. The EPA would have to either re-register or cancel up to 50,000 products that had been allowed to remain on the market (grandfathered in) under the 1972 amendments.

President G.H.W. Bush

In 1989 Bush unveiled a major overhaul of the nation's food safety laws with a plan to make it easier to remove carcinogenic pesticides from the market. The proposal would give the EPA more flexibility to declare a pesticide an imminent hazard and enable the agency to remove it from the market within two to three years rather than four to eight years. However, environmentalists and some members of Congress argued that the proposals might end up weakening standards for determining cancer risks. They maintained that the EPA already had the authority under FIFRA to undertake many of Bush's proposals for streamlining the pesticide removal process.[32]

1991–92: 102nd Congress

Efforts to rewrite FIFRA failed to advance in the 102nd Congress. Both environmental and consumer protection groups wanted to make it easier to take pesticides off the market quickly if they proved to be dangerous.[33] For example, the Senate considered S 638, a bill to require the EPA director to maintain a facility for the biological testing of pesticides. The director would be required to conduct tests at the facility to verify data submitted by pesticide registration applicants and to test marketed pesticides to ensure compliance with the law. Another proposal to amend FIFRA was S 3304. This proposal would include nitrogen stabilizers in the definition of *pesticide.* This did not pass.

President Clinton

Clinton sought to strengthen pesticide laws in the country in an effort to keep people safe.[34] In a weekly radio address, President Clinton noted the link between cancer and possible exposure to pesticides.[35] He sought to reform the approval process for new pesticides, and make it easier for newly developed, safer pesticides to be put on the market to replace the older and possibly more dangerous ones.[36]

To serve as an example of how landscaping could be accomplished with reduced pesticide use, Clinton promised to implement landscaping practices on federal properties that used the least amount of pesticide as possible.[37]

When it came to Congressional action on pesticides, Clinton complained that Congress cut appropriations for enforcement of environmental rules, possibly leading to more pesticides in the food supply.[38] Toward the end of his administration, Clinton signed the Food Quality Protection Act that required supermarkets to provide information about the pesticides used on the food in the store.[39]

1993–94: 103rd Congress

Pressure grew in 1993 to rewrite the nation's major pesticide laws, including FIFRA.[40] Congressional members passed a new FIFRA bill (S 1913/PL 103-231) that extended the effective date of certain EPA regulations, mostly relating to administrative requirements, but it did not postpone pesticide safety protections for farmworkers. The bill was intended to give the EPA more time to provide necessary training, education, and compliance information to farmers and regulators.[41] In 1994, Congress passed HR 967, which was aimed at keeping less profitable minor-use pesticides on the market. But it did not pass the Senate.[42]

1995–96: 104th Congress

The House Agriculture Committee approved sections of a bill, HR 1627, that were aimed at expediting government decision making on pesticide use. Called the Food Quality Protection Act, the bill would have amended federal pesticide regulations in FIFRA, and it proposed to restrict the EPA's review process for a potentially dangerous pesticide to 450 days. It also proposed to streamline the registration of several classifications of chemicals. Congress tried to strike a balance between consumers and the chemical industry in the bill by allowing the continued use of pesticides while setting limits on residues of cancer-causing chemicals in both raw and processed foods, specifically designed to protect children. Clinton signed the bill (PL 104-170).[43]

In 1996 Congress ended a stalemate on pesticide policy when they passed the Food Quality Protection Act. The law ended up including a major revision of the nation's pesticide law by requiring the EPA to use a new, uniform, reasonable-risk approach to regulating pesticides used on food, fiber, and other crops. It required that special attention should be given to the many ways in which people of all ages are exposed to such chemicals. The act passed through Congress with no opposition because the food industry was fully behind it.[44]

1997–98: 105th Congress

The Senate considered, but did not pass, S 2652, the Circle of Poison Prevention Act of 1998. The bill would have amended FIFRA to require pesticide producers to inform the administrator of the EPA of the types and quantities of pesticides, and active ingredients used in producing pesticides, that were exported to a foreign country as well as the date the chemical was exported.

1999–2000: 106th Congress

In HR 1913, the Pesticide Registration Harmonization Act of 1999, the House requested that the president direct the U.S. representative to the United States-Canada Technical Working Group to urge the group to make harmonization of pesticide registrations the highest priority. The concern was over those pesticides that were registered in Canada for use on agricultural crops but registered in the United States for use on food crops. It would also direct the administrator of the EPA to accept a request from a Canadian registrant for same-use U.S. registration unless substantial evidence existed precluding such acceptance. This did not become law.

President G. W. Bush

In the beginning of the Bush administration, the President and administrator of the EPA issued a rule about limiting the use of pesticides as a way to improve the health of the country.[45] Throughout his term in office, Bush continued to support new, scientific rules to limit the use of dangerous pesticides.[46] He believed that the United States needed to play an important role in the international community with regard to the management of pesticides. He sent the Stockholm Convention on Persistent Organic Pollutants to the Senate for their approval. This convention was intended to stop the production and use of certain dangerous pesticides in other countries.[47] Bush also reported to the American public that the United Nations was supporting a program of biotechnology that could improve crop production in developing countries while allowing them to use fewer pesticides.[48]

2001–02: 107th Congress

The Arsenic-Lumber Child Protection Act (HR 2727) was proposed but not passed in this session. The bill would direct the EPA administrator to regulate lumber treated with inorganic arsenicals or chromated copper arsenicals (CCA lumber) under FIFRA. The proposal would require that such regulation deem CCA lumber as misbranded unless its labeling disclosed that it had been treated with arsenic. Further, the proposal would make the manufacture or commercial distribution of a product constructed with such lumber a violation of federal law if the product was playground equipment or any item intended for use primarily by children.

In another bill, the House debated HR 111, the School Environment Protection Act of 2001. This bill would amend FIFRA to require the administrator of the EPA to establish a National School Integrated Pest Management Advisory System to develop and update uniform standards and criteria for implementing integrated pest management systems in schools. The EPA administrator would have to provide grants to local educational agencies to develop such systems.

The proposal would prohibit the application of a pesticide when a school is occupied or in use. It would also prohibit the use of an area or room that has been treated by a pesticide, other than a least-toxic pesticide, for a 24-hour period beginning at the end of the treatment. Violations of the act would be punishable with civil penalties.

In HR 1084, the House debated legislation to amend FIFRA to allow a state to register a Canadian pesticide for distribution and use within that state. This did not pass.

2003–04: 108th Congress

In S 337, the Arsenic-Treated Residential-Use Wood Prohibition Act, the Senate members debated legislation to amend FIFRA and list CCA-treated wood as a hazardous waste. The proposal would also require disposal of discarded CCA wood in a lined landfill with a leachate system and groundwater-monitoring system. The director of the EPA would be asked to conduct an assessment of the risks of CCA-treated wood production and processing and to suggest prohibiting treated wood. This did not pass.

In the House, members debated the School Environment Protection Act of 2003 (HR 121). The provisions of the bill were essentially similar to those of the 2001 bill of the same name. This proposal did not have enough support to pass through Congress and did not become law.

2005–06: 109th Congress

Once again, members of the House proposed the School Environment Protection Act, now of 2005 (HR 110/S 1619). The bill's provisions were again essentially the same as in the 2001 and 2003 bills. One new provision concerned the establishment of the National School Integrated Pest Management Advisory Board. The bill once again did not have enough support to pass Congress.

2007–08: 110th Congress

During this session, Congress was able to pass a new law (S 1983/PL 110-94) called the Pesticide Registration Improvement Renewal Act. The new law amended FIFRA to repeal a requirement that the administrator of the EPA (EPA) review applications to register a pesticide within 45 days of receiving the application. The administrator would complete an initial review of each pesticide and then report, annually, on pesticide registration. Registrants of the pesticides would be required to pay an annual fee.

A bill considered by the House, but not passed, was HR 3290, the School Environment Protection Act of 2007. This law was primarily the same as others proposed in previous sessions concerning the use of pesticides in a school or on school grounds. Under HR 3399, the use, production, sale, importation, or exportation of any pesticide containing atrazine would be prohibited. The bill was not passed. The Congress passed S 2571 (PL 110-193) to make technical corrections to FIFRA. The bill amended FIFRA to authorize the administrator of the EPA to exempt a manufacturer from, or waive a portion of, the registration service fee for an application for minor uses for a pesticide.

President Obama

The Obama administration chose to reevaluate the use of the popular herbicide atrazine because of findings that it might cause cancer. He cited new studies indicating that the herbicide has dangerous effects on humans and that it can disrupt hormones in amphibians, so that a frog might become both male and female.[49]

2009–2010: 111th Congress

The School Environment Protection Act was proposed again (now as HR 4159), and it still did not pass. The proposed law would amend FIFRA and require school districts to implement an integrated pest management program and establish a coordinator to create an integrated pest management plan for addressing any pest problems in the school buildings or grounds. Schools would not be allowed to use any pesticides other than nontoxic or least-toxic types, or any synthetic fertilizers. A school could use the least-toxic pesticide only as a last resort and only if the area or room treated was unoccupied or not in use. This did not become law.

Conclusion

After concerns were raised about the quality and safety of pesticides, Congress passed FIFRA and then amended it in later years. The original law set the tone for later legislation designed to keep the dangers from pesticides to a minimum. As we learn more about pesticides and their impact, it is clear that more regulation may be necessary. We will undoubtedly continue to learn more about pesticides and create more pesticides, and Congress will continue to pass laws to keep citizens from feeling the possible effects of pesticides, despite the controversy such laws create.

Chapter 6

The Politics of Solid Waste

In the 1970s, environmentalists announced that Americans would be swimming in garbage if it was not brought under control quickly. They pronounced that new and alternative methods to reduce the amount of trash through recycling and other methods were needed. Environmentalists pointed to vanishing landfills that were not only costly and dangerous to people and animals but also increasingly more clogged with the byproducts of a consumer society.[1] As disposal costs rose and landfills reached their capacity, bitter interstate battles erupted over what to do with waste products. Table 6.1 indicates the amount of solid waste produced in the United States from 1960 to 2008. Not only does this show how many tons of waste are produced each year but also the increase over time. Overall, the amount of solid waste in the United States has more than tripled from 1960 to the present.

The issue of what to do with municipal solid waste, ranging from household garbage to nonhazardous industrial wastes, has become a problem not only for environmentalists and city managers but also for members of Congress. This chapter describes the attention given to solid waste by Congress and the president.

President Johnson

President Johnson was concerned with the problem of solid waste disposal, and he called for legislation to aid states in developing programs for solid waste disposal.[2] On March 8, 1968, he gave a message to Congress on conservation, called "To Renew a Nation." In it, he recommended that the Congress approve "a national planning, research and development program to find ways to dispose of the annual discard of solid wastes—millions of tons of garbage and rubbish, old automobile hulks, abandoned refrigerators,

TABLE 6.1. Solid Waste Production in the United States, 1960–2008, in thousands of tons

Materials	1960	1970	1980	1990	2000	2005	2008
Paper and Paperboard	29,990	44,310	55,160	72,730	87,740	84,840	77,420
Glass	6,720	12,740	15,13	13,100	12,760	12,540	12,150
Total Metals	10,820	13,830	15,510	16,550	18,910	20,060	20,850
Plastics	390	2,900	6,830	17,130	25,540	29,240	30,050
Rubber and Leather	1,840	2,970	4,200	5,790	6,710	7,360	7,410
Textiles	1,760	2,040	2,530	5,810	9,440	11,380	12,370
Wood	3,030	3,720	7,010	12,210	13,110	14,080	16,390
Other	70	770	2,520	3,190	4,000	4,170	4,500
Total Solid Waste	54,620	83,280	108,890	146,510	178,210	183,670	181,140

Source: U.S. Environmental Protection Agency, Office of Resource Conservation and Recovery, "Municipal Solid Waste Generation, Recycling, and Disposal in the United States, Detailed Tables and Figures for 2008" November 2009. Available at http://www.epa.gov/osw/nonhas/municipal/msw99.htm; accessed 10/27/2010.

slaughterhouse refuse. This waste—enough to fill the Panama Canal four times over—mars the landscapes in cities, suburbia and countryside alike. It breeds disease-carrying insects and rodents, and much of its find its way into the air and water."[3]

Johnson then went on to recommend a one-year extension of the Solid Waste Disposal Act, and he instructed the director of the Office of Science and Technology to undertake a comprehensive review of solid waste disposal technology. He believed that the research would indicate the best way to reduce the costs of solid waste disposal and to improve research and development in this field.[4]

1965–66: 89th Congress

Congress passed the Solid Waste Disposal Act of 1965 (HR 8248/PL 89-272), which established $92.5 million during fiscal years 1966–69 for federal research to explore better methods of solid waste disposal; this would be overseen by the Department of Health, Education and Welfare (HEW) and the Department of the Interior. The new law left the states and local governments with the primary responsibility for the collection and disposal of waste. The law also established the Bureau of Solid Waste Management in the Department of HEW.[5]

The intent of the act was to start a national research and development program to create new methods for the disposal of solid waste. It was also intended to provide both technical and financial aid to states and local governments so they could develop, establish, and conduct their own solid waste disposal programs. The bill covered up to 50 percent of the costs of state surveys of disposal practices and provided grants for up to two-thirds of the costs of any demonstration projects states and local governments wanted to try.[6] Most of the funds were to be used for grants to or contracts with public and private agencies for the completion of studies, projects, surveys, demonstrations, and the construction of facilities.[7]

President Nixon

Nixon often addressed the problem of solid waste. He noted that packaging methods for products, especially nonreturnable bottles and cartons, had created an increasing amount of waste and refuse. He described the traditional method of dealing with the problem as continuing to spend money on collection and disposal of wastes, which "amounts to a public subsidy of waste pollution."[8]

To deal with the problem, Nixon proposed an extension of the Solid Waste Disposal Act with a redirection of research to place greater emphasis on techniques for recycling materials, and on development and use of packaging and other materials which will degrade after use.[9] Nixon wanted to develop a program of incentives, regulations, and research for reducing the amount of wastes by reusing a large proportion of waste materials. He also asked for a system that would "promote the prompt scrapping of all junk automobiles," which he considered to be an eyesore.[10]

Finally, Nixon understood the need to address the problem of paper waste. He said, "The Nation's solid waste problem is both costly and damaging to the environment. Paper, which accounts for about one-half of all municipal solid waste, can be reprocessed to produce a high quality product. Yet the percentage the Nation recycles has been declining steadily." He then reported that the General Services Administration (GSA) found many prohibitions in federal purchasing policies that contradicted the use of paper with recycled content. Because of that, the GSA changed its policies to require a minimum of 30 to 50 percent recycled content in over $35 million per year of paper purchases. Nixon then directed that state governors also review their purchasing policies and where possible revise them to require recycled material.[11]

1969–70: 91st Congress

During this session, the Congress proposed a bill (HR 11833/PL 91-512) called the Resource Recovery Act of 1969, which extended the Solid Waste Disposal Act of 1965 by authorizing more federal funds for the construction of waste treatment facilities and for additional research and planning in solid waste disposal. The Congress extended the research program for one year by authorizing appropriations of $32 million in fiscal year 1970.[12] The law also established major research programs on the problem of waste and expanded federal grants to state and local agencies to develop their own resource recovery and solid waste disposal systems. The secretary of HEW was to focus on the health and welfare effects of waste and on ways to reduce the amount of waste. The legislation provided for three-year, $462 million extension grants for solid waste disposal programs. Whereas earlier federal programs had been concerned with disposing of solid wastes, this bill created a demonstration program and construction grants for innovative solid waste management systems. The legislation also emphasized the need for more recycling efforts. President Nixon signed the bill into law on October 26, 1970.[13]

1973–74: 93rd Congress

During this session, Congress passed a bill (HR 5446/PL 93-14) that authorized $238.5 million for fiscal year 1974 for a one-year extension for the waste recycling program originally funded under the Solid Waste Disposal Act of 1970. It continued funding for the program at fiscal year 1973 authorization levels.[14]

The Senate also considered S 2062, intended to ban interstate shipment of nonreturnable beverage containers and to require a deposit on all bottled and canned beverages and outlaw the sale of cans with detachable openers. Supporters argued that the provisions would save energy, reduce litter, and cut consumer costs. The bill was considered in the Senate Commerce Subcommittee on Environment, but no further action was taken.[15]

Congress passed a bill (HR 16045/PL 93-611) that authorized $76 million for federal solid waste disposal and recycling programs in fiscal year 1975. The authorizations for the program, first established in the Solid Waste Disposal Act of 1965 and the Resource Recovery Act of 1970, expired on June 30, 1974, but were funded under a continuing appropriations resolution through September 30, 1975. The one-year extension of authorizations was intended to give Congress more time to consider a major overhaul of the programs.[16]

President Ford

Ford signed S 3894, a law to provide loan guarantees for construction of municipal wastewater treatment plants.[17]

1975–76: 94th Congress

A major bill (S 2150/PL 94-580), called the Resource Conservation and Recovery Act (RCRA) of 1976, passed Congress during this session. The law expanded the federal solid waste program and provided funds for state and local efforts to cope with garbage and sludge. The bill authorized $365.9 million for solid waste programs, mostly for fiscal years 1978–79. That included $80 million for general use by the Environmental Protection Agency (EPA), $70 million to finance state solid waste management programs, $50 million for state hazardous waste programs, and $35 million to help finance demonstration projects on new methods of recycling, extracting resources from, or disposing of solid wastes. The bill also directed the EPA to establish an Office of Solid Waste to administer the law and to advise state and local governments on solid waste programs. Additionally, the bill created a new federal regulatory program for hazardous waste, placed a ban on open dumping within five years, and authorizations for state and local solid waste management and research programs.[18]

With regards to hazardous waste, the EPA was required to issue regulations defining the term *hazardous waste* and to list the regulations. All people and organizations that handled hazardous waste would need to notify the EPA of their operations. The EPA would be required to issue regulations that would set safety standards for those who produced, transported, or stored hazardous waste. The EPA would work with states to issue guidelines for hazardous waste programs, and those states meeting standards would take over the administration of permits in the future. Premises that handled hazardous wastes would be inspected, as would their records. The law established a civil penalty of up to $25,000 a day for hazardous waste violations that occurred after state or federal compliance deadlines were set. Criminal penalties were set at up to $25,000 a day and/or one year in prison for knowing violations of the law.[19]

President Carter

Carter sought to address two principal causes of the solid waste problem: excessive packaging and inadequate use of recycled materials. He asked the

EPA to give him recommendations to address the use of solid waste disposal charges. He also encouraged resource conservation within the White House, including the use of recycled paper.[20]

1979–80: 96th Congress

Congress heard testimony that a billion gallons of used oil were produced annually in the United States, about half from automobile use and half from industrial use. They were told that if all used lubricating oil were recycled, 42,000 barrels of oil a day would be saved. So Congress passed S 2412, the Used Oil Recycling Act (PL 96-463), which was aimed at encouraging the use of recycled motor oil for automobiles.

Specifically, the bill authorized $10 million in grants to states ($5 million in fiscal 1982 and $5 million in 1983) to encourage the recovery and recycling of lubricating oil. It voided a Federal Trade Commission (FTC) requirement that recycled motor oil had to be labeled "used." Third, the bill required that all containers of lubricating oil bear the words, "DON'T POLLUTE—CONSERVE RESOURCES; RETURN USED OIL TO COLLECTION CENTERS." Finally, the bill required the federal government to set health and environmental standards for used oil recycling.[21]

S 1156, the Solid Waste Disposal Act Amendments of 1980 (HR 3994/PL 96-482), authorized funds for fiscal years 1980–82 for programs to help deal with solid waste. Much of the act revolved around coal mining waste. The law directed the EPA administrator to review regulations concerning coal mining wastes and to suggest revisions of the regulations to the secretary of the interior. The EPA administrator would also be required to conduct a study of the adverse effects of coal ash and other fossil fuel wastes, uranium mining waste overburden, phosphate and other mining wastes, and cement kiln dust wastes on the environment.

The secretary of the interior would be responsible for carrying out the provisions of the new law. A new committee, the Interagency Coordinating Committee on Federal Resource Conservation and Recovery Activities, was established that would coordinate all activities dealing with conservation and recovery of solid waste.

President Reagan

President Reagan did not give many statements concerning solid waste. He proclaimed June 1998 as National Recycling Month in Proclamation 5830.[22] He also noted to farmers in Iowa about the need to recycle to save energy.[23]

1981–82: 97th Congress

During this session, the Congress passed a new law (HR 5288/PL 97-278) that granted congressional approval to a compact between New Hampshire and Vermont for construction and operation of a facility that would burn solid wastes to produce electricity for communities in both states. The agreement was the result of new technology that made it possible to convert waste into electric power, while separation of solid wastes facilitated recovery of metals for reuse. For the compact to become official, congressional approval was required. The legislatures of the two states had already approved the compact.[24]

President G.H.W. Bush

President Bush thought it was necessary to "make every effort to stem the rising tide of garbage and industrial waste through a more aggressive use of waste minimization and recycling practices."[25] He wanted to see households begin to minimize waste in their homes and to begin recycling practices. But he also wanted to see companies do the same, and announced that the 3M Corporation had begun a recycling program and had saved 388,000 tons of solid waste from being put into a landfill.[26]

Like President Reagan, President Bush declared National Recycling Month. This month would be a time to inform the public about the benefits of recycling solid waste and encourage state and local governments to do so.[27]

1989–90: 101st Congress

After several states filed suit against Amtrak for flushing human waste onto railroad tracks, Congress gave the rail system six years to find an alternative. The bill, called the National and Community Service Act, S 1430, was signed by President Bush (PL 101-610). Under federal law, the unregulated public dumping of human waste was outlawed, but Amtrak was exempted in 1976. Even though a study commissioned found no epidemiological link between such waste disposal and the spread of disease, states still filed against the company. In one case in Florida in 1989, an Amtrak train emptied its retention tanks over a bridge in Putnam County, Fla. The liquefied waste sprayed down on William and Mary Trammeil, two residents were fishing on the St. Johns River with another couple from Alaska. In 1989, a jury in that state convicted Amtrak of four counts of commercial littering. Based on that case, other states were considering prohibiting Amtrak from operating in their states.[28]

Another bill (S 1140/HR 1056) was a proposal that would have allowed states to levy fines against federal facilities that failed to comply with hazardous and solid waste laws; it did not pass. The Congress wanted to establish the power of the EPA and state governments to enforce compliance with the RCRA at federal facilities. RCRA was the 1976 law that regulated hazardous and solid waste.[29]

1991–92: 102nd Congress

Two bills, HR 2194 and S 596, known as the Federal Facilities Compliance Act, gave the EPA and state governments the right to fine or penalize federal facilities that violated provisions of the RCRA, which governed the disposal of solid wastes.[30] The House passed HR 2194, and the Senate followed suit within hours. It was signed by the president and became law (PL 102-386).

In some states, particularly Ohio, Maine, and Washington, hazardous and nuclear wastes had been dumped and had escaped into the air and contaminated groundwater supplies. The federal government polluting facilities had claimed sovereign immunity. A proposed bill (HR 2194) removed sovereign immunity claimed by federal agencies as a shield against prosecution and fines for violating general solid and hazardous waste laws. The federal government had long been among the country's worst polluters, with the prime offenders being the Energy and Defense departments' nuclear weapons and energy complexes. The bill, known as the Federal Facilities Compliance Act, was signed by the president, becoming law (PL 102-386).[31]

Three bills (HR 3865, S 976, and S 2877) were proposals to reauthorize the RCRA and regulate the growing problem of out-of-state waste. There was debate about proposals to handle garbage, particularly by some Midwestern states that complained of being dumping grounds for garbage from the eastern states. The House Energy and Commerce Committee proposed a small reauthorization bill (HR 3865) that sidestepped regulation of industrial, oil, and gas wastes to concentrate on municipal and household garbage. The Senate Environment and Public Works Committee drafted a broader bill (S 976), but neither version advanced. The Senate passed a narrower bill (S 2877) that dealt only with the issue of interstate garbage. The proposal would allow state governors to ban or limit garbage imports, subject to a request by local authorities. The bill allowed landfills that were already receiving out-of-state garbage to continue to do so under a grandfather clause.[32] None of the proposals was passed into law.

President Clinton

As part of an executive order on waste prevention, President Clinton ordered that each Executive agency will incorporate waste prevention and recycling into its operations.[33] He also asked that government agencies purchase only recycled paper.[34] Clinton, in a similar fashion to Bush and Reagan, announced a America Recycles Day to bring attention to the issues of solid waste and pollution in general.[35]

1993–94: 103rd Congress

Two bills (HR 4779, S 2345) that would have given states the authority to control their receipt of out-of-state garbage died at the end of the session. The bills were called the State and Local Government Interstate Waste Control Act of 1994. Another bill, to allow local governments to direct waste to publicly operated waste disposal facilities (HR 4683), also did not pass.

1995–96: 104th Congress

The Waste Disposal Bill, S 534, was aimed at giving states new authority to keep out unwanted garbage while at the same time attempting to accommodate the needs of states searching for somewhere to put their waste. The bill also sought to provide relief to local governments that had issued bonds to build incinerators, landfills, and other disposal sites. If a governor requested it, out-of-state waste could be banned at facilities that had not received such waste in the past. States were allowed to charge a fee of $1 per ton for the disposal or processing of out-of-state trash, if they had imposed such a special fee on out-of-state trash previously. S 534 was not passed in 1996.[36]

HR 2024 was a proposal to phase out the use of mercury in batteries and establish new labeling requirements for rechargeable batteries. When the batteries were disposed of in a landfill, the mercury broke down and polluted groundwater; if incinerated, it produced toxic emissions. Rechargeable batteries contained heavy metal nickel-cadmium or lead, also known pollutants. The bill phased out the use of mercury in most batteries manufactured one year after its enactment. It also required manufacturers to label rechargeable batteries as recyclable. After one year, the bill prohibited the sale of rechargeable batteries that did not conform to the standard.[37] In the end, the bill was signed by the president and became law (PL 104-142).

1997–98: 105th Congress

HR 942 was a proposal to amend the Solid Waste Disposal Act to authorize states to limit the interstate transportation of municipal solid waste. Title I of the bill had to do with the interstate transportation of municipal solid waste. It would authorize state governors, if requested by a local government, to prohibit the disposal of out-of-state municipal solid waste in landfills or incinerators subject to their jurisdiction. It would permit governors to limit the quantity of out-of-state waste received at landfills and incinerators to an annual amount equal to or greater than the quantity received during 1993. The proposal did not make it out of the subcommittee.

In another bill, the State and Local Government Interstate Waste Control Act of 1997 (HR 1346), amendments were made to the Solid Waste Disposal Act. The proposal would prohibit a landfill or incinerator from receiving out-of-state municipal solid waste for disposal or incineration unless the waste was received pursuant to a new or existing host-community agreement. Under the law, states could establish limits on the amount of out-of-state waste received annually for disposal at each facility. The bill did not pass.

HR 1358, the Interstate Transportation of Municipal Solid Waste Act of 1997, would amend the Solid Waste Disposal Act to authorize a state governor, if requested by an affected local government, to prohibit the disposal of out-of-state municipal waste in any landfill or incinerator subject to the jurisdiction of the governor or the local government. This proposal did not become law.

The National Beverage Container Reuse and Recycling Act of 1997 (HR 1586) was a proposal to amend the Solid Waste Disposal Act. It would prohibit the sale of beer, mineral water, soda water, wine coolers, or carbonated soft drinks in beverage containers by retailers and distributors unless such containers carried a refund value of 10 cents. The proposal did not have enough support in Congress and did not pass.

1999–2000: 106th Congress

The debate over states accepting waste from other places generated many proposals during this Congress. For example, HR 378 was a proposal in Congress to authorize individual states to regulate solid waste from other states. The proposal would amend the Solid Waste Disposal Act and authorize state officials to enforce laws regulating, and collecting fees for, the treatment and disposal of solid waste that was generated in another state. It also would give congressional approval to agreements or compacts entered into by two or more states for cooperative efforts and mutual assistance for solid waste management. The bill did not become law. A similar proposal, HR 379, was a bill to permit states to prohibit the disposal of solid waste imported from other nations. The proposed law would amend the Solid Waste Disposal Act to allow any state to prohibit disposal of any foreign solid waste. This did not pass.

HR 891 was also called the Solid Waste Compact Act. It would amend the Solid Waste Disposal Act to allow states with approved solid waste management plans to prohibit the importation of solid waste from outside the state. This proposal did not have enough support in Congress to become law. Neither did the Solid Waste Interstate Transportation and Local Authority Act of 1999 (HR 1190/S 663). This proposal was an attempt to amend the Solid Waste Interstate Transportation and Local Authority Act of 1999 to prohibit a landfill or incinerator from being forced to accept municipal waste from another state.

HR 2676 was the National Beverage Container Reuse and Recycling Act of 1999. The provisions of the bill were the same as in the 1997 bill of the same name. Again, the proposal did not pass.

President Bush

Like other presidents, President G. W. Bush did not discuss the issues of solid waste frequently. He declared November 14, 2002 and November 15, 2004 as America Recycles Day. These days would provide opportunities to educate the public about the need to recycle in order to reduce pollution.[38]

2001–02: 107th Congress

Many proposals that did not pass during the 106th Congress were reintroduced into Congress during this session. HR 667 was the Solid Waste Compact Act, intended to amend the Solid Waste Disposal Act to allow states with approved solid waste management plans to prohibit the importation of solid waste from outside the state. This again did not pass.

Another proposal that did not pass during either this session or the previous one was HR 845/HR 1667, the National Beverage Container Reuse and Recycling Act of 2001. The law would amend the Solid Waste Disposal Act to prohibit the sale of certain beverages unless the containers carried a refund value of 10 cents.

Also reintroduced was HR 1213, the Solid Waste Interstate Transportation Act of 2001. The proposal would amend the Solid Waste Disposal Act to prohibit a landfill or incinerator (facility) from receiving out-of-state municipal solid waste for disposal or incineration unless the waste was received pursuant to a new or existing host-community agreement or an exemption from this prohibition (which could be limited by the state). This again did not pass.

HR 1548 was a proposal to phase out the incineration of solid waste. The proposal would require each state to adopt and submit to the administrator of the EPA a three-year implementation plan to increase recycling by at least 75 percent and to reduce water source pollution. The bill would also restrict landfill dumping to materials that are not recyclable or compostable as well as phase out solid waste incineration within four years and six months after the enactment of the law. Overall, the proposal sought a waste reduction rate of 10 percent and an increase in composting of 10 percent. The law did not pass Congress.

HR 1927, the Solid Waste International Transportation Act of 2001, would amend the Solid Waste Disposal Act to authorize states to enact laws prohibiting or limiting the receipt and disposal of municipal solid waste generated outside the United States. This did not pass.

2003–04: 108th Congress

HR 382, the Solid Waste International Transportation Act of 2003, was a proposal originally introduced in the previous congressional session to amend the Solid Waste Disposal Act to authorize states to enact laws prohibiting or limiting the receipt and disposal of municipal solid waste generated outside the United States. This did not pass. Similarly, HR 411 was a proposal that would direct the EPA administrator to carry out provisions of an agreement made with Canada about the importation of municipal solid waste. The proposal would amend the Solid Waste Disposal Act to prohibit any person from importing, transporting, or exporting municipal solid waste for final disposal or incineration. This did not pass.

Laws were considered about waste from other states. HR 418, the Solid Waste Compact Act, was a proposal to allow states with approved solid waste management plans to prohibit the importation of solid waste from outside the state. This did not pass. HR 1730 was the Solid Waste Interstate Transportation Act of 2003. This proposal would amend the Solid Waste Disposal Act to prohibit a landfill or incinerator (facility) from receiving out-of-state municipal solid waste for disposal or incineration unless the waste was received pursuant to a new or existing host-community agreement or an exemption from this prohibition.

HR 1122 was an proposal to amend the Solid Waste Disposal Act to provide funding for the cleanup of methyl tertiary butyl ether (MTBE) contamination from underground storage tanks. This did not pass. Likewise, the Stop Solid Waste Incineration Act of 2003 (HR 2827) did not pass. The proposal would require each state to adopt and submit to the EPA a three-year implementation plan to achieve (1) an increase in recycling of at least 75 percent; (2) water source pollution reduction; (3) restriction of landfill dumping to materials that are not recyclable or compostable; (4) the phasing out of solid waste incineration within four years and six months after this act's enactment; (5) a waste reduction rate of 10 percent; and (6) an increase in composting of 10 percent. It would also provide grants to states that phase out the incineration of solid waste prior to the deadline established under this act.

2005–06: 109th Congress

Whether states had to accept waste from other states or not was once again a topic of debate during this session. HR 70, the State Waste Empowerment and Enforcement Provision Act of 2005, was not passed during this session. The proposal would have amended the Solid Waste Disposal Act to authorize a state to limit, place restrictions on, or otherwise regulate out-of-state municipal solid waste that was received or disposed of annually at each landfill or incinerator in the state.

HR 274, the Solid Waste Interstate Transportation Act of 2005, did not pass. The proposal would amend the Solid Waste Disposal Act to prohibit a landfill or incinerator (facility) from receiving out-of-state municipal solid waste for disposal or incineration unless the waste is received pursuant to a new or existing host-community agreement or an exemption from this prohibition. A state could establish limits on the amount waste it received each year from other states. A similar proposal, HR 553, the Solid Waste Compact Act, did not become law. It would have amended the Solid Waste Disposal Act to allow states with approved solid waste management plans to prohibit the importation of solid waste from outside the state. The director of the EPA would have identified alternative solid waste disposal methods and established and published technical guidance regarding their implementation.

Congress also debated receiving waste from other countries in HR 593, HR 2109, and S 346, none of which passed. The proposals were amendments to the Solid Waste Disposal Act to authorize states to restrict receipt of foreign municipal solid waste. HR 2491 and S 1198 were proposals that were not enacted. Called the International Solid Waste Importation and

Management Act of 2006, the proposals were amendments to the Solid Waste Disposal Act to authorize states to enact laws or issue regulations or orders restricting the receipt and disposal of foreign municipal solid waste. This was called the transboundary movement of hazardous waste between the United States and Canada.

Recycling of waste as an option to deal with solid waste was discussed in the House. HR 320, the Tax Incentives to Encourage Recycling Act of 2005, was a proposal to allow manufacturers of certain computer, cell phone, and television equipment a business tax credit for the disposal and recycling of such equipment. This did not pass. Another bill on recycling was HR 1723, a proposal to amend the Internal Revenue Code of 1986 to allow a credit against income tax for recycling or remanufacturing equipment. This failed to become law.

HR 1958, the Stop Solid Waste Incineration Act of 2005, did not pass. The provisions of the proposal were the same as those in the 2003 act of the same name. The law would have provided for federal implementation plans for states that failed to meet plan submission and approval requirements.

Another proposal, (S 510/HR 4316), the Electronic Waste Recycling Promotion and Consumer Protection Act, did not pass. It would have amended the Internal Revenue Code to give waste recyclers and individual consumers a tax credit for recycling qualified electronic waste. The law would have banned the disposal of electronic waste without recycling three years after its enactment.

2007–08: 110th Congress

The House passed a bill to reauthorize the popular Clean Water State Revolving Fund, the largest source of low-interest loans for construction of wastewater treatment facilities and other water pollution abatement projects. The White House threatened to veto the bill because it included a plan to apply federal wage rules to projects funded by the loans; the Senate did not consider similar legislation.[39]

The question of receiving waste from other states continued during this session of Congress. HR 70, the State Waste Empowerment and Enforcement Provision Act of 2007, was a proposal to amend the Solid Waste Disposal Act to authorize a state to limit, place restrictions on, or otherwise regulate out-of-state municipal solid waste received or disposed of annually at each landfill or incinerator to the extent authorized by a host state agreement (between the owner or operator of a landfill or incinerator and the affected local government). This did not pass.

HR 274 and HR 6166 were also proposals that did not pass. The Solid Waste Interstate Transportation Act of 2007 would prohibit a landfill or incinerator (facility) from receiving out-of-state municipal solid waste for disposal or incineration unless the waste was received pursuant to a new or existing host-community agreement or an exemption from this prohibition (which could be limited by the state). Additionally, HR 387, the Solid Waste Compact Act, did not pass. It would have allowed states with approved solid waste management plans to prohibit the importation of solid waste from outside the state. It would have directed the administrator of the EPA to identify alternative solid waste disposal methods and establish and publish technical guidance regarding their implementation. HR 518, the International Solid Waste Importation and Management Act of 2007, likewise did not pass. The law would have authorized states to enact laws or issue regulations to restrict the receipt and disposal of foreign municipal solid waste within their borders.

2009–2010: 111th Congress

Two bills on solid waste were proposed in this session, but neither became law. The first was HR 274, the Solid Waste Interstate Transportation Act of 2009. The proposal would have prohibited a landfill or incinerator from receiving out-of-state municipal solid waste for disposal or incineration unless the waste was received pursuant to an existing host-community agreement. It would allow states to establish limits on the amount of out-of-state waste received annually for disposal at each facility. The second proposal that was denied passage was HR 1481, the Solid Waste Compact Act. The proposal would have allowed states with approved state solid waste management plans to prohibit the importation of solid waste from outside the state.

Conclusion

Over the years, many laws have been proposed in Congress to address the problem of solid waste, and some have been passed into law. The concerns held by states regarding accepting waste from other states, or even from other countries, were often debated by congressional members. Recycling of solid waste was also often debated. As solid waste continues to increase in most cities and states, the problem will continue to be on the policy agenda of Congress.

Chapter 7

The Politics of Toxic and Hazardous Waste

Beginning in the 1970s, a series of accidents involving toxic waste illustrated the potential harm it poses to humans and the environment. One of those events occurred in the late 1970s, in a suburb of Niagara Falls, New York, called Love Canal. The residents of the suburb had to be evacuated from their homes because highly toxic industrial waste was found in the basement of homes, and at least one five-year-old boy was experiencing seizures connected with his exposure to the chemicals. It turns out that in the 1940s and 1950s, a company called the Hooker Chemical Corporation had buried 22,000 tons of hazardous waste in barrels and in liquid form. Years later, an elementary school, playground, and homes were built on top of the buried hazardous waste. Very few of the residents of Love Canal even knew the waste was there until the 1970s, when citizens began to notice bad odors and have health issues. Residents demanded that the area be cleaned up and that the government reimburse them for the value of their homes, which were deemed valueless.

Then on March 29, 1979, a valve stuck open on one of two reactors at the Three Mile Island nuclear power plant on the Susquehanna River near Harrisburg, Pennsylvania, causing a loss of cooling water. The nuclear fuel core overheated and, at least partially, melted. Residents of the community had to evacuate after it was thought that a hydrogen explosion could cause the reactor building to rip apart. The majority of the radioactive water was removed from the reactor building and the adjoining building. But the hardest part of the cleanup was to decontaminate the reactor building and deal with the reactor core. The owner of the plant, General Public Utilities Corp (GPU), estimated the cleanup would cost more than $1 billion. Insurance payments covered only $300 million of that cost, so when the insurance money was spent, GPU slowed the cleanup. Politicians and residents feared that, over time, the plant would begin leaking radioactive material and wanted the cleanup to continue.[1]

In early 1993 in Glen Avon, California, a civil trial began that had 3,700 plaintiffs, all of whom were residents of Glen Avon. The case had 13 defendants, including the state of California and major corporations such as Rockwell International, Northrop, McDonnell Douglas, and Montrose Chemicals. At issue was liability for injuries alleged to have been inflicted on the Glen Avon residents from exposure to more than 200 chemicals in 34 million gallons of waste dumped into the Stringfellow Canyon between 1956 and 1972. The defendants' injury claims exceeded $800 million.[2]

In the northern part of New Orleans, Louisiana, residents live in an area nicknamed Cancer Alley. The community was built over a buried dump that contains approximately 150 toxic chemicals, including lead, which can cause physical and mental developmental problems in children and kidney and nervous system damage in adults. The contaminants were found in the soil around residents' homes. The area was relatively isolated, and the residents mostly poor, having no political influence with legislators. Nonetheless, the residents went to Washington to ask for help, and the Environmental Protection Agency (EPA) agreed to clean up their neighborhood. However, it decided to simply dig up two feet of the contaminated soil, lay down a layer of synthetic matting, and place two feet of uncontaminated soil over the mat. The plan angered many residents, who did not believe the EPA's actions would protect them from future exposure to the contaminated soil. Most residents wanted to move off the site entirely and were stunned that their health was not important to anybody in Washington.

These events are clear examples of how some materials can cause or significantly contribute to serious illness or death in humans if not properly managed. These materials are considered to be hazardous materials. Hazardous waste is any garbage (liquid or solid) that threatens human health or the environment if it is carelessly thrown away or handled improperly, stored improperly, or disposed of carelessly. Both hazardous materials and waste can pose a substantial threat to human health and the environment if not kept isolated. Toxic waste also poses a problem. Although most chemicals are safe, those categorized as toxic are usually defined as a subset of hazardous substances that produce adverse effects in living organisms.[3] The primary concern for hazardous materials and wastes is leakage from corroded containers or unlined or leaking landfills, ponds, and lagoons, which can contaminate groundwater.[4]

To be classified as hazardous waste, a material must meet one of the following criteria:

a. Flammability: wastes that pose a fire hazard during routine handling
b. Corrosivity: wastes requiring special containers or segregation from other materials
c. Reactivity: wastes that react spontaneously when heated, shaken, or exposed to air and/or water
d. Toxicity: wastes that pose a substantial hazard to human health and the environment[5]

The disposal of high-level radioactive wastes from commercial nuclear power plants is a significant problem.[6] It became clear to environmental groups and scientists after the Love Canal and the Three Mile Island accident that grim health dangers could result from nuclear energy. The public's concern about the potential health effects of hazardous waste rose quickly on the national agenda at the end of the 1970s. Concurrently, legislation about hazardous and toxic waste grew rapidly as well. Hazardous waste then became one of the major environmental issues of the 1980s.[7]

The emergence of toxic chemicals as an environmental issue energized the chemical industry to become a leader in the anti-environmental movement. Their involvement was pushed even deeper when Congress passed the Toxic Substance Control Act of 1976 and then increased attention to hazardous waste sites. The chemical industry created the American Industrial Health Council, which became a major political group representing the chemical industry. It fought against proposed laws to regulate chemicals used in industry. The chemical industry also created the Chemical Industry Institute of Toxicology, responsible for carrying out research studies on toxic chemicals. The industry also provided information that could be used to combat scientific research concerning the effects of toxic chemicals, especially in terms of human health.[8]

Many laws have been passed by Congress to minimize the effects of hazardous materials and wastes on humans and the environment. The following chapter presents a discussion of that information.

1946–47: 79th Congress

Congress passed PL 79-585, the Atomic Energy Act, in 1946. This was one of the first pieces of federal legislation in Congress concerning toxic waste. The Atomic Energy Act established a five-member Atomic Energy Commission

(AEC) and a Congressional Joint Committee on Atomic Energy that would provide federal supervision over nuclear technology.

1953–54: 83rd Congress

In 1954, Congress passed the Civilian Nuclear Power Act. This law charged the AEC with developing peacetime applications of nuclear power.[9] It outlined the policy for both civilian and military uses of nuclear materials. For civilian use, the bill designated the building of demonstration reactors and facilities in the United States. All nuclear materials and facilities had to be licensed, and the Nuclear Regulatory Commission (NRC) was supposed to enforce any standards it developed.

President Johnson

Johnson realized and discussed the problems of toxic waste pollution. He gave the Department of Health, Education and Welfare (HEW) the responsibility for developing a computer-based file on toxic chemicals, and he called for legislation to tighten control over the manufacture and use of agricultural chemicals.[10]

President Nixon

On February 8, 1971, Nixon gave a speech to Congress on the environment. In it, he said, "As we have become increasingly dependent on many chemicals and metals, we have become acutely aware of the potential toxicity of the materials entering our environment. Each year hundreds of new chemicals are commercially marketed and some of these chemicals may pose serious potential threats. Many existing chemicals and metals, such as PCB's (polychlorinated biphenyls) and mercury, also represent a hazard. It is essential that we take steps to prevent chemical substances from becoming environmental hazards. Unless we develop better methods to assure adequate testing of chemicals, we will be inviting the environmental crises of the future."[11]

Nixon then went on to propose that the EPA administrator be given the ability to restrict the use or distribution of any substance that he found to be dangerous to human health or the environment. He also suggested that the EPA administrator be granted the ability to seek immediate injunctive relief when use of a substance could be hazardous to human health or the environment. Finally, Nixon suggested that the administrator should set minimum standard tests for hazardous substances.[12]

1971–72: 92nd Congress

The Senate failed to complete a bill called the Toxic Substances Control Act (S 1478) in this session. The proposal would have established a federal regulatory system to protect the public and the environment from potentially hazardous chemicals.[13] This did not pass.

In 1972, the organochlorine pesticide DDT was banned for most uses. Since its development during World War II, the chemical had been used to control insects on crops. After a three-year study, the EPA administrator reported that the chemical posed an unacceptable risk to the environment and to human health.

1973–74: 93rd Congress

Both the Senate and House passed different versions of a bill (S 426/HR 5356) that would establish a federal regulatory system to protect human health and the environment from potentially hazardous chemicals. One issue of contention between the two bills was the extent to which the EPA administrator would be required to use existing laws rather than the law under consideration to regulate potentially dangerous chemicals. The legislation was originally proposed by Nixon in his 1971 environmental message to Congress. The bill was still in conference when Congress adjourned, so it did not become law.[14]

President Ford

According to Ford, the Toxic Substances Control Act of 1976 provided authority to regulate any of the chemicals in commerce, which was particularly important since only a few of these chemicals had been tested for their long-term effects on human health or the environment.[15]

1975–76: 94th Congress

In this session, Congress passed the Toxic Substances Control Act of 1976 (S 3149/PL 94-469). The law gave the EPA broad powers to regulate chemicals in all phases of their production and use, from acquisition or manufacture to disposal. In doing so, the law expanded federal regulation of industrial and commercial chemicals and required premarket testing for potentially dangerous chemicals. The bill directed the EPA to require that chemical manufacturers test products that might pose a risk to human health or the environment. Those companies planning to produce a new chemical or to market an existing substance for a new purpose would have to notify the agency 90 days in advance. The EPA was given the power to hold up marketing of that chemical while more testing was done or even to ban a chemical in an extreme case.[16]

The new law required the EPA to screen new chemicals to assess their safety before they could be marketed. The manufacturers would be required to notify the EPA of their intent to produce a new chemical substance (excluding pesticides, drugs, chemicals in cosmetics, food additives, and radioactive materials that were already regulated by another statute) or use an established chemical in a new way or for a new purpose. The chemical manufacturer had to present information on the nature, use, toxicity, and possible carcinogenic effects of the chemical to the EPA at least 90 days before its manufacture. Based on that information, the EPA would then determine whether the chemical was safe for use or whether it should be prohibited from manufacture, sale, use, or disposal if it presented an unreasonable risk to health or the environment. The EPA could also limit the amount of the chemical that could be manufactured and used, or the manner in which the chemical could be used. In cases where the EPA determined there was a "reasonable risk" of injury to human health or the environment, it could prevent manufacture for 180 days. Manufacture of the chemical could be prevented for longer than 180 days by court injunction if there was evidence of a risk to health or the environment. Once an injunction was issued, the manufacturer had to prove that the compound was safe before it could be produced.[17]

When it came to chemicals already on the market, the new law created the Toxic Substances Program, which would be administered by the EPA. A committee would be created to develop a priority list of chemical substances to be tested for possible harmful effects. The committee could name up to 50 chemicals that could then be tested within one year. However, the act also required the EPA to test about 55,000 chemical substances already in commerce. If the EPA found that a chemical could present an unreasonable risk to citizens or the environment, it could then impose virtually any type of controls on those chemicals.[18] It could prohibit the manufacture, sale, use, or disposal of the chemical, as well as limit the amount of the chemical that could be manufactured and used or the manner in which the chemical could be used.[19] If the EPA decided that there were insufficient data or that the data were inadequate to predict the effects of the substance, the EPA would require testing. In addition, the law regulated the labeling and disposal of polychlorinated biphenyls (PCBs) and prohibited their production and distribution after July 1979.

The second major law Congress passed was S 2150 (PL 94-580), the Resource Conservation and Recovery Act of 1976 (RCRA). The provisions targeted those who generated, transported, treated, stored, or disposed of hazardous solid waste and outlined goals for hazardous waste management. Some have called this one of the most important laws passed to protect the environment from hazardous pollutants because it provided a program aimed at solving one of the highest-priority environmental problems confronting the nation: the disposal of hazardous waste. It was the first federal law focused on the disposal of solid hazardous waste, and it set basic rules for the handling, storage, and disposal of hazardous wastes.[20]

Specifically, the RCRA called for the development of criteria for distinguishing hazardous wastes. The act clearly defined the responsibilities of the generator, transporter, and disposer of hazardous materials. The "waste generator" was required to keep track of the amount of hazardous wastes that were generated and report this amount to the government. A manifest had to be prepared for all wastes identified by the EPA. The "transporter" of qualifying wastes had to ensure that the waste was properly packaged and labeled and that the manifest accompanied the wastes from the point of generation to the point of disposal. Under RCRA, industries were required to track waste through its final disposal, from "cradle to grave." If hazardous wastes were sent offsite for disposal, only approved transportation companies could be used to transport the wastes, and only specially licensed "treatment, storage and disposal" (TSD) facilities could accept the wastes. If hazardous wastes were kept onsite, the industrial plant itself became a type of hazardous waste treatment, storage, and disposal facility under RCRA and was therefore subject to licensing and other special requirements. Land disposal, burying wastes in a local landfill, is stated to be the "least favored method for managing hazardous wastes."[21] No matter what was done with the wastes, RCRA required that paperwork be filled out along the way, allowing the government to track hazardous waste from its creation to its disposal. The law developed a permit system based on EPA standards to treat, store, and dispose of hazardous wastes.

The law also provided standards for state waste management programs. A state could operate its own hazardous waste program in lieu of the federal program if it met the EPA standards. The law required financial capability to clean up any spills. Minimal criminal penalties were established for violations. Failure to comply with the provisions of the law could result in civil action and, in certain cases, fines and imprisonment.[22]

Over time, the RCRA became the primary tool of the federal government for controlling hazardous wastes. It put limits on the dumping of hazardous wastes and supported more cautious handling of hazardous waste. At the same time, the requirements for handling of, and disposing of, hazardous waste added significant costs for manufacturers, which in turn created an incentive for companies to generate less waste.[23] Thus, the law provided an economic incentive for the recycling and recovery of waste materials. RCRA also gave the government a better understanding of the hazardous waste problem by allowing it to track the types and quantities of waste being generated by industries in the United States, which had not been done before.

Another law passed during this session of Congress was S 2498 (PL 94-305) on small business assistance. This included a grant program to permit small businesses to finance pollution control equipment through the sale of tax-exempt industrial revenue bonds.[24] Ford signed the bill, even though he opposed the combination of a federal guarantee and a tax-exempt security. The formal name of the bill was the Small Business Export Development Act.[25]

President Carter

In 1979, an accident at Three Mile Island in Pennsylvania caused the nuclear reactor to melt down. It was a costly near disaster. Calls for new U.S. nuclear reactors ended. During the first year of his presidency, Carter said, "The presence of toxic chemicals in our environment is one of the grimmest discoveries of the industrial era. Rather than coping with these hazards after they have escaped into our environment, our primary objective must be to prevent them from entering the environment at all."[26] He asked the Council on Environmental Quality to develop a program to collect data on toxic chemicals and coordinate federal research in the area. He also ordered the EPA to give highest priority to developing industrial effluent standards to control toxic pollutants and to set standards to limit human exposure to toxic substances in drinking water.[27]

In his 1980 State of the Union address, Carter promised to pursue policies with proper regard for environmental values. He specifically mentioned that he wanted to create a hazardous waste management program.[28] In August 1978, Carter declared the Love Canal area in Niagara Falls, New York, a national disaster because of past hazardous waste dumping. With this action, Carter authorized the use of federal disaster relief aid. In addition, he proposed legislation called the Oil, Hazardous Substances and Hazardous Waste Response, Liability and Compensation Act to provide the first comprehensive program to address releases of oil and hazardous substances from spills and from inactive and abandoned sites into navigable waters, groundwater, land, and air.[29] He said that the legislation would allow the federal government, along with state and local governments and industry, to identify abandoned hazardous dump sites across the nation, establish a uniform system of reporting spills and releases, and provide emergency government response and containment to clean up pollution without delay. The law would also provide vigorous investigation of spills or other releases, stronger authority to compel the responsible parties to clean up dangerous sites, and compensation for damage to property resulting from spills. The financing for these actions would come from a national fund of appropriations from a fee on the oil and chemical industries.[30]

President Carter said that the country needed the "development and implementation of a strong, responsible program to manage and dispose of nuclear wastes."[31] Carter proposed legislation for nuclear waste that included plans to permanently establish a state planning council to deal with nuclear waste problems. The licensing authority of the NRC would be extended to cover low-level waste storage. The bill included assistance to aid states in managing commercial low-level waste. Finally, the bill would clarify the federal responsibility for continued management of abandoned federal or federally utilized facilities.[32]

On January 21, 1980, Carter addressed Congress and said, "One of the most important environmental and public health issues facing our Nation is the threat caused by the improper disposal of hazardous substances. Accidents like those at Love Canal and Valley of the Drums have highlighted the inadequacy of the existing laws and inability of governments at all levels to respond quickly and efficiently to these dangerous incidents."[33]

Carter followed that speech with another on February 12, 1980, in which he announced the "nation's first comprehensive radioactive waste management program" to manage nuclear wastes and protect the health and safety of all Americans. Using an executive order, Carter established a State Planning Council to "advise the Executive Branch and work with the Congress to address radioactive waste management issues, such as planning and siting, construction, and operation of facilities." During that speech, Carter asked Congress to pass legislation he proposed that would authorize the Department of Energy to design, acquire or construct, and operate one or more away-from-reactor storage facilities. These facilities would also accept domestic spent fuel, and a small amount of foreign spent fuel, for storage, until permanent disposal facilities were available.[34]

1977–78: 95th Congress

S 1531, approved by the Senate, would amend the 1976 Toxic Substances Control Act to allow victims of chemical spills to be compensated for their out-of-pocket losses.[35] The bill was not passed by the House. Another bill, HR 13650, was signed by the president (PL 95-604). The law, called the Uranium Mill Tailings Radiation Control Act, set up a program to clean up 25 million tons of potentially hazardous uranium waste. It mandated the cleanup and also provided for stricter control on future handling and disposal of wastes from the processing of uranium.[36]

1979–80: 96th Congress

On March 29, 1979, the cleanup began at the Three Mile Island power plant. The majority of the radioactive water was removed from the reactor building, but decontaminating the reactor building and dealing with the reactor core were more complicated. When the insurance money did not pay for the entire cost of the cleanup, the plant owner, GPU, stopped working on it. Many community members believed that, in time, radioactive material would start leaking, so they demanded that the cleanup be completed. To pay for the costs of the cleanup, the Senate Energy Committee approved a bill (S 1606) to force the nation's utilities to pay part of the cost of cleaning up the reactor. But the bill never made it to the Senate floor and did not become law.[37]

In 1980, the Comprehensive Environmental Response, Compensation, and Liability Act (CERCLA), also called Superfund, was passed (HR 7020/S 1480/PL 96-510). This was an attempt by the federal government to expedite the cleanup of contaminated hazardous waste dump sites and remedy problems created in the past rather than reduce pollution as it is created.[38] The law was signed by Carter as one of his last acts as president.

CERCLA established a $1.6 billion trust fund for the emergency cleanup of hazardous substance spills and for the cleanup of inactive hazardous waste sites for which responsible parties could not be identified or if the responsible party was unable to clean it up.[39] The Superfund fund allowed the government to set aside money that could be used to pay for emergencies arising from hazardous waste contamination.[40] The money for Superfund did not come from taxpayers but from special taxes on the chemical industry. A tax on oil and certain other chemicals provided 87.5 percent of the funding, with the remaining 12.5 percent appropriated by the federal government. The fund was necessary because the regulatory program of the RCRA did not address abandoned and inactive dump sites.[41]

The bill also required anyone in charge of a facility where hazardous substances were stored or disposed to notify the federal government of the existence of that facility and notify the government if there were any likely discharges. If the facility was found to have released a hazardous substance, it was to immediately notify the government. If they did not, the punishment could be a fine of up to $10,000 and/or imprisonment for up to one year.[42]

Strict joint liability for all parties responsible for contamination was a key part of the law. All those who had used the contamination sites were jointly and individually liable for the costs of cleanup and damage. The EPA was given the authority to identify the parties responsible for waste sites and to force them to clean them up. If the responsible parties refused, the government could sue them for reimbursements to the Superfund for cleanup costs and other expenses.[43]

As part of the law, the EPA was required to create a National Priority List to identify the worst hazardous waste sites in the country. The EPA was to assess the risks from the sites and rank them according to the hazards they presented. This would become the National Priority List of the most dangerous sites. The EPA could then take emergency action to control any immediate hazards.[44] Although the bill did not address oil spills, as Carter had proposed, he signed the bill. Some members of the House were also not pleased with the bill and tried to kill it. Nonetheless, it was passed and became law.[45]

Another bill (S 1156) gave the EPA tougher enforcement powers to control "midnight dumpers" of hazardous wastes. Under existing law, violators could be charged only with a misdemeanor. Called the Solid Waste Disposal Act Amendments of 1980, the proposed bill made it a felony for anyone to "knowingly" endanger human life when disposing of hazardous wastes. The bill also allowed EPA to go to court to seek a cleanup order if a hazardous waste might present an imminent or substantial danger to health or the environment. It also expanded some of EPA's toxic waste enforcement powers. The proposal was signed by the president (PL 96-482).[46] Congress also proposed legislation on toxic substances. A bill (S 1147/HR 6995) to extend the Toxic Substances Control Act, first passed in 1976, was passed in the Senate but not acted on in the House.[47]

In S 685, passed by the Senate Energy Committee, the Department of Energy would be required to come up with a design and site for a long-term nuclear waste storage facility within a year. In doing so, the Congress was trying to force action on the long-standing problem of what to do with nuclear waste. They wanted to force the government to design a giant concrete vault-like facility from which the nuclear waste could be retrieved and placed in permanent storage. The waste would be monitored and could be stored there for decades or longer. In the meantime, the government would continue researching plans for permanent storage of the waste, probably in an underground site where radioactive waste could be isolated and stored for centuries.[48]

The Senate cleared legislation on December 13, 1980 (S 2189/PL 96-573) that gave states the responsibility for burying low-level radioactive waste. In doing so, the Congress postponed deciding on a solution to the more difficult problem of disposing of high-level nuclear waste and storing spent fuel from nuclear reactors. Finding a solution to the problem of low-level waste was critical because a large amount of it was being generated in every state, but only three states had dumps to hold it. These states were threatening to stop taking it all. The new law, called the Nuclear Waste Policy Act, made the disposal of commercial low-level waste a state responsibility. States could build their own dump sites or, if they chose, could form regional compacts with other states to establish a mutual burial site. After 1985, these regional groups, which had to be approved by Congress, could refuse waste from states that chose not to join the compact. Waste generated by the federal government would not be the responsibility of the states, nor could the compacts make policies affecting federal waste, which came mostly from military weapons production. The bill also required the Department of Energy to evaluate existing burial sites and transportation problems.[49]

Congress addressed the problem related to recycling of used oil in S 2412 (PL 96-463). In the Used Oil Recycling Act of 1980, Congress established a program to recycle used lubricating oil. The law required that lubricating oil be labeled with a statement concerning the recycling of used oil. The EPA administrator would make grants available to those states with an approved or proposed solid waste plan that encouraged the use of recycled oil while discouraging the use of hazardous oil as a way to protect public health and the environment. There were calls for informing the public on the uses of recycled oil and establishing a program for the collection and disposal of oil in a safe manner. The bill also authorized the administrator to provide technical assistance to states in removing impediments to the recycling of used oil.

President Reagan

Reagan was concerned that future generations of Americans be protected from the adverse health effects of toxic chemicals and hazardous wastes.[50] He spoke about accelerating the efforts to put the Superfund to good use to clean up hazardous dumps that might pose a danger to human health.[51]

When the Superfund was set to expire in 1985, Reagan asked Congress to extend it to allow for more time to clean up dump sites.[52] He proposed legislation to triple the size of the Superfund program to $5.3 billion. The fund would be partially comprised of a tax imposed on the manufacture of certain chemicals and a new fee on the disposal and treatment of hazardous waste.[53]

1981–82: 97th Congress

S 1211/HR 3495 (PL 97-129) was passed to extend the Toxic Substances Control Act for one year. HR 3809, the Nuclear Waste Policy Act (PL 97-425), established a national plan for the disposal of highly radioactive nuclear waste. It required that the secretary of energy issue guidelines for the recommendation of sites for nuclear waste repositories and that an environmental assessment of each potential repository site be completed. The bill also contained provisions for spent fuel storage, retrievable storage, and an evaluation of a possible location for an underground storage facility.[54] The new law gave the Department of Energy until 1998 to open a permanent underground repository for high-level nuclear waste. The measure also established the Nuclear Waste Fund, made up of fines imposed on nuclear utilities that produced electricity.

Two separate committees (Energy and Environment) passed different versions of a bill (S 1662) that set a timetable for locating and approving construction of a permanent nuclear waste repository; this was to occur by the year 2000. The House committee passed HR 5016, which called for construction of a waste repository by 1990. This was important because no policy had been established for disposing of spent fuel from nuclear reactor plants, which was piling up at reactor sites around the country, or the millions of gallons of highly radioactive liquid waste that had been generated, mostly in weapons production. It was being stored in tanks in Washington State, Idaho, South Carolina, and New York until a waste program was developed.

However, some key issues were controversial. One concerned the best way to permanently dispose of nuclear waste, either in deep underground geological formations or in monitored, man-made vaults near the Earth's surface. Also of concern was the role of the states. Third was radioactive waste from the nuclear weapons program, which constituted more than 90 percent of the nation's total volume of nuclear waste.[55]

The House passed HR 6307 to reauthorize and tighten the nation's principal law governing the management and disposal of hazardous wastes, but it died in the Senate. Much of the support for the House bill came from members' concern that the Reagan administration was not enforcing hazardous waste disposal laws. The RCRA was the basis for management of the nation's hazardous waste and for encouraging conservation and recovery of valuable materials and energy from solid wastes. Some House members thought the EPA's efforts to enforce sections of the legislation were weak.[56]

1983–84: 98th Congress

This session, Congress passed HR 2867 (S 757/PL 98-616), the Hazardous and Solid Waste Amendments of 1984, the most significant pollution control legislation enacted in 1984. The law tightened the original law, passed in 1976, by removing enforcement discretion from the EPA after some members claimed the agency had abused it. The newly created law banned the disposal of hazardous liquids in landfills and brought small businesses under requirements for safe disposal of the hazardous wastes they produced.[57] The EPA was required to issue regulations on ways to minimize the disposal of containerized liquid hazardous waste. If waste was stored underground, the owners would be required to notify state agencies if the tanks were used to store hazardous substances. Any new or expanding land disposal facilities for hazardous waste would be required to have two liners and to monitor groundwater for leaks.[58]

More specifically, the amendments strengthened the nation's main law regulating the handling of hazardous waste by banning the disposal of all liquid and some solid hazardous wastes on land. To reduce the chance of groundwater contamination, Congress enacted a rule called the Land Ban, which prohibited the disposal of hazardous waste in landfills unless the waste was treated first or unless its disposal involved no risk to public health. The Land Ban went into effect on May 8, 1990.

It also set up the first federal program regulating the design of underground tanks that stored gasoline and other hazardous liquids. All hazardous waste landfills now needed to have a double liner along their bottom and sides. The liners are typically made of synthetic materials or clay, and they help hold the hazardous waste in place. A pump must be placed between the liners so that if the first liner cracks and hazardous waste begins to leak out, the waste can be pumped back into the landfill before it breaches the second liner.[59]

The amendments set new deadlines for EPA rulemaking on control of hazardous chemical wastes. It also directed the EPA to regulate new classes of facilities, including generators of small quantities of hazardous wastes and underground petroleum and chemical storage tanks. In other words, the law applied hazardous waste requirements to "small-quantity generators," and it required cleanup and replacement of all underground storage tanks that might be subject to leakage.

In the HSWA, Congress gave little attention to the costs and economic impacts of the program it was directing EPA to implement. Because the RCRA does not explicitly allow EPA to consider costs in setting standards, the HSWA rules are among the most expensive ones the EPA has issued.[60]

In December 1984, the Union Carbide pesticide plant in Bhopal, India, released toxic gas, killing or injuring 3,000 people. It was the worst industrial disaster in history, focusing public and official attention on emergency releases of toxics. In response, the Congress passed a community-right-to-know law in the reauthorized Superfund law of 1984 and created an accident investigation board.[61] Under the proposal, citizens would have the right to know what chemicals were being stored in a manufacturing facility and their potential danger to the community. The proposal, the Superfund Expansion and Protection Act of 1984 (HR 5640), was an attempt to renew and expand the 1980 Superfund for the cleanup of hazardous waste dumps. The bill died in the Senate.[62]

The Department of Energy chose three sites for geologic exploration to determine which should house the nation's first repository for high-level radioactive waste; the naming of the sites was a step in the procedure set up by a 1982 law (PL 97-425) to establish a deep, underground permanent storage area for highly radioactive nuclear waste by 1998. The sites being considered were Yucca Mountain in Nevada; Deaf Smith County, Texas; and Hanford, Washington.[63]

A proposed bill passed that reauthorized funding for programs designed to regulate shipments of hazardous materials (S 2706/HR 5530/PL 98-559). The law, called the Hazardous Materials Transportation Act Amendments of 1984, authorized $7.5 million for fiscal year 1985 and $8 million for fiscal year 1986 for programs under the Hazardous Materials Transportation Act, housed within the Department of Transportation. The legislation also authorized the secretary of transportation to enter into agreements with private entities to report hazardous material transportation accidents to a central system. Further, the bill required the secretary and the director of the Federal Emergency Management Agency to jointly evaluate training and planning for responding to incidents involving the transportation of hazardous materials and to report their findings to Congress.[64]

1985–86: 99th Congress

In this session of Congress, the members agreed to pass the Low-Level Radioactive Waste Policy Amendments Act of 1985 (HR 1083/PL 99-240), that would impose strict deadlines for states or regions to set up disposal facilities for low-level radioactive wastes. The measure made states responsible for disposal of nonfederal low-level radioactive wastes generated within their borders. Further, the states would be legally liable for any damages incurred by those businesses that had to shut down because their states did not provide for disposal of wastes with low levels of radioactivity.[65] Three states, South Carolina, Washington, and Nevada, would continue to accept waste from other states. The legislation set a deadline of January 1, 1988, by which each compact commission had to have identified a "host state" for its disposal facility, and each host state had to have a plan to establish the location for a facility. States that were not members of a compact also had to develop plans for choosing facility sites within their own borders.[66]

Also during this session of Congress, members voted to strengthen Superfund in the Superfund Amendments and Reauthorization Act (SARA; HR 2005). President Reagan supported the bill and signed it (PL 99–499), making it law.[67] The new legislation set strict standards for cleaning up hazardous dump sites and required the EPA to begin work to clean up 375 sites within five years. Congress also provided for a five-fold increase in the money available in the trust fund to pay for emergency actions and for cleanups when the companies responsible could not pay the entire costs. The fund was increased from $1.6 to $8.5 billion. The money for the fund was to come from taxes on crude oil and certain chemicals and a special environmental tax on businesses. Additionally, Congress tightened cleanup requirements by giving the president and the EPA broader powers to force companies to pay for cleanups. At the same time, the law gave victims of toxic dumping a longer opportunity to sue those deemed responsible for the dumping.

The Congress also added a separate title to the act, called the Emergency Planning and Community Right-to-Know Act (EPCRA). This was a new program that mandated nationwide reporting for toxic and hazardous chemicals produced, used, or stored in communities, resulting in the Toxics Release Inventory, as well as state and local emergency planning for accidental chemical releases. The community right-to-know provisions required firms to maintain and make available data about harmful chemicals used or stored on a site and to report the annual emission of such chemicals. The Toxics Release Inventory has become a major source of data on industrial emissions.[68] The act also established a special program within the Department of Defense for restoration of contaminated lands, somewhat similar to the Superfund under CERCLA.

SARA included three key provisions for when a hazardous substance is handled and when an actual release has occurred. Even before an emergency occurs, facility owners and operators must provide information pertaining to any regulated substance held at the facility to state and local government officials as well as the general public. Three types of information were to be reported:

1. Material safety data sheets (MSDSs), which were to be prepared by the manufacturer of any hazardous chemical and should be kept by the facility owner or operator. They would contain information on a hazardous chemical.
2. Emergency and hazardous chemical inventory forms, which were to be submitted to the state and local authorities each year. The form would include the maximum amount of a hazardous chemical that could be present at any time during the year and the average daily amount present during the year prior. Also included is the general location of the hazardous chemicals. This information would be made available to the general public upon request.
3. Toxic chemical release reporting, which provides general information about effluents and emissions of any toxic chemicals.

If a hazardous substance was released, a facility owner or operator had to notify the authorities. They would need to identify the chemical involved, the amounts released, the time and duration, the possible environmental impact, and the suggested action. An emergency plan on the state and local level was to be established.

This bill was needed because, under PL 96-573 (passed in 1980), states were encouraged to form regional compacts for joint disposal of low-level radioactive wastes. In the following few years, the states made little progress toward creating disposal sites. Only three states (South Carolina, Washington, and Nevada) had disposal sites. Under the law, the regional compacts to which those states belonged had the power to exclude waste from other states. Congress did not ratify the proposed interstate compacts because states with low-level radioactive waste were fearful that the waste-receiving states would not allow them to dispose of their waste.[69]

Congress amended the Toxic Substances Control Act of 1976 with the Asbestos Hazard Emergency Response Act (HR 4311). This required inspection for, and removal of, asbestos in all local schools, as well as notification of parents. The act also called for a study of the extent of, and responses to, the public health danger posed by asbestos in public and commercial buildings. The proposal did not have enough votes to pass the House.

Finally, in HR 3917, hazardous waste disposal facilities that had to close because they could not meet a November 8 deadline for obtaining mandatory insurance coverage would be required to remain open. The bill passed the House but not the Senate before Congress adjourned. The bill would have extended deadlines for disposal facilities to comply with the RCRA for one year, under which they were licensed to operate. The 1976 law authorized the EPA to set requirements for safe disposal of hazardous waste. Companies already operating hazardous waste disposal facilities in 1976 could continue to operate under interim permits, without meeting all safety requirements, until the EPA decided whether they should receive final permits.[70]

1987–88: 100th Congress

House and Senate negotiators agreed on a plan to put all of the nation's nuclear waste in Nevada, even though representatives from that state protested. The plan was added to the fiscal year 1988 Omnibus Budget Reconciliation Act (HR 3545/PL 100-203), which Reagan signed. The plan reversed a decision made in Congress in the 1982 Nuclear Waste Policy Act, which declared that selection of a site would be based purely on science and safety rather than politics. Officials in the Department of Energy narrowed down the search for the first nuclear waste repository to three Western states and named Yucca Mountain as the site where the department was to drill the first exploratory shaft for a permanent repository for the nation's highly radioactive waste.[71]

The House passed HR 3070 to close loopholes in existing hazardous waste law. PCBs, chemicals used in electrical transformers, were largely banned in 1976 because they were proven to cause cancer and disorders of the nervous, gastric, and reproductive systems. Under the original law, only those who ultimately disposed of PCBs were regulated. The generators of the used PCBs, as well as the middlemen who transported and stored the chemicals before disposal, did not need special permits. The new law changed that.[72] The bill was referred to the Committee on the Environment and Public Works, but got no further.

President G.H.W. Bush

The president's budget proposed $315 million to pay for the cleanup of toxic waste sites through the Superfund program. He reported that his administration had opposed congressional efforts to cut the Superfund budget.[73]

He also asked for a billion dollars in the federal government's effort to clean up the environmental effects of federal nuclear weapons plants. He signed a new bill into law that waived federal immunity for violation of laws related to hazardous waste.[74]

1989–90: 101st Congress

The House voted to require federal agencies to comply with solid and hazardous waste regulations. HR 1056 (S 1140) would reaffirm the power of the EPA and state governments to enforce compliance with the RCRA. The application of RCRA regulations to federal facilities, particularly plants producing nuclear weapons, had been challenged in court by the Defense and Energy departments.[75] The bill did not pass during this session.

1991–92: 102nd Congress

Congress considered a bill to reauthorize RCRA in 1991, but it didn't pass. HR 6199, the Voluntary Environmental Response Act of 1992, applied to any facility where there had been a release or threat of release of a hazardous substance into the environment. It required that an action plan for responding to a release of a hazardous substance had to be filed with the EPA administrator or the state.

A proposal by the Department of Energy to open a billion-dollar facility in New Mexico to store defense-generated nuclear waste passed the Senate and gained approval from three House committees in this session but not the entire House. The bill (HR 2637, S 1671) would have been the first step toward the opening of the Waste Isolation Pilot Plant, the Department of Energy's underground nuclear waste storage facility near Carlsbad. The Department of Energy became frustrated when Congress failed to act, and it attempted to obtain control of the site from the Interior through an administrative land transfer. Officials from New Mexico filed a lawsuit and won. This forced Congress to act on legislation to allow the site to open.

In 1991, the House and Senate moved forward on plans to build a high-level nuclear waste dump site at Yucca Mountain, Nevada. The two chambers gave permission to the Department of Energy to study the site, even though officials from Nevada argued the area was unsuitable. The Department of Energy was allowed to begin construction of a temporary waste dump, known as a monitored retrievable storage facility, at Yucca Mountain. Additionally, the Senate Energy Committee approved a bill (S 1138), the Nuclear Waste Policy Amendment, that would allow the federal government to preempt state permit laws and move ahead with the project. This did not pass. Instead, Congress passed HR 776 (PL 102-486), the Energy Policy Act, which directed the EPA administrator to create public health and safety standards to oversee releases of radioactive materials stored or disposed of in the Yucca Mountain repository site. The law directed the administrator to work with the National Academy of Sciences to conduct a study to analyze recommendations on reasonable public health and safety standards.[76]

In 1992, Congress cleared legislation (Waste Isolation Pilot Plant Land Withdrawal Act) that paved the way for the Department of Energy to begin storing defense-related nuclear waste at the Waste Isolation Pilot Plant near Carlsbad, New Mexico. President Bush signed the bill, making it a law (HR 2637/PL 102-579). The site, designed to store plutonium-tainted waste from the nation's nuclear weapons factories, had been ready to accept the waste, but until now, the Department of Energy had lacked the necessary approval from Congress to begin storing waste there. The new legislation transferred the New Mexico dump site to the Department of Energy and set certain restrictions that had to be met before testing could begin, including oversight by the EPA. The law also authorized giving $20 million a year in federal aid to the state of New Mexico for 15 years.[77]

Because lead exposure in children leads to lower IQs and can lead to cardiovascular, reproductive, and testicular disorders, as well as osteoporosis and miscarriages, in adults, the Congress passed a law that directed the EPA and other agencies to take actions to reduce risks from lead paint, especially for children in urban areas. The Residential Lead-Based Paint Hazard Reduction Act was proposed in the Senate but not turned into law (S 2341).[78]

A bill that would give states and the EPA greater power to require federal agencies to obey hazardous waste laws passed both chambers and became law (HR 2194/S 596/PL 102-386). The federal government was considered one of the worst polluters, with the main culprits being the Department of Energy's nuclear weapons plants and the Department of Defense's military installations. However, none of the federal agencies was exempt from antipollution laws. The proposed bills gave the EPA additional authority to force federal agencies to comply with RCRA. The legislation that eventually became law, the Federal Facilities Compliance Act, also clarified states' authority to force compliance with state environmental laws. Such authority had been hindered because some courts had ruled that federal agencies were immune from state regulation.[79]

President Clinton

President Clinton was concerned that too many kids in America were growing up near toxic waste dump sites,[80] especially since he realized the possible link between toxic waste and cancer.[81] He wanted to clean up the waste sites, or brownfields, and claimed to have cleaned up more toxic waste sited in 3 years than the previous administrations did in 12.[82]

In 1994 Clinton proposed changes to Superfund to allow for faster and quicker cleanups because the toxic land posed such serious dangers to the health of people living near them.[83] He blamed the Congress for blocking legislation to clean up toxic waste dumps,[84] reporting that the Congress filibustered the bill to prevent a vote on it.[85]

The president supported a Community Right-to-Know Act that would force factories to disclose what chemicals are stored in their facility and what harm they might cause if accidentally released into the community.[86]

Clinton wanted to make the use of nuclear energy safer, but he knew the spread of nuclear material was a serious global concern.[87] He had a joint agreement with the Japanese concerning nuclear waste disposal.[88]

1993–94: 103rd Congress

Two bills (HR 3800, S 1834) that were proposals to overhaul the nation's hazardous waste cleanup program, otherwise known as Superfund, died in the final weeks of the 103rd Congress, derailing one of the Clinton administration's top environmental initiatives. Clinton had made passage of a Superfund bill one of his top three environmental priorities for the year. The proposals did not make it to the floor of either chamber.[89]

HR 540/HR 541, the Toxic Pollution Responsibility Act of 1993, would amend CERCLA to absolve municipalities or persons that generate or transport municipal solid waste from liability under that act. This did not pass.

1995–96: 104th Congress

In 1995, Congress suspended the special taxes on crude oil and certain chemicals and the special environmental taxes on business that financed the Superfund trust fund. In a short time, the money available through Superfund dwindled. Some members of Congress even discussed retiring Superfund as it ran out of money. Nonetheless, a proposal to overhaul Superfund was debated during this session. The bill, entitled Reform of Superfund Act (HR 2500/S 1285), would have allowed the government to take legal action against a single polluter to recover the cost of cleaning up a hazardous waste site. A controversial issue was "retroactive liability" that allowed a private business to be held responsible for the cleanup of waste that had been dumped legally before the Superfund law was enacted in 1980.[90] This proposal did not pass. Provisions in an appropriations bill barred the EPA from adding any toxic waste sites to the Superfund cleanup list unless requested by the governor of that state. Further, it denied selected disaster relief benefits to illegal aliens.[91]

In 1995, the House membership rejected a bill (HR 558) that would have given congressional consent to an agreement among Maine, Vermont, and Texas to deposit their low-level radioactive waste at a yet-to-be-built facility in Texas. A Senate version of the bill (S 419) won voice-vote approval from the Judiciary Committee but went no further. Maine and Vermont would each have paid Texas $25 million for hosting the disposal facility. Under the compact, Texas would have been able to exclude radioactive waste from other states. Under 1985 amendments to the Low-Level Radioactive Waste Policy Act, Congress required states to open their own waste-disposal sites or make arrangements to use facilities in other states. The regional compacts that resulted from these negotiations were required to have the approval of Congress.[92]

In this session, the Senate voted to construct a temporary dump for the nation's high-level radioactive waste at Yucca Mountain in Nevada. The House considered the Nuclear Waste Policy Act (HR 1020/S 1271) concerning storage of nuclear waste at Yucca Mountain in Nevada.[93] The battle, which had been ongoing for 14 years, was between Nevadan officials who opposed the plan to store nuclear waste in their state and those in the government who believed the site to be an appropriate place for a storage facility.[94] Nevada's members managed to stall the legislation long enough to block final action on it, and the bill did not pass.

In 1996, the House Commerce Committee approved legislation (HR 1663) designed to speed up the opening of the Waste Isolation Pilot Plant in New Mexico that would store nuclear waste from the nation's weapons industry. But the House did not take up the bill before adjourning, and the measure died at the end of the 104th Congress.[95]

In 1996, a proposal (HR 3391) to expedite federal funding of state efforts to clean up leaking underground storage tanks did not make it through the Senate. The proposed bill would increase the amount of money disbursed by the federal government to the states from the Leaking Underground Storage Tank (LUST) Trust Fund. The LUST fund had been established in 1986 to ensure a source of federal assistance to clean up underground tanks that were leaking or had leaked gasoline and other toxic substances. Such leakage could contaminate groundwater and soil, potentially causing substantial health risks.[96]

1997–98: 105th Congress

S 8, HR 2727, and HR 3000 were bills to revise Superfund. Congressional members disagreed over how to clean up sites, who should pay the costs and how to remedy damage done by Superfund sites to rivers, streams, and other natural resources. There was also disagreement about how far back to go in scaling back existing retroactive liability provisions that allowed the government to hold a company responsible for the costs of cleaning up toxic waste dumped before the Superfund law was enacted.[97] Congress could not work out their disagreements in this session.[98]

A new proposal to require construction of an interim nuclear waste storage site at Yucca Mountain, Nevada, was once again introduced into the House and Senate (HR 1270/S 104). The bills passed their individual chambers, but the Senate vote fell short of the two-thirds needed to counter a threatened veto, and conference action was postponed indefinitely. The legislation would have required the Department of Energy to open a temporary site for nuclear waste near Yucca Mountain. The plan was intended to sidestep the lengthy process of building a permanent dump at the mountain site. But the Clinton administration opposed the temporary solution, because of fears the solution would kill efforts to create a permanent system for long-term underground waste disposal. They argued that a temporary location would bias scientific studies of the waste site.[99]

During this session, a bill (HR 629) concerning a three-state compact for disposing of low-level radioactive waste in western Texas won House approval, and the Senate Judiciary Committee approved a companion bill (S 270) in March. The legislation provided a new disposal site in Texas that would accept low-level radioactive waste from Maine and Vermont. Maine and Vermont would be required to ensure the safe transport of the waste and pay $25 million each to build and operate the facility. The compact had been endorsed by the governors and state legislatures of the three states, as required by law. The proposal had enough support by members of Congress to clear both chambers, and the president signed it (PL 105-236).[100]

The House passed a bill (HR 688) aimed at hastening the cleanup of leaking underground storage tanks by assuring states that they would receive federal funding to help pay for the task. Called the Leaking Underground Storage Tank Trust Fund Act, the bill would require the EPA to send the states at least 85 percent of the federal money appropriated to the agency each year from the LUST Trust Fund. The bill also sought to provide states more flexibility in determining how to spend the money. The bill did not become law.[101]

1999–2000: 106th Congress

Senators from Nevada again prevented the Senate from passing legislation on storing nuclear waste (HR 45/S 1287: the Nuclear Waste Storage Bill) in the first year of the session.[102] But in 2000, Congress designated Yucca Mountain as the likely permanent burial site for nuclear waste even though questions remained about its safety. The site was not expected to begin receiving waste before 2010. The bill would authorize the Department of Energy to offer utilities a combination of money and storage casks for short-term storage. Clinton vetoed the bill, and a Senate attempt to override the veto failed.[103] Upon vetoing the bill, Clinton explained that the bill would "do nothing to advance the scientific program at Yucca Mountain or promote public confidence in the decision of whether or not to recommend the site for a repository in 2001." He also said that the bill would limit the EPA's authority to issue radiation standards to protect human health and the environment and that it would do little to minimize the potential for continued claims against the federal government for damages that occurred as a result of the delay in accepting spent fuel from utilities. Clinton said, further, that the bill would impose new requirements on the Department of Energy without establishing sufficient funding mechanisms to meet those obligations, creating new unfunded liabilities for the department.[104]

HR 1300/S 1090, bills to overhaul Superfund, stalled in 1999.[105] Called the Recycle America's Land Act of 1999, the bill would amend CERCLA and would direct the president to establish a program to provide grants to states to inventory and assess the brownfield facilities in their jurisdiction. The proposal would also require the president to provide for meaningful public participation.

S 2962 was a bill to ban MTBE (methyl tertiary butyl ether), which was found to be leaking from underground storage tanks in 30 states and was a suspected carcinogen. Despite the danger, the legislation died at the end of the session.[106]

President G. W. Bush

President Bush gained congressional approval of important legislation to reclaim brownfields, which refers to a typically abandoned industrial site in a city. The measure involved a compromise[107] between Republicans in the House, who had wanted to reduce the liability for small businesses under the Superfund legislation, and Democrats, who preferred to see contaminated and abandoned industrial sites in urban areas cleaned up. The final bill authorized $250 million a year for five years to help states clean up and redevelop contaminated industrial sites.[108] Bush noted that between 2001 and 2003, his administration helped to clean up over 1,000 brownfields to usable condition.

President Bush also recommended the Yucca Mountain Site for the disposal of nuclear fuel and nuclear waste. He wanted to use this site to isolate highly radioactive materials that were scattered throughout the nation into one repository at a remote location.[109]

2001–02: 107th Congress

The Nuclear Liability Limit Law, HR 2983, would have extended for 15 years an earlier law to permanently limit the liability of owners of existing nuclear reactors in case of a catastrophic accident. The Senate refused to debate the bill.[110]

Congress endorsed a decision by President Bush to make Yucca Mountain in Nevada the permanent repository for high-level radioactive waste from the nation's nuclear power plants. The joint resolution, which Bush signed in 2002 (HJ Res 87/PL 107-200), overrode a veto of the project by the governor of Nevada. Nevada state officials and environmentalists had several lawsuits pending in federal court and planned to file more.[111]

A bill (HR 2869) combined with other proposals (S 350 and HR 1831) to become the Small Business Liability Relief and Brownfields Revitalization Act (PL 107-118). The law, which increased funding for the cleanup and redevelopment of polluted industrial sites (brownfields), cleared Congress shortly before it adjourned for the year. Under the new law, a total of $1.25 billion would be provided to states to assess sites for cleanup and then $50 million a year in grants would be available for the actual cleanup. The bill, which also extended superfund liability protection for small businesses and landowners, was signed into law. This was one of President Bush's top environmental priorities.[112]

A proposal in the House (HR 2941) would have made it easier for communities to use federal programs to clean up brownfields, but the Senate did not act, and the measure died at the end of the session. It would have formally authorized the Department of Housing and Urban Development Brownfields Redevelopment Grant Program and eased requirements that discouraged smaller communities from applying for the grants. The bill died in session.[113]

2003–04: 108th Congress

In the 108th Congress, the House considered HR 2926, the Nuclear Waste Terrorist Threat Assessment and Protection Act, but it did not have enough support to become law. The proposal would have directed the secretary of homeland security to develop and carry out an interagency plan to prepare for and defend against terrorist crimes targeting the Yucca Mountain project.

2005–06: 109th Congress

House members again considered the Nuclear Waste Terrorist Threat Assessment and Protection Act (HR 895). Similar to the bill in the previous session, this bill directed the secretary of homeland security to develop and carry out a plan to prepare for and defend against terrorist crimes targeting the Yucca Mountain project. It did not pass.

HR 433 was a proposal in the House to amend the Residential Lead-Based Paint Hazard Reduction Act of 1992 to provide assistance for residential properties designated as Superfund sites. It did not pass. The proposal would have required the secretary of housing and urban development, when awarding grants for lead-based paint hazard reduction, to give priority to those residential properties that are included on the National Priorities List established by CERCLA.

There were many bills concerning Superfund during this session. One of those was the Superfund Enhancement and Prioritization Act (HR 434), a proposal to transfer funds to the EPA for cleaning up identified sites. The proposal would have authorized additional appropriations for such cleanup in fiscal years 2006 through 2009 but would have limited the amount of appropriations that could be used for management and administration costs.

The Superfund Polluter Pays Act (HR 3584) was a proposal to amend the Internal Revenue Code by reinstating the Hazardous Substance Superfund financing rate and the corporate environmental income tax. This did not pass. A similar proposal, HR 4199, was the Superfund Revenue Reinstatement Act of 2005. In this proposal, the Internal Revenue Code would be amended to extend the environmental tax on corporate income. Superfund was also the topic of another bill, HR 4341, a proposal to amend CERCLA of 1980 to provide that manure is not considered a hazardous substance, pollutant, or contaminant under that act. It did not pass.

2007–08: 110th Congress

During this session, the House proposed a bill similar to one proposed in the previous session to amend Superfund. HR 1887, the Superfund Equity and Megasite Remediation Act of 2007, would amend the Internal Revenue Code to reinstate the Hazardous Substance Superfund financing rate and the corporate environmental income tax, except in cases in which the unobligated balance in the Hazardous Substance Superfund exceeded $5.7 billion. In a similar vein, House members proposed HR 3636, the Superfund Reinvestment Act of 2007, which would amend the Internal Revenue Code by reinstating the Hazardous Substance Superfund financing rate and the corporate environmental income tax. Neither bill passed. In HR 3962/S 3544, the Superfund Polluter Pays Act, the Internal Revenue Code would be amended by reinstating the Hazardous Substance Superfund financing rate and the corporate environmental income tax. This did not pass.

In HR 5336, the Brownfields Reauthorization Act of 2008, the CERCLA of 1980 would be amended to require the inspector general of the EPA to report to Congress on the management of the brownfields program every four years. This did not pass.

President Obama

President Obama declared during his presidential campaign that Yucca Mountain was "not an option" for nuclear waste storage. He followed through with that statement by pressing for a new solution that could easily take decades.[114] In S 3635, the fiscal year 2011 energy-water appropriations bill, the Senate Appropriations Committee left funding for the Yucca Mountain nuclear waste site out of the bill. In essence, the Obama administration cut off funding for the Yucca Mountain program in 2009.[115]

2009–10: 111th Congress

The debate over the proposed nuclear waste storage facility site continued during this session of Congress. There were many proposals concerning different options, but none passed. One of those was HR 515, which would give the NRC the authority to deny a plan by Utah-based Energy Solutions Inc. to import up to 20,000 tons of waste. The House Energy and

Commerce subcommittee approved the bill in November 2009. The waste would come from Italy's nuclear power program and move through ports in South Carolina and Louisiana. Italy does not have facilities for disposing of the waste. S 232, the companion bill, was introduced in the Senate.[116]

Another bill denied passage was a proposal in the House (HR 2868) to protect chemical plants and water systems from acts of terrorism. It would extend and modify the authority of the Department of Homeland Security and the EPA to enhance security at chemical facilities and wastewater treatment and drinking water plants.[117]

In S 1397, the Electronic Device Recycling Research and Development Act, Congress would authorize grants for programs aimed at researching ways to prevent improper disposal of electronic devices and mitigate the environmental effect of such waste. The Environment and Public Works Committee approved the bill on December 10, 2009, but it went no further.[118]

In HR 564, the Superfund Reinvestment Act of 2009, the Internal Revenue Code would be amended by reinstating the Hazardous Substance Superfund financing rate and the corporate environmental income tax. In another bill, House members again considered the Superfund Polluter Pays Act (HR 832/S 3125). The bill would amend the Internal Revenue Code by reinstating the Hazardous Substance Superfund financing rate and the corporate environmental income tax. Neither of these proposals passed.

Conclusion

We depend on the use of chemicals for many purposes in our daily lives, and these chemicals can pose serious risks to public health and the environment.[119] Congress has passed laws to clean up the effects of past underground chemical storage and has established new policies to prevent future brownfield areas. Nonetheless, the dangers of toxic waste disposal will continue to be debated. Future policies from Congress will continue to focus on new laws to manage toxic waste in a way that will protect the environment and the public's health.

Chapter 8

The Politics of Land

The ongoing dispute between different groups over drilling for oil in the Arctic National Wildlife Refuge (ANWR) is a clear example of the controversy between environmentalists who seek to protect the land and wildlife in that region and developers who seek to drill for oil to reduce America's reliance on foreign oil or develop the land in some way. The refuge is made up of 100 million acres of land in the state of Alaska. It is the largest protected wilderness area in the United States, created in the Alaska National Interest Lands Conservation Act of 1980, signed by President Carter. The law allowed for oil drilling but only with the approval of Congress. In November 1986, the Fish and Wildlife Service recommended that drilling should be allowed off the coastal plain of the refuge because the oil was needed to boost the country's economy. Environmental groups argued that the drilling might threaten the existence of the Porcupine caribou and harm the tundra environment.

In 1990, President George H. W. Bush banned coastal oil exploration in Alaska, but a few years later, in 1996, the Republican Congress passed a bill to allow drilling in ANWR. This proposal was vetoed by President Clinton. Environmentalists asked Clinton to appoint the area as a national monument and close the area to oil drilling permanently, but he chose not do this. President G. W. Bush supported oil drilling in the area and asked Congress to remove the ban, explaining that it could help the economy and create jobs.

Controversy remains as to whether oil drilling should be allowed in ANWR. Those who favor protecting the environment are strongly opposed to the action, but others believe the land should be explored further. This is not the first time the two sides have clashed. Many new laws have been passed to protect the land, beginning in 1920 with the Mineral Leasing Act. This law established a leasing system for fuel-mineral extraction on public lands as an alternative to the sale of lands themselves. The law did not grant free and open rights to mine. The Wilderness Act of 1964 established a national wilderness system on federal lands. Millions of acres of land were eventually set aside for designation as wilderness and protection from development. Congress also ordered a review of all public lands that might qualify for wilderness designation.[1] The law was the first federal endangered species law. It was an important, albeit early, piece of legislation and a forerunner for much of the environmental laws that would follow.[2]

1947–48: 80th Congress

Congress took no action on some proposals (HR 4150/S 1621) that would transfer the Soil Conservation Service to the Extension Service of the Department of Agriculture. Another proposal (HR 4417) would put all soil conservation work under the Soil Conservation Service. This also did not pass.[3]

1949–50: 81st Congress

A new law passed in 1949 (S 979/PL 81-128) created a program to complete a national survey of forest resources across the country. The survey would help to determine the area and location of forest land, volume of timber by species and types, the rate of growth of timber, the rate of depletion, and losses from fire, insects, and disease.[4]

President Eisenhower

On July 31, 1953, President Eisenhower sent a message to Congress dealing with conservation of natural resources. In it, he noted a "growing recognition on the part of land users and the public generally of the need to strengthen conservation in our upstream watersheds and to minimize flood damage."[5]

President Kennedy

President Kennedy gave a message on conservation to the Congress on March 1, 1962. As part of the speech he focused on outdoor recreation resources, saying that adequate outdoor recreational facilities are among the basic requirements of a sound national conservation program. He said he would appoint an Outdoor Recreation Advisory Council that would be responsible for recreation policy and promised to create a Bureau of Outdoor Recreation in the Department of the Interior. He urged Congress to pass legislation to fund grants for the development of state plans for outdoor recreational programs.

President Kennedy also supported protecting land for recreation. He said that "we must move forward with an affirmative program of land acquisition for recreational purposes. For with each passing year, prime areas for outdoor recreation and fish and wildlife are pre-empted for suburban growth, industrial development or other uses." He proposed a "Land Conservation Fund" to help fund the purchase of land by the federal government for that purpose. Kennedy urged congressional action on legislation to create Point Reyes National Seashore in California, Great Basin National Park in Nevada, Ozark Rivers National Monument in Missouri, Sagamore Hill National Historic Site in New York, Canyonlands National Park in Utah, Sleeping Bear Dunes National Lakeshore in Michigan, Prairie National Park in Kansas, Padre Island National Seashore in Texas, and a National Lakeshore Area in northern Indiana.[6]

1961–62: 87th Congress

A bill passed by the Senate (S 174) would create the National Wilderness Preservation System. The proposal had the strong support of the administration and conservationists but was opposed by economic interests in the Western states who were afraid that the bill would prevent development of the land. The bill would have designated approximately 6.7 million acres in 44 different national forests throughout the county as the original portion of the wilderness. The areas had already been reviewed and classified as wilderness by the secretary of agriculture. The bill also designated about 45 million additional acres, including "primitive" national forest area, roadless national park areas, and wildlife preserves and game areas, mostly under the jurisdiction of the Department of the Interior, for inclusion in the wilderness system, subject to review. The House took no action on this bill, and the proposal was not passed.[7]

In HR 7391 (PL 87-383), the Congress authorized $105 million over a seven-year period to accelerate the Department of the Interior's program for purchasing wetlands. Until then, the department had been purchasing wetlands to preserve the habitats of migratory waterfowl and other wild birds. The land purchases were financed by a special fund set up through the sale of duck hunting stamps. Any hunter across the country who purchased a license to hunt ducks had to also purchase a stamp, and that money went to the fund to purchase wetlands. In order to accelerate acquisition of land, the Interior proposed that the Congress provide $200 million over a 10-year period for wetlands purchases. They presented evidence that some wetland areas were being destroyed and needed to be protected while still available.[8]

1963–64: 88th Congress

In 1964, the administration backed a Senate bill (S 4/HR 9070/PL 88-577) that established a National Wilderness Preservation System. Under the bill, 8.2 million acres of federally owned land that had already been classified as "wilderness, "wild," and "canoe" areas would be added to the wilderness system. The lands had not yet been developed and were still in a wild, natural state. An additional 57.2 million acres of public lands would be conditionally added into the system, subject to future review. Over a 10-year period, the departments of Agriculture and the Interior would review "primitive" national forest areas, roadless national park areas, wildlife preserves, and game ranges and recommend to the president whether the land should be included in the system.[9] Any land placed in the wilderness system would be permanently safeguarded against future commercial use.[10]

Another new law (HR 3846/PL 88-578) created this session was the Land and Water Conservation Fund Act. This bill, requested by both presidents Kennedy and Johnson, was designed to meet the public demand for outdoor recreation facilities. The new law established a special federal fund to help state, local, and federal agencies pay for the purchase of land and water areas that could be used as public parks and outdoor recreation areas. It was thought that there was a need to purchase land quickly, particularly in the East, because land was becoming increasingly scarce as a result of urban expansion, commercial development, and highway construction. The fund would receive allocations from the sale of federal surplus property, from the federal motorboat fuel tax, and also from a new system of entrance and recreation user fees at recreation areas on federal lands that were managed by the National Park Service, Forest Service, and other federal agencies. The funds allocated to federal agencies could be used only for land acquisition, not for construction of facilities or other development activities.[11]

Congress passed an administration-supported bill (HR 8070/PL 88-606) creating a 19-member Public Land Law Review Commission to study existing public land laws and policies and make recommendations to the president. The Congress wanted to modernize and overhaul the land management and land disposal laws concerning the federal government's public landholdings.[12]

President Johnson

On March 8, 1968, Johnson gave a speech to Congress on conservation, called "To Renew a Nation." In it, he talked about surface mining. He said that "stripping machines have torn coal and other minerals from the surface of the land, leaving two million acres of this Nation sterile and destroyed. The unsightly scars of strip mining blight the beauty of entire areas, and erosion of the damaged land pours silt and acid into our streams." He also said that "America needs a nationwide system to assure that all lands disturbed by surface mining in the future will be reclaimed." The president then proposed the Surface Mining Reclamation Act of 1968, which would establish criteria that would be used in developing regulatory plans. The states would develop their own plans within two years and submit them to the secretary of the interior for review and approval. If the states did not submit any plans, or their plans were inadequate, the secretary would have the authority to impose federal standards on them.[13]

Another topic for Johnson was the system of national parks. He said that his administration had authorized more than 2.2 million acres to be added to the nation's park system, so that people were not forced to drive many miles to enjoy the parks. But he had plans to do more. He urged the House to pass a redwood bill to preserve the tallest trees in nature. He also recommended that the House pass two other bills, to protect the North Cascades National Park and the Apostle Islands National Lakeshore in Wisconsin. He then urged the Congress to authorize the development of the Potomac National River, a uniquely historic area.[14] Also in that same speech, Johnson asked Congress to provide $130 million for the Land and Water Conservation Fund to provide for more recreation acreage to serve the people.

Johnson again talked about strip mining in a speech on July 3, 1967. He discussed a report from the Department of the Interior in which a national program for regulating surface mining operations and for reclaiming lands already damaged by unwise surface mining techniques was proposed. The study was originally authorized by the Appalachian Regional Development Act of 1965 (PL 89-4). The report recommended that new federal standards be developed to prevent future damage to land and water resources, that damage lands be returned to their natural state with the help of cost sharing agreements between federal, state, and local governments; that privately owned lands be acquired in an effort to promote conservation efforts; and that research into surface mining be supported.[15]

In another speech, on natural beauty, President Johnson outlined a massive program to beautify the nation. He proposed that cities create new programs for development of open spaces and also establish community extension programs to focus the resources of universities on community problems. Johnson also discussed beautifying the countryside, through initiation of a "Parks for America Decade." Highways could be beautified through the control of billboards and junkyards and through expanded programs of landscape development. Rivers could be beautified through preservation of specified streams in their natural state. Overall, we could beautify the total environment through new controls over air and water pollution.[16]

In October 1968, one of the final months of his administration, President Johnson signed into law many acts that protected the environment. These included the Wild and Scenic Rivers Act, the National Trails System Act, and the acts to establish the North Cascades National Park and the Redwood National Park.[17]

1967–68: 90th Congress

In this session, Congress added more than 800,000 acres of federally owned land to the National Wilderness Preservation System. One of the laws involved was S 889 (PL 90-271) to designate the San Rafael Wilderness in Los Padres National Forest, California, as the first addition to the National Wilderness System. The area was 42,722 acres that included wooded, brushy, and mountainous terrain suitable for camping, riding, hiking, fishing, and studying nature. President Johnson requested the legislation in a letter to leaders in the House and Senate.[18] Also passed was legislation adding San Gabriel in California (S 2531/PL 90-318) and Mount Jefferson in Oregon (S 2751/PL 90-548). Also added were the 520,000-acre Pasayten Wilderness and a 10,000-acre addition to the existing Glacier Peak Wilderness (S 1321/PL 90-544). Another Senate proposal (S 2515/PL 90-545) established Redwood National Park in California.

Three other bills were referred to the House Interior and Insular Affairs Committee but not acted on. They included bills on Pelican Island in Florida (S 3343), Monomoy Island in Massachusetts (S 3425), and Moosehorn in Maine; the Huron Islands, the Michigan Islands, and Seney in Michigan; and Green Bay and the Gravel Islands in Wisconsin (S 3502).

In HR 25, Congress completed action on a bill supported by the administration that would authorize a program to preserve the nation's estuaries. The areas to be preserved included coastal marshlands, bays, sounds, seaward areas, lagoons, and the land and waters of the Great Lakes. The bill authorized appropriations of $500,000 in fiscal years 1969–70 for a study and report by the Department of the Interior on how best to preserve estuaries and to determine whether a national system of estuarine areas should be established. A second part of the bill authorized the Secretary of the Interior to enter into agreements with any state or locality for the permanent protection of estuarine areas already publicly owned. The preservation of estuaries was considered important both because of their natural beauty and recreational value and because of their function as breeding grounds for many species of fish and birds.[19]

A new law passed this session (HR 12121/PL 90-213) extended the life of the Public Land Law Review Commission for another 18 months. It also authorized additional appropriations of $3.2 million to complete the commission's work.

The Senate Committee on Interior and Insular Affairs, Subcommittee on Minerals, Materials and Fuels held hearings on S 217, S 3126, and S 3132 for the regulation and reclamation of surface mining areas. In S 217, also known as the Surface Mining Reclamation Act of 1968, states would be required to submit plans for reclamation of current and future (but not past) surface mining areas for approval by the secretary of the interior. The secretary would determine whether state plans were being enforced. If they were not, the secretary could withdraw approval of the state standards and issue federal regulations. S 3126 would provide federal grants of up to 50 percent the help states pay for the costs of developing and enforcing their plans. The last bill, S 3132, would authorize the secretary to conduct research in surface mining reclamation to aid in carrying out the provisions of S 3132. None of these bills passed.[20]

President Nixon

President Nixon named three members of the newly created Council on Environmental Quality. They issued a report to Congress in August. One major recommendation was the formulation of a national land use policy to direct the type, location, and timing of future land development.[21]

On February 8, 1970, Nixon gave a speech on the environment in which he said that the use of our land not only affects the natural environment but also shapes the pattern of our daily lives. He called on Congress to adopt a national land use policy. He also proposed initiatives that would bring parks to the people so that everyone would have access to nearby recreational areas. He wanted to expand the wilderness system, to restore and preserve historic and older buildings, to provide an orderly system for power plant siting, and to prevent environmental degradation from mining.

He promised to submit to the Congress several bills that would be part of a comprehensive effort to preserve our natural environment and to provide more open spaces and parks in urban areas where they can be scarce.

He also proposed legislation to establish a national land use policy that would encourage state and local governments to plan for and regulate major developments affecting growth and the use of critical land areas. He proposed a new "Legacy of Parks" program to help states and local governments provide parks and recreation areas that would serve today's Americans but tomorrow's as well. By setting aside and developing such recreation areas now, we can ensure that they will be available for future generations. He requested $200 million to begin a new program for the acquisition and development of additional park lands in urban areas. This would be administered by the Department of Housing and Urban Development. He also asked that the appropriation for the Land and Water Conservation Fund be increased to $380 million.

When Nixon talked about wilderness areas, he said that "there must still be places where nature thrives and man enters only as a visitor. These wilderness areas are an important part of a comprehensive open space system. We must continue to expand our wilderness preservation system, in order to save for all time those magnificent areas of America where nature still predominates. I will soon be recommending to the Congress a number of specific proposals for a major enlargement of our wilderness preservation system by the addition of a wide spectrum of natural areas spread across the entire continent."[22]

In a speech on the environment on February 10, 1970, Nixon spoke about parks and public recreation areas. He proposed more money for additional park and recreational facilities, with more locations that could be easily reached by citizens. He also established a Property Review Board to review reports and recommend what properties should be converted or sold.[23]

In 1972, Nixon proposed 16 new wilderness protection areas that would add 3.5 million acres to the system. They included the Brigantine National Wildlife Refuge in New Jersey, the Blackboard Island National Wildlife Refuge in Georgia, the Chassahowitzka National Wildlife Refuge in Florida, and the Lostwood National Wildlife Refuge and the Chase Lake National Wildlife Refuge in North Dakota. A sixth area would be within the Cumberland Gap National Historical Park on the borders of Tennessee, Virginia, and Kentucky. The areas also included parts of Yellowstone National Park, the Grand Canyon, and Yosemite National Park and the Grand Teton National Park, as well as adding land to the Great Sand Dunes National Monument in Colorado, the Theodore Roosevelt National Memorial Park in North Dakota, the Badlands National Monument in South Dakota, the Guadalupe Mountains National Park in Texas, the Carlsbad Caverns National Park in New Mexico, and the Haleakala National Park in Hawaii.[24]

In 1972 Nixon gave a lengthy speech to Congress and outlined more plans to build on what he had already done. He planned to tighten pollution control, make technology an environmental ally, protect the natural heritage, expand international cooperation on the environment, and protect children from lead-based paint. He also proposed amendments to the National Land Use Policy Act to provide federal assistance to encourage the states, in cooperation with local governments, to protect lands that are of critical environmental concern. The proposed amendments would provide that any state that had not established an acceptable land use program by 1975 would be subject to annual reductions of certain federal funds. Nixon proposed legislation to establish a Golden Gate National Area around San Francisco Bay. He proposed converting more federal land to be used for parks and recreational use and proposed more wilderness areas to add 1.3 million acres to the wilderness system.[25]

1969–70: 91st Congress

New legislation (S 3014/PL 91-504) designated 23 new wilderness areas in 12 states, for a total of 201,212 acres, including national parks, national monuments, wildlife refuges, national forests, and other lands already owned by the federal government.[26] It added more than 200,000 acres of land to the national wilderness system.

Two laws that did not pass were HR 12025, the National Forest Timber Conservation and Management Act, and S 812, a bill to establish the Florissant Fossil Beds National Monument. This consisted of 6,000 acres near Pikes Peak, Colorado.

During this session, Congress moved closer to establishing a national land use policy. On June 23, 1970, the Public Land Law Review Commission, established in 1964, released a report after a six-year study of land use practices. The commission made more than 350 recommendations and guidelines to help Congress overhaul outmoded laws and make sense of the often-conflicting regulations that govern the use and sale of federal lands. Congress did not act on the recommendations in this session. They recommended enhancement and maintenance of the environment, and rehabilitation where necessary, as well as the establishment federal standards for environmental quality. The Commission also recommended that Congress establish policies for increased access and new fees for use of federal land.

They proposed a merger of the Department of the Interior and the Department of Agriculture's Forest Service.

In the final weeks of the 91st Congress, the Senate Interior and Insular Affairs Committee reported legislation (S 3354) that would create a cabinet-level agency called the Land and Water Resources Council that would plan land use on a national level. The bill would have provided financial aid for states to conduct similar planning activities. There was no further action on S 3354 in this session.[27]

The Senate held hearings on a bill (S 719) to replace the Mining Act of 1872 and establish a more modern national mining and minerals policy. This was needed because the 1872 law remained the basic mining law for the public domain. The new bill called for a new mineral leasing system and other wise updated the outdated law. This proposal became law (PL 91-631).[28]

Many bills were proposed to add land to the national wilderness system. These included S 713 (PL 91-82), which designated the 63,469-acre Desolation Wilderness Area of Eldorado National Forest, and S 714 (91-58), which added the 94,728-acre Ventana Wilderness of the Los Padres National Forest. Another bill, S 2564 (PL 91-88), authorized funds for the acquisition of an additional 6,640 acres for the Everglades National Park, and HR 11069 (PL 91-42) authorized appropriations of approximately $4 million to complete the acquisition of land for the Padre Island National Seashore in Texas.[29] S 3014 (PL 91-504) added more than 200,000 acres of land to the national wilderness system.

The Senate passed a bill (S 1830) called the Alaska Native Claims Settlement Act of 1970; it authorized payment of $1 billion and a grant of up to 10 million acres of land to the natives to cancel their claim to land in Alaska. The House took no action.[30]

1971–72: 92nd Congress

A House bill (HR 7211) was proposed to establish a comprehensive national policy and program for federal and nonfederal land use in the United States. The bill, the National Land Policy, Planning and Management Act of 1972, would have established a public land policy and provide guidelines for its administration. The Senate passed S 632 as a companion bill, but the House took no action on it.

Congress cleared HR 736 (PL 92-364) to designate certain lands in Cedar Keys National Wildlife Refuge in Florida as wilderness land.[31] Other bills that added land were HR 10655 (PL 92-510), providing for the creation and administration of a 78,982-acre wilderness area in the Lassen Volcanic National Park in California;[32] S 141 (PL 92-537), adding the Fossil Butte National Monument in Wyoming; S 1852 (PL 92-592), adding Gateway National Recreation Area in New York and New Jersey; HR 5838 (PL 92-493), adding a 28,460-acre wilderness area in the Lava Beds National Monument in California; S 3129 (PL 92-475), establishing the Longfellow National Historic Site in Massachusetts; HR 15597 (PL 92-533), adding Piscatawy National Park in Maryland; and S 1497 (PL 92-501), adding the Sitka National Monument in Alaska.[33] A House bill was approved by the president (HR 6957/PL 92-400) that established the Sawtooth National Recreation Area in Idaho.[34] Other new lands added to the National Wilderness Preservation System included the Coconino Kaibab and Prescott National Forests in Arizona, established as the Sycamore Canyon Wilderness (S 960/PL 92-241); the Gulf Islands National Seashore in Florida and Mississippi (S 3153/PL 92-275); Minmam River Canyon in Oregon (S 493/PL 92-521); the Glen Canyon National Recreation Area in Arizona and Utah (S 27/PL 92-593); the Indian Peaks Area of Colorado (S 1198/PL 92-528); the Golden Gate National Urban Recreation Area in San Francisco and Marin County (HR 16444/PL 92-589); and Helena, Lewis and Clark, and Lolo National Forests (S 484/PL 92-395).[35]

Bills passed in both the House and Senate (S 3699 and HR 14392) would establish a system of wild areas in the eastern United States within the national forest system. S 3255 would establish a southeastern wild areas preservation system, but none of these became law.[36]

A proposal to regulate strip mining of coal was debated by members of the House (HR 6482), but action was not completed on it. The bill would require mine owners to obtain permits for strip mining operations and to submit reclamation plans and purchase reclamation bonds. No strip mining would be permitted on slopes of greater than 20 degrees unless the mine owner could prove that environmental damage would be prevented and the land reclaimed. The secretary of the interior would administer the federal regulatory and permit program, but states would have been able to run their own programs if they met federal standards. Those state programs that did not comply would have been taken over by the Department of the Interior. The bill was passed by the House, but the Senate did not take action on the companion bill, S 630. The House bill was endorsed by most environmental groups, including the Coalition Against Strip Mining, but was opposed by the National Coal Association and the American Mining Congress.[37]

1973–74: 93rd Congress

In proposed legislation (S 268/HR 2942), the Senate approved a strong federal land use program designed to establish federal control over how the nation's urban, suburban, and rural lands are developed. However, the House killed it, and it went no further.[38]

In S 424, Senate members passed a proposal to consolidate and strengthen the authority of the Bureau of Land Management (BLM) to manage land under its jurisdiction. The legislation reflected a basic change in federal land management policy away from disposal of federal lands toward their retention under federal control. The House did not act on the bill.[39]

A ban on strip mining on lands where the federal government owned the mineral rights but not the surface rights was proposed in a bill in the Senate (S 425), The bill, which was opposed by the administration and the coal industry, would have regulated surface mining of coal by establishing minimum federal standards for the reclamation of mined lands. Strip mining had caused severe damage on lands in Appalachia and the Midwest by the 1970s and was moving into the northern Great Plains and the Southwest. But the nation needed more coal. Strip-mined coal could be extracted quickly, easily, and cheaply. Among the standards set in the bill was a requirement that land stripped in the future would have to be returned to its original contours unless a better use could be established; it also would establish a reclamation program for land that had already been stripped and abandoned. It would provide a mechanism by which land that could not be reclaimed would be designated unsuitable for strip mining, provide for public participation in the development of state programs, and provide protections for the rights of surface owners whose land lay on top of federally owned coal. Neither environmentalists nor the mining industry were satisfied with the version passed by the Senate.[40] In 1974, the legislation (S 425) passed both the House and Senate and was sent to the president. Ford pocket vetoed the bill without giving Congress a chance to override the veto.[41]

In 1974, Congress proposed the Forest and Range Renewable Resources Planning Act (RPA), which required 5- and 10-year management plans for timber harvested from the national forests, accompanied by budget estimates. It also provided support for professional forest planning and required special justification for any deviation from the planned budget estimates. This law did not pass.[42]

Two other bills that did not pass were S 22 (the National Forest and wild Areas Act) and S 316/HR 1758. If passed, the bills would have created wilderness areas in national forests that had been disturbed and then allowed to return to their natural state. Both versions would immediately designate a number of eastern areas as wilderness and provide for the study of more areas for possible inclusion in the National Wilderness Preservation System.[43]

Many new bills created additional wilderness areas this session. Designation as a wilderness area brought the land under protective laws prohibiting timber cutting, construction of power facilities or reservoirs, and other activities that could destroy its natural state. One law that emerged from the Senate (S 3433/PL 93-622), the Eastern Wilderness Areas Act, created 16 additional wilderness areas in the eastern United States, totaling 206,988 acres. It also designated 17 additional areas for study for possible inclusion in the national wilderness system.[44] Congress cleared HR 6395 (PL 93-429), which designated certain lands in the Okefenokee National Wildlife Refuge as wilderness. The bill added over 340,000 acres of remote swampland in Georgia to the growing number of wilderness areas in the eastern United States.[45]

A bill establishing the 9,910-acre Cascade Head Scenic Research Area in and adjacent to the Siuslaw National Forest in Oregon was passed during this session (HR 8352/PL 93-535). Proponents of the bill said it was needed to protect the area from the adverse effects of commercial and residential development. The bill also extended the boundaries of the Siuslaw National Forest by adding to it 5,325 acres, of which 4,945 had been privately owned.[46]

President Ford

Ford noted his commitment to preserve America's wilderness areas, which he called America's irreplaceable heritage, in a proposal to the speaker of the House and president of the Senate on establishing new national wilderness areas.[47] He later said, "The preservation of wilderness areas across the country today enables us to recapture a vital part of the national experience: like our forebears, we can journey into primeval, unspoiled land. The Nation as a whole is enriched by the availability of the wilderness experience to those who are able and willing to seek it. Wilderness preservation insures that a central facet of our Nation can still be realized, not just remembered."[48] In this speech to Congress, Ford proposed the creation of 37 additional wilderness areas.[49] However, Ford made it clear that economic issues needed to be considered when designating land as wilderness area. He said, "The Administration has strongly and consistently urged the Congress not to designate National Forest areas as wilderness where the evidence of man's activity is clearly apparent....More careful consideration must be given to these proposals if we are to maintain a high-quality Wilderness System while protecting many other important management opportunities for these lands."[50] In 1976 Ford proposed a Bicentennial Land Heritage Act that would establish a 10-year commitment to double America's national parks, recreation areas, wildlife refuges, urban parks, and historic sites.[51]

However, President Ford overturned the environmentalists' victory of 1974 by pocket vetoing a bill to impose strong controls on the strip mining of coal. The coal industry lobbied against the bill. Ford said he refused to sign it because "the adverse impact of this bill on domestic coal production is unacceptable at a time when the national can ill afford significant losses from this critical energy resource."[52]

1975–76: 94th Congress

In 1975, the Senate Interior Committee reported the National Resource Lands Management Act, and it was passed by Congress (S 507/PL 94–579). Under the law, 60 percent of all federally owned property would be administered as "national resource lands" by the secretary of the interior through the BLM.[53] The act set forth for the first time in a single statute the authority for management of millions of acres of federal lands. The BLM in the Department of the Interior would supervise the care and use of most of this land. Until passage of S 507, its mandate came only from thousands of individual public land laws.[54]

In the House, a bill (HR 25) would set minimum federal standards for control of strip mining of coal and for reclamation of strip-mined lands. Known as the Surface Mining Control and Reclamation Act of 1975, it would have been the major new environmental legislation approved in 1975. The provisions of the bill would have set minimum federal standards to be followed by the states in drawing up mandatory strip mining control programs. If states did not adopt the standards, a federal program would be imposed.[55] President Ford vetoed the bill, and the House failed by three votes to override the veto. In vetoing the bill, Ford cited the economic and energy-related costs of the proposal, saying that it would cost more in lost jobs, lost coal production, and higher electricity bills than the American economy could withstand.

HR 7792 was a proposal to create a wilderness area in the Central Cascade Mountains in Washington State. The president signed it into law (PL 94-357). Environmentalists wanted to preserve the area for its natural beauty, while logging to companies and labor resisted the loss of valuable timber lands.[56]

The National Forest Management Act (S 3091/HR 12503/HR 15069/PL 94-588) in 1976 authorized clear-cutting as a legitimate forest management practice but also established restrictive guidelines for its use.[57] The act essentially justified the multiple-use policies that the Forest Service and BLM had been pursuing. However, the act created widespread resentment among ranchers, miners, and others who for decades had used public land with little supervision.[58]

The House passed a law (HR 3130/PL 94-83) to amend the National Environmental Policy Act. The new law made it clear that the environmental impact statement required by the law for major federal actions was not legally insufficient just because it had been prepared by a state agency or official—if the state official had statewide jurisdiction and responsibility for the action dealt with by the impact statement and if the responsible federal official guided and participated in such preparation and independently evaluated the statement before it was approved and adopted. The second part of the law required that the responsible federal official should notify and seek the views of any other state or federal land management body on any action or any alternative action that might have significant impact on that body. If there was disagreement over the impact, a written assessment should be incorporated into the impact statement. Third, the law stipulated that this amendment did not relieve the federal official of responsibility for the impact statement or affect the legal sufficiency of impact statements prepared by state agencies that did not have statewide jurisdiction.[59]

President Carter

President Carter supported many policies to protect the land and wilderness areas. He proposed an accelerated program to develop new and existing parks, as well as $759 million to help pay for that. He wanted to develop a National Heritage Trust that would protect places of cultural, historic, and ecological value. Carter proposed designating major additions to the Park, Forest, Wildlife Refuge and Wild and Scenic River System in Alaska.[60]

In 1978, President Carter restricted the use of 56 million acres of federally owned wilderness in Alaska. This act protected millions of acres of Alaskan wilderness from commercial development. Carter then used his authority under the Antiquities Act of 1906 to create 17 new national monuments, which prevented mining, logging, and other commercial development. Carter also signed the Endangered American Wilderness Act

(HR 3453/PL 95-237) to add about 1.3 million acres of Western state land to wilderness areas.[61] His decisions more than doubled the size of the national park system.[62]

In a speech to Congress on January 21, 1980, Carter said his highest environmental priority was the passage of legislation that adequately resolved the allocation of federal lands in Alaska. His proposal would "assure that Alaska's great national treasures can be preserved, while providing for increased domestic energy production and for the economic needs of all Alaskans."[63]

1977–78: 95th Congress

In 1977, the Surface Mining Control and Reclamation Act passed Congress and became law (HR 2/PL 95-87). The bill found that surface mining operations adversely affect commerce and the public welfare by destroying land, polluting water, and damaging natural beauty and habitats. As a result, Congress established environmental controls over strip mining and put limits on mining on farmland, alluvial valleys, and slopes. It also required restoration of land to its original contours. The provisions were directed primarily at strip mining for coal, as other types of strip mining were largely untouched. The Office of Surface Mining Reclamation and Enforcement was established to maintain an Information and Data Center on Surface Coal Mining, Reclamation and Surface Impacts of Underground Mining to provide information to the public and other agencies. Each state would be required to establish a state mining and mineral resources research institute that would conduct research and train scientists.[64]

The House and Senate passed different versions of a bill entitled the Endangered American Wilderness Act (HR 3454/PL 95-237) that added substantial acreage to the National Wilderness Preservation System. A large portion of the debate in the Senate revolved around an unrelated amendment to authorize the sale of 30,000 tons of tin from stockpiles and use the proceeds to purchase up to 250,000 tons of copper. There was also disagreement about the specific areas to be included as wilderness. The bills designated some different areas in Oregon. The Senate bill included land in Washington and Idaho, but the House bill did not. Instead, the House bill included new wilderness acres in Montana and Wyoming. While the Senate added one area in Arizona, the House added two A compromise version was worked out and the president signed it into law.[65]

1979–80: 96th Congress

A new law that passed this term (HR 2676/PL 96-229), the Environmental Research, Development, and Demonstration Authorization Act, authorized $379.5 million in fiscal year 1980 funds for the environmental research programs of the Environmental Protection Agency (EPA). The bill authorized funds for several new EPA programs, including a project to find less expensive ways to control hazardous waste dumps and groundwater contamination from those dumps. The bill also authorized funds for the National Bureau of Standards.[66]

Congress completed action on a bill (HR 39/PL 96-487), the Alaska National Interest Lands Conservation Act, which designated 102 millions of acres of federal land in Alaska as national parks, wildlife refuges, and other protected areas. The designation would sharply limit the extent of commercial and recreational development that could occur on those lands. The issue revolved around the amount of Alaska's federal lands that would be preserved in their pristine state and how much would be opened up to development, such as logging, mining, and oil and gas exploration, and to sport hunting. One of the most controversial points was the potentially oil-rich ANWR, which was also a calving ground for caribou.[67] The final version of the bill set aside 104.3 million acres divided into conservation units that imposed varying degrees of restrictions on exploration for oil, minerals, and timber. The bill also

- Set aside 43.6 million acres for new national parks;
- Set aside 53.8 million acres for new wildlife refuges;
- Provided 56.7 million acres of wilderness protection in various conservation units;
- Allowed seismic exploration for oil and gas—but no drilling—on 90,000 acres in the western coastal plain of the William O. Douglas Arctic Wildlife Range on Alaska's North Slope;
- Established a 2.3 million–acre national monument in the Misty Fjords area of southeastern Alaska, all of which was designated as wilderness, except for a 149,000-acre area surrounding the U.S. Borax moldenum mine; and
- Established a 921,000-acre national monument in the Admiralty Island area of southeastern Alaska, all of which was designated as wilderness.[68]

HR 2043 (PL 96-182) was a new law created to authorize funding for the preservation of up to 20 million additional acres of wooded and swamp wetlands in several southern states. In addition to increasing the types of lowlands eligible for funding under the national wetlands conservation program, the bill tripled the amount available for conserving those wetlands, beginning in fiscal year 1981. The new funds would expand the program to cover the newly eligible areas in the South. Other bills set aside land as national wilderness land. For example, HR 39 (PL 96-487) set aside millions of acres designated as wilderness in the Alaska lands bill. Another Senate bill (S 2009/PL 96-312) set aside land in central Idaho as the River of No Return Wilderness. This created the largest wilderness in the United States outside of Alaska. Another 1.4 million acres of wilderness were set aside in Colorado, Missouri, Louisiana, South Dakota, and South Carolina in the Colorado Wilderness bill (HR 5487/PL 96-560). A House bill set aside 610,000 acres in New Mexico as wilderness area (HR 8298/PL 96-550). Other bills to propose new national wilderness lands did not pass, one of these was S 95, which would have set aside 2.2 million acres as the River of No Return Wilderness in Idaho. Also not passing was a Florida wilderness bill (HR 6218), which passed the House only. It would have set aside six parcels of wilderness totaling 36,290 acres in the Ocala and Apalachicola National Forests.[69]

In 1979, the Senate passed legislation (S 1403) to allow states to design their strip mining plans without following the pages of federal regulations written by the Department of the Interior's Office of Surface Mining. The administration and environmentalists opposed the bill, claiming it would "gut" the landmark strip mining law just passed. However, proponents said the bill would only free states from rules that were "voluminous and suffocating."[70] The House membership refused to act on the Senate proposal. Instead, they proposed a bill (HR 4728) that would also exempt states from being forced to comply with the regulations written to implement the original act. Coal-state senators said the federal regulations were arbitrary and inflexible and did not allow for different topographic and geographic conditions in states.[71] None of the bills passed.

President Reagan

Reagan supported adding to the National Wild and Scenic Rivers system. He said, "We must protect wilderness areas and wild, free-flowing rivers...for future generations to enjoy in their natural, undeveloped state."[72] In 1983, Ronald Reagan proposed to the Congress 57 areas, for a total of 2.7 million acres of land, as wilderness land.[73] However, Reagan did not add to the system freely. On January 14, 1982, Reagan pocket vetoed HR 9, which designated federal wilderness areas in Florida. In doing so, he said, "I support the designation of additions to the National Wilderness Preservation System in the State of Florida....Although HR9 is intended to resolve an issue that has been in contention during three prior Administrations, it does so in a way that is unnecessarily costly to the Federal taxpayer."[74]

1981–82: 97th Congress

Congress continued to pass bills that would add land to the national wilderness areas. One of those was S 1119 (PL 97-250), which adjusted the boundary of Crater Lake National Park in Oregon and designated certain lands in the Cumberland Island National Seashore in Georgia as wilderness land.[75] Other bills to add to the national wilderness land did not have enough support to pass. HR 7702, a bill to designate 2.1 million acres as wilderness, passed the House but not the Senate. HR 4083, to designate millions of acres in California as protected wilderness, also passed the House but not the Senate.[76] Finally, HR 9 was a proposal to designate 49,150 acres in three national forests in Florida as wilderness and to resolve a long-standing dispute over phosphate mining in the Osceola National Forest.[77]

The 97th Congress cleared RARE II bills (Roadless Area Review and Evaluation) covering five states—Florida, West Virginia, Missouri, Indiana, and Alabama. But Reagan vetoed the Florida measure (HR 9) because he opposed provisions calling for compensation of companies with pending applications to strip-mine phosphate from some of the lands covered by the bill. Other RARE II bills that died in the 97th Congress included measures for California (HR 4083), Montana (S 2110), Wyoming (S 2118), and Oregon (HR 7340).[78]

In 1982, the House passed HR 861 designating three new national scenic trails and authorizing the study of six other routes for possible inclusion in the national trail system. It designated three new scenic trails: the Potomac Heritage, Natchez Trace, and Florida trails. The bill forbade entrance fees for any national recreational area, national wild and scenic river area, or any portion of the national trail system. In addition, it prohibited entrance fees at national parks where the cost of collecting them would exceed the revenue from such fees. This provision was opposed by the Reagan administration and did not pass the Senate.[79]

HR 396 was a proposal in the House to make amendments to the Surface Mining Control and Reclamation Act of 1977. The bill, which did not pass, would have provided that performance standards for the reclamation of surface coal mining operations should not require restoration of lands to the approximate original contour if the surface owner specified a different reclamation standard.

Bills to protect lands did not pass during this session. One of those, HR 9/HR 4474, the Florida Wilderness Act of 1982, would have designated certain lands in Florida as part of the National Wilderness Preservation System. The bill was pocket vetoed by the president. Another bill (HR 856) would have designated land in California as national park wilderness. Also failing in Congress was HR 1135, a bill to add about 25 acres to the Sandia Mountain Wilderness system in New Mexico. Another failing bill was HR 1214, a bill to designate national forest system lands in Texas as components of the National Wilderness Preservation System. Finally, S 1117, a bill to designate lands in the Shasta Trinity National Forest, California, as the Mt. Shasta Wilderness did not pass.

A bill to prohibit Congress from designating more than 3.5 million cumulative acres in the state of Idaho as wilderness (HR 293) did not pass. The law would have declared that wilderness land created by Congress in excess of 3.5 million acres must be offset by statutory withdrawal from the National Wilderness Preservation System in Idaho. This was called the Idaho Wilderness Ceiling Act.

1983–84: 98th Congress

During this session, Senate members passed a bill (S 626) that designated 6,670 acres of the Aravajpa Canyon and adjoining lands in southern Arizona as wilderness. The House Interior and Insular Affairs Committee took no action, and the bill did not pass.[80] But Congress cleared more wilderness legislation in 1984 than in any other year since passing the 1964 Wilderness Act. Twenty separate bills were sent to President Reagan during the year that would provide wilderness protection to lands in Arizona, Arkansas, California, Florida, Georgia, Mississippi, Missouri, New Hampshire, New Mexico, North Carolina, Oregon, Pennsylvania, Tennessee, Texas, Utah, Vermont, Virginia, Washington, Wisconsin, and Wyoming. The total amount of land set aside was more than 8.3 million acres. These bills were S 2125 (PL 98-508), HR 4707 (PL 98-406), HR 1473 (PL 98-425), HR 9 (PL 98-430), S 2773 (PL 98-514), S 2808 (PL 98-515), S 64 (PL 98-289), HR 3921 (PL 98-323), HR 6296 (PL 98-603), HR 3960 (PL 98-324), HR 1149 (PL 98-328), HR 5076 (PL 98-585), HR 4263 (PL 98-578), HR 3788 (PL 98-574), S 2155 (PL 98-428), HR 4198 (PL 98-322), HR 5121 (PL 98-586), S 837 (PL 98-339), HR 3578 (PL 98-321), and S 543 (PL 98-550).[81]

The House introduced a bill (HR 1621: the Abandoned Mine Reclamation Improvements Act) to direct the secretary of the interior to invest part of the Abandoned Mine Reclamation Fund; the interest earned from the investments would be deposited back into the fund. In order for a state to collect a reclamation fee payment, it would be required to have a state reclamation plan.

Congress attempted to create a trust fund for the reclamation of underground mines and surface mines by amending the Surface Mining Control and Reclamation Act. In the bill (S 1755), the Abandoned Mine Reclamation Fund would be abolished and replaced with two new trust funds, the Abandoned Surface Mine Reclamation Fund and the Abandoned Underground Mine Reclamation Fund. This did not pass.

1985–86: 99th Congress

Most of the action by Congress in this session was to choose land to be designated as wilderness land. For example, the Senate cleared a new law (HR 5496/PL 99-555) that more than doubled the amount of federal wilderness in Georgia. The bill set aside five areas totaling 42,258 acres in the Chattahoochee National Forest, Raven Cliffs Wilderness, Brasstown Wilderness, Tray-Mountain Wilderness, Rich Mountain Wilderness, and an addition to the Cohutta Wilderness. The bill was supported by the Reagan administration.[82]

The Nebraska Wilderness bill was also passed (S 816/PL 99-504). This new law established the 8,100-acre Soldier Creek National Wilderness and the 6,600-acre Pine Ridge Recreation Area, both in the Nebraska National Forest.[83] The Congress established six wilderness areas totaling 33,735 acres in the Cherokee National Forest in Tennessee in HR 5166/PL 99-490.[84] The House passed a bill (HR 5508) to add 15,900 acres of national forest land to the existing Sipsey Wilderness in Alabama, but the Senate did not act on the bill.[85] And, finally, S 2506/PL 99-565, an administration-backed bill, established the 44,000-acre Great Basin National Park in eastern Nevada.

During this session of Congress, the members of the Senate once again proposed a bill (S 478) to amend the Surface Mining Control and Reclamation Act of 1977 to create a trust fund for the reclamation of underground mines and a trust fund for surface mines. As before, the Abandoned Mine Reclamation Fund would be abolished and replaced with two new funds, the Abandoned Surface Mine Reclamation Fund and the Abandoned Underground Mine Reclamation Fund. Again, this did not pass.

1987–88: 100th Congress

In 1988 President Reagan announced a pocket veto of a bill (S 2751) that would have created additional Montana wilderness reserves. The proposal would have made about 1.4 million acres of National Forest System land in Montana off-limits to logging, road building, and other development. Reagan said the bill "could cost jobs and eliminate vast mineral development opportunities."[86]

A bill in the House, the Recreation and Public Purposes Amendment Act (HR 4362/PL 100-648), was an attempt to prevent Western towns from accepting gifts of federal land, filling them with hazardous waste materials, and then giving them back to the federal government. This had happened under the Recreation and Public Purposes Act of 1926, which authorized the BLM, which oversaw the vast tracts of empty rangeland in the West, to give tracts to municipalities that could put them to use. Often, these tracts became city dumps. The law contained a clause specifying that if the municipality misused or abandoned the land, the title would revert to the federal government. In cases where hazardous wastes were dumped in such landfills, that would mean that liability for cleanup costs also reverted to the government.[87]

The House also passed a law (HR 1963/PL 100-34) to prevent abuses of the 1977 Surface Mining Control and Reclamation Act. The new law repealed an exemption in the original law whereby a two-acre mine would not have to comply with environmental standards imposed on larger mines. The new bill also allowed states to set aside up to 10 percent of their abandoned- mine/land reclamation grants for use after the taxes that supported the grants expired in 1992.[88]

Members of the Congress tried again to create a trust fund by amending the Surface Mining Control and Reclamation Act of 1977 in HR 1434/S 643. This would allow states to create a special trust fund with up to 10 percent of the annual state allocation from the Abandoned Mine Land Reclamation fund for expenditures to reclaim abandoned mines in the future. This did not pass.

A bill (HR 14) to designate certain river segments in New Jersey as rivers that could be included in the National Wild and Scenic Rivers System became law (PL 100-33). A bill to designate certain lands in Michigan as part of the National Wilderness Preservation System became law (PL 100-184). The bill, HR 148/S 1036, was known as the Michigan Wilderness Act of 1987.

Some laws did not pass. For example, a bill (HR 39) to designate certain lands in Alaska as part of the National Wilderness Preservation System did not pass. Neither did HR 371, the California Desert Protection Act of 1987, to designate new additions from California as wilderness. HR 708/HR 2142, the Nevada Wilderness Act of 1987, would have designated land in Nevada as part of the National Wilderness Preservation System, but it did not pass.

President G.H.W. Bush

President Bush believed that protecting our forests is not a political issue, nor a partisan issue. Instead, he described it as a practical issue that we must come together and solve.[89] Bush proposed a global convention on forests that would emphasize an international cooperative effort to help address possible threats to the world's forests. The president outlined many areas of action, including research and monitoring; education, training, and technical assistance; reforestation and rehabilitation; tropical forestry action plan reform; reduction of air pollution; bilateral and multilateral assistance programs; a debt-for-nature swap; and removal of harmful subsidies.[90]

In 1991, Bush submitted to Congress the California Public Lands Wilderness Act. This proposal would direct the secretary of the interior to review the possibility of adding 62 areas of land to the wilderness program.[91] That year, he also announced a comprehensive plan to improve the protection of the nation's wetlands because they provide an important habitat for many animals. The plan included ways to protect current wetlands and restore those that had been damaged.[92]

In the last year of his presidency, Bush signed the Marsh-Billings National Historical Park Establishment Act, which protected that area from future development.[93] He also signed legislation that designated area in Colorado as wilderness area and a new law to establish the Keweenaw National Historical Park.[94] He sent proposals to Congress to designate land in Arizona, Idaho, Colorado, and Montana as wilderness.[95]

1989–90: 101st Congress

In 1989 the House cleared and sent to President Bush a proposal called the Nevada Wilderness Protection Act of 1990 (S 974/HR 2066/HR 2320/PL 101-195), which would create 733,000 acres of wilderness in Nevada. Over the protests of Western Republicans who argued that the bill would lock up valuable deposits of minerals, oil, and gas, the president signed it.[96] More land in Arizona was designated as wilderness in 1990. In HR 2570/HR 2571/HR 4398, 39 areas of different sizes were set aside as wilderness in Arizona.[97] This law was known as the Arizona Desert Wilderness Act of 1990 (PL 101-628).

In HR 987 (PL 101-626), timber cutting was prohibited on more than 1 million additional acres of the Alaskan Tongass National Forest. This law designated 296,000 acres of the forest as wilderness, meaning no logging, mining, or road building could be done; on another 722,000 acres, some mining and road building could occur.[98]

1991–92: 102nd Congress

In 1991, the House passed HR 2929, the California Desert Protection Act, to protect millions of acres of California's desert. The proposal was contentious because the desert holds lucrative gold and mineral deposits, multiple archaeological sites, and acres of shifting sand dunes that beckoned off-road-vehicle enthusiasts. The bill would turn the land into 73 national wilderness areas to limit activities such as mining and off-road-vehicle access. The bill also sought to expand the 3.3 million–acre Death Valley National Park and 800,000-acre Joshua Tree National Monument into similarly protected national parks. President Bush did not agree with the bill, which stalled in the Senate.[99]

That same year, the Senate passed a bill (S 1029) to add to federal wilderness areas in Colorado, but the measure stalled in the House because of controversy over water rights. The proposal designated about 700,000 acres of federal land as a protected wilderness. But one provision retained jurisdiction for Colorado over water rights within wilderness areas created in the bill. House leaders opposed the water-rights language because it differed from wilderness laws that gave the federal government the right to water found within wilderness areas.[100]

In 1992, a proposal that would have clearly defined the borders of Montana's federally protected wilderness failed, leaving Montana as one of only two Western states, along with Idaho, without a statewide federal wilderness plan. Backers of S 1696 argued the bill would help the state balance its development needs while conserving its wilderness, but opponents said the bill would block needed development, drive business away, and give the U.S. Forest Service the power to make lasting land management decisions that would not be subject to judicial review.[101]

A 1992 proposal in the Congress, the Mineral Exploration and Development Act (HR 918), was aimed at updating the 1872 mining law. Proponents said the original law was passed when the settlement of the West was supposed to be a national priority. Some problems with the old law included a transfer of federal land to private use for as little as $2.50 an acre, improper use of mining claims for real estate development, a legacy of abandoned mines left to scar the landscape and pollute water supplies, and mining companies' practice of extracting billions of dollars in minerals without paying royalties to the federal Treasury. The bill would have imposed an annual $100 fee on miners to keep their mining claims active and eliminated the requirement that miners perform $100 worth of mining-related work each year to keep a claim active.[102] This bill did not pass.

President Clinton

Clinton's goal concerning natural sites was to improve our national parks and preserve them[103] because, he said, they are the envy of the world.[104] During his administration he signed many bills to expand the National Wilderness Preservation System, including the Colorado Wilderness Act, the California Desert Act,[105] the Mollie Beattie Wilderness Area Act,[106] and a bill to protect the Black Canyon of the Gunnison National Park.[107] Clinton also created the largest national park south of Alaska in the Mojave Desert in California.[108] He also signed legislation to protect Forest Roadless areas that remain largely untouched by humans.

In order to encourage educating people about the natural environments, historical events, and cultural resources in our country, Clinton proclaimed May 22–28 as National Park Week.[109] He also ordered a study of the possible harm done by automobiles and other forms of transportation on the parks.[110]

1993–94: 103rd Congress

Legislation (S 21/PL 103-433) to protect huge stretches of the California desert was signed into law. Called the California Desert Protection Act, the new law created 71 separate wilderness areas and protected about 6.4 million acres of desert land. It also created 1.2 million-acre Mojave National Park. The proposal e expanded and upgraded the nearly 800,000-acre Joshua Tree and over 3 million–acre Death Valley National Monuments to national parks.[111]

HR 631 (PL 103-77) was a proposal to protect more than 600,000 acres of Colorado wilderness. Specifically, the bill designated 611,730 acres as wilderness and set aside 174,510 acres in a less protected management area. HR 631 did not assert any federal water rights in the protection area. It did not include any mention of the controversial issue, but it did not allow anyone to claim a reserved water right in court or through any administrative proceeding. The bill also prohibited construction of new or expanded water projects on the lands.[112]

In 1993, Congress took up a host of park and wilderness bills, three of which passed. The first, HR 698, was a new law to protect the nation's deepest cave from oil and mineral exploration. The cave was intended to protect Lechuguilla Cave in Carlsbad Caverns National Park in New Mexico. The bill was signed by the president (PL 103-169). Another bill signed by the president was HR 38, a bill to establish the Jemez National Recreation Area in New Mexico. The new law protected one of the nation's richest ancient Indian settlements and created a 57,000 acre recreation area. This became PL 103-104. Finally, HR 236 was a bill to set aside nearly half a million acres around Idaho's Snake River as a federally protected conservation area. This was signed by the president and became law (PL 103-64). One bill that did not pass was HR 704/S 291, to allow fishing to continue at its existing level.[113]

The Congress in 1994 considered a bill (HR 2473/S 2137), the Montana Wilderness Act of 1994, that would designate land in Montana as components of the National Wilderness Preservation System. Disagreements between preservationists and development advocates led to the bill's death.[114] Another bill that did not pass was HR 3732, which would have classified 1.36 million acres in five national forests in Idaho as protected wilderness; this barred harvesting of timber, mining, and recreational development. Idaho was one of only two states that did not have a congressionally approved statewide wilderness plan, even though the state had the largest amount of roadless national forest land in the nation outside Alaska. The House passed a separate bill (S 2100) to authorize the Forest Service to continue for one year a pilot project in the Idaho Panhandle National Forest, but it also did not pass.[115]

1995–96: 104th Congress

The Forest Health Act (S 391) would have mandated salvage logging in national forests, but it did not pass. HR 1745 and S 884 were bills to designate 1.8 million acres in Utah as wilderness. Under the bill, the areas would be off-limits to all but the most restricted uses. These bills did not pass.[116]

Some bills that did not pass (HR 1500/HR 1745) would have designated land in Utah as wilderness. This was called America's Red Rock Wilderness Act of 1995. Another bill in the House (HR 1825/HR 1538/HR 2292/S 879) that failed was a proposal to limit acquisition of land on the 39-mile headwaters segment of the Missouri River in Nebraska and South Dakota; this had been designated as a recreational river.

1997–98: 105th Congress

In 1997, the House and Senate panels approved versions of a bill (S 738/HR 1739) to ease motorboat and truck restrictions in the Boundary Waters Canoe Area Wilderness in northern Minnesota, but no final action was taken on it, so it did not become law.[117] A House proposal (HR 302) would have designated land in the Rocky Mountain National Park in Colorado as a component of the National Wilderness Preservation System, but it also did not pass. Another proposal that did not have enough support to pass was HR 1423, the Northern Rockies Ecosystem Protection Act of 1997. This proposal would have designated land in Idaho as wilderness areas.

However, many other bills did become law. Two bills (HR 449 and S 94) comprising the Southern Nevada Public Land Management Act of 1998 became law (PL 105-263). Under the new law, the secretary of the interior would be required to dispose of certain federal lands in Nevada that were under the jurisdiction of the director of the BLM. State officials could choose to purchase the land for local public purposes under the Recreation and Public Purposes Act (RPPA).

1999–2000: 106th Congress

HR 701, to distribute federal offshore drilling revenues to the states for conservation projects, was approved by the House Resources Committee but was not passed by the entire Congress.[118] In HR 15 (PL 106-145), the Congress designated specified public lands in the California Desert District of the BLM as the Otay Mountain Wilderness. Further, the law recognized that because of the area's proximity to the U.S.-Mexico international border, drug interdiction, border operations, and wildland fire management operations needed to continue.

A bill that did not pass was HR 488, the Northern Rockies Ecosystem Protection Act of 1999. If the bill had passed, land in Idaho, Montana, Oregon, Washington, and Wyoming would have been declared wilderness areas and components of the National Wilderness Preservation System. Similarly, HR 1165, a bill to establish the Black Canyon National Park in Colorado as part of the wilderness system, was not passed.

President G. W. Bush

President Bush reiterated that the wildlife refuges across the country had played a vital role in conserving and recovering our country's wildlife and natural resources. To protect these sites, Bush announced a National Parks Legacy Project that he hoped would be a major investment in our national parks. He proposed to spend $5 billion to clean up the backlog in maintenance and make parks more inviting and accessible to all citizens. He planned to protect nearly 4,000 miles of river alongside that.[119]

In other speeches, Bush maintained that our failure to protect our forests had had dangerous and devastating consequences. He said that the "uncontrolled growth left by years of neglect chokes off nutrients from trees and provides a breeding ground for insects and diseases.... Such policy creates the conditions for devastating wildfires." To prevent such damage, Bush announced the Healthy Forests Restoration Act, a program to thin trees and clean out the underbrush to restore our forests.[120]

In 2004, Bush announced a new policy to protect wetlands. He noted that wetlands are the habitat for thousands of species of wildlife, including some threatened and endangered species. He also noted that wetlands trap pollution. His new policy was aimed at restoring wetlands that had been damaged in the past and improving the quality of existing wetlands. To do this, he proposed to spend $349 million in federal money.[121] As part of this, he proposed a program to protect the Everglades and restore it to be a vibrant, alive, and available place for future generations.[122]

2001–02: 107th Congress

One bill passed during this session was HR 451/S 1205/S 1894 (PL 107-334), a proposal to make adjustments to the boundary of the Mount Nebo Wilderness Area. In another bill (HR 1576/ S 1711/ PL 107-216), Congress passed the James Peak Wilderness and Protection Area Act. This proposal was an idea to amend the Colorado Wilderness Act of 1993 to designate specified lands in the Arapaho/Roosevelt National Forest as the James Peak Wilderness Area. Congress also debated protecting more land in HR 2908, the Alaska Rainforest Conservation Act of 2001. This proposal would have designated public lands in the Chugach National Forest as wilderness as well as land in the Tongass National Forest, but it did not become law.

2003–04: 108th Congress

In 2003, the Congress approved the Healthy Forests Initiative (HR 1904), one of the Bush administration's environmental priorities. The program was designed to allow increased logging in specific national forests in an effort to reduce the risk of wildfires. The bill reduced the number of environmental reviews that would be required for logging projects and sped up judicial review of any legal challenges to the projects. Environmental groups opposed the legislation, but concern over communities at risk from wildfires overrode their concerns, and the bill was passed into law (PL 108-148).[123]

HR 353, the Federal Lands Improvement Act of 2003, did not pass. The proposal was an idea to have the secretary of the interior dispose of all public lands administered by the BLM that were identified for disposal under the Federal Land Policy and Management Act of 1976. It would require the disposal of at least one-third of the land within three years, two-thirds within five years, and all of the land within seven years. However, land located within wilderness areas would not be eliminated. The proceeds would be placed into an account, with some of the money given to the county in which the land was located and some going to the Treasury's general fund.

HR 640, the Rocky Mountain National Park Wilderness Act, also did not pass. This was a proposal to designate specific lands in the Rocky Mountain National Park, Colorado, as components of the National Wilderness Preservation System. A similar proposal, HR 979, was a bill to designate land in the Chugach National Forest in Alaska to be part of the National Wilderness Preservation System. This was not signed into law.

2005–06: 109th Congress

In the 109th Congress, a bill called the Brownfields Redevelopment Enhancement Act (HR 280/S 3620) failed to pass. This bill was a proposal to amend the Housing and Community Development Act of 1974 and allow the secretary of housing and urban development to make grants to different public entities and Indian tribes to assist in the environmental cleanup of brownfield sites and of land that had been affected by mines.

A bill passed into law was HR 233/S 128 (PL 109-362) to designate land in California as wilderness areas and as components of the National Wilderness Preservation System. Another bill (HR 539/S 272) that was also passed into law (PL 109-118) made land in Puerto Rico a component of the National Wilderness Preservation System.

2007–08: 110th Congress

During this session, Congress considered but did not pass laws to increase the lands protected in the National Wilderness Preservation System. One of those (HR 488), the Northern Rockies Ecosystem Protection Act of 2001, would have designated land in Idaho, Montana, Oregon, Washington, and Wyoming as components of the National Wilderness Preservation System. Another, HR 944, would have designated land in Colorado as wilderness. Finally, S 2316 was introduced to designate part of the Arctic National Wildlife Refuge as part of the wilderness system. This did not pass.

President Obama

Upon signing the Omnibus Public Land Management Act of 2009, Obama noted that there were places that we needed to preserve and set aside so that we could guard their sanctity and allow everyone to share them.[124] Obama proclaimed September 2009 and September 2010 as National Wilderness months and asked Americans to learn more about our wilderness areas and what they can do to help protect them.[125] April 18–26, 2009, was proclaimed as National Park Week, and people were urged to enjoy the park system during this time.[126]

2009–10: 111th Congress

During this session, the Senate passed an omnibus land bill (S 22/HR 146/PL 111-11), the Omnibus Public Land Management Act of 2009. The final bill was actually a series of smaller bills that were proposed in both the House and Senate (including S 22, HR 146, HR 169, HR 170, HR 222, HR 328, HR 867, S 109, S 135, and others). The new law would set aside more than 2 million acres of newly protected wilderness nationwide and establish other protections for federal lands, including protections for wild and scenic rivers and national historic sites.[127] The protected land included 517,000 acres in Idaho, almost 256,000 acres in Utah, and almost 250,000 acres in Colorado.[128] The bill established new programs in the Department of the Interior (on water reclamation and the effects of climate change on water availability), as well as new programs in the National Oceanic and Atmospheric Administration (on ocean exploration, undersea research, and ocean acidification).[129]

Members of the Republican Party accused the Obama administration of plans to take over more than 10 million acres of land in western states by designating new national monuments via the Antiquities Act of 1906 (PL 59-209). This would prevent mining, energy exploration, and other activities that Republicans felt were needed.[130] Finally, HR 167, the Rio Grande Wild and Scenic River Extension Act of 2009, did not pass. The law would have amended the Wild and Scenic Rivers Act to modify the boundary of the Rio Grande Wild and Scenic River.

Conclusion

Congressional action to protect natural land in the United States has ranged from protecting significant land areas as wilderness to reclaiming land ruined by strip mining. Environmentalists and those who use the land for recreation support such legislation; however, there are often opponents who successfully block those bills. There is no doubt that debate over what lands, if any, should be protected, and to what extent, will continue long into the future, and those elected to Congress will continue to debate legislation over the protection of the land for years to come.

Chapter 9

The Politics of Endangered Species and Wildlife

At one time in the United States, passenger pigeons flew in flocks so dense they blocked the sun. Some of the flocks contained more than 2 billion birds. But by 1914, target shooters and market hunters had reduced the species to a single individual, Martha, who lived at the Cincinnati Zoo until she died. The disappearance of the passenger pigeon and the decline of others led to more interest in protecting wildlife during the 1960s.[1]

In the mid-1970s, controversy erupted between environmentalists and developers over the Tellico Dam in Loudon County, Tennessee, then under construction by the Tennessee Valley Authority. When the dam was nearly complete, a species of endangered fish, the snail darter, was discovered in the river that fed the dam. Environmentalists quickly brought a lawsuit to prevent completion of the dam and ensuing harm to the fish species. The case, *Tennessee Valley Authority v. Hill* (437 U.S. 153 (1978)), traveled all the way to the U.S. Supreme Court, which stopped final construction on the dam. The justices argued that the construction project violated the Endangered Species Act, which provided that the habitat of any endangered species should not be disrupted. The justices pointed out that the dam would destroy the habitat of the snail darter, causing its extinction. A committee was established to consider a waiver, but the committee members refused to grant one, further preventing the completion of the dam. In the end, the Congress finally granted an exemption to the dam in an unrelated bill, and the dam was finally completed in 1979. Later, other schools of the snail darter fish were found in nearby streams.

The debate over the use of wilderness areas and protection of wildlife has been a contentious issue in environmental politics for many years. Two schools of thought exist: conservationism, which supports a multiple-use policy governing the national parks, and preservationism, which wants national parks to be put aside for preservation only.[2]

One of the earliest legislative acts to protect endangered species and wildlife was the Lacey Act, passed in 1900. This law prohibited interstate commerce in illegally captured wildlife. The law effectively ended the commercial hunting industry. Three years later, the first national wildlife refuge was created at Pelican Island, Florida, by Theodore Roosevelt. In 1911, the Fur Seal treaty was signed, and in 1918, the Migratory Bird Treaty Act authorized federal management of ducks and other migratory waterfowl across state lines. Since these early years, Congress has continued to pass legislation to protect wildlife, as outlined in the following chapter.

1965–66: 89th Congress

The Endangered Species Conservation Act was passed in 1966 (HR 9424/PL 89-669). The new law allowed the secretary of the interior to take special action and develop a program to conserve, protect, restore, and propagate selected species of fish and wildlife that were in danger of extinction. That included actions to protect the habitats of threatened species. The Congress authorized $15 million from the Land and Water Conservation Fund for acquisition of lands and waters to preserve threatened species. The bill, backed by the administration, was designed to protect about 35 types of mammals and 30 to 40 species of birds. The bill was supported by various conservation groups including the National Audubon Society, National Wildlife Federation, Wildlife Management Institute, Sport Fishing Institute, and Defenders of Wilderness, Inc.[3]

1969–70: 91st Congress

In this session, the Endangered Species Conservation Act of 1969 (HR 11363/PL 91-135) was passed to remedy some limitations of the earlier law. The new act expanded the original 1966 Endangered Species Conservation Act (PL 89-669) and was designed to strengthen laws that limited the importation and sale of animals, fish, and amphibians that were in danger of becoming extinct. The secretary of the interior was given the authority to develop and publish a list of species that were threatened with worldwide extinction and to prohibit their import into the United States.[4] The law barred importation of endangered species of fish or wildlife and prohibited shipment of any amphibians, reptiles, mollusks, or crustaceans in interstate commerce if they were obtained in violation of any federal, state, or foreign law. The new law included more specific definitions of types of protected wildlife and included procedures for identifying animals as endangered. The new law also gave the secretary of the interior the authority to protect reptiles, amphibians, mollusks, and crustaceans.[5] Finally, the government's authority to acquire land for species protection and preservation was also expanded.[6]

In 1970, the Congress passed a bill (HR 12475/PL 91-503) that revised and clarified the Federal Aid for Fish and Wildlife Restoration Act. The bill made numerous changes in both the fish and wildlife restoration programs and gave states the alternative of using comprehensive long-range plans rather than plans made on a project-to-project basis. It also amended the wildlife program by applying certain firearms excise taxes directly to the wildlife restoration fund and gave states the option of using up to three-quarters of their share of the fund for hunter safety programs. The administration opposed the provision that transferred revenues from certain firearms taxes from the general fund of the Treasury to the wildlife restoration fund, but Nixon chose to sign it despite that.[7]

1971–72: 92nd Congress

In this session the Congress created the National Marine Mammal Protection Act (HR 10420/PL 92–522). Although an early version of the bill allowed for a permit system for killing of certain marine mammals, the final act provided for a permanent moratorium on killing of ocean mammals and on importation of their products. It applied to seals, sea lions, whales, porpoises, dolphins, sea otters, manatees, walruses, and polar bears, some of which were in danger of extinction because of commercial slaughtering or recreational hunting.[8]

A proposal passed by the House (HR 13152), originally from President Nixon, was designed to restrict the use of poisons in predator control programs. The bill was not passed by the Senate and did not become law.[9] A related bill, HR 14163, sought to indemnify ranchers whose domestic animals were killed by predators. In the end, the House killed that bill as well. Other bills that did not pass were the Endangered Species Conservation Act of 1972 (S 3199); the Nature Protection Act of 1972 (S 249); and bills to prohibit poisoning of predators (S 2083), to provide a study of predator control methods (S 2821), to establish a national policy and program on predators (S 273), and to prohibit the killing or capture of endangered species by making it a federal crime (S 3818).[10]

Another bill that passed was HR 11091 (PL 92-558), in which additional funds for wildlife restoration projects and hunter safety programs were approved by Congress. The bill provided for an 11 percent sales tax on bows and arrows and other archery accessories. Revenues from this tax, which would not apply to bows and arrows manufactured by Indians on reservations, would be earmarked for the Wildlife Restoration Fund.[11]

The House members debated a bill (HR 12143/PL 92-330) that established a 21,662-acre San Francisco Bay National Wildlife Refuge in California that would help to preserve and enhance the wildlife habitat of South San Francisco Bay and protect the migratory waterfowl that frequented the area. Also living within the proposed refuge was a colony of harbor seals. The law directed the secretary of the interior to establish the refuge and authorized him to modify the boundaries where necessary, providing the total land area did not exceed 23,000 acres. The bill authorized up to $9 million in funds to carry out the provisions of the legislation.[12]

Congress completed action on two bills designed to protect wildlife from commercial and recreational hunters. The first, HR 5060 (PL 92-159), made it a federal crime to shoot, harass, or hunt any bird, fish, or other animal from an airplane. The bill set maximum penalties of up to $5,000 in fines and a year in prison for violations of the new laws.[13] The other established federal protection over the 9,000 to 10,000 wild horses and 10,000 wild burros roaming free on public lands in the West.[14]

HR 12186 (PL 92-535) strengthened the penalties imposed for violations of the Bald Eagle Protection Act of 1940. The 1940 act prohibited the shooting or molesting of bald and golden eagles. The new bill increased the maximum fines and prison sentences, as well as authorized rewards for information leading to conviction of offenders.[15] However, Congress did not pass a bill to call for a moratorium on killing of polar bears (HJ Res 1268) or a proposed international treaty to illegalize the harvesting of seals in Antarctic waters. Conversely, they successfully passed a bill (HR 16074/PL 92-604) to provide for the control or elimination of jellyfish and other ocean pests.[16]

1973–74: 93rd Congress

Congress cleared, at the president's request, a bill (S 1983/HR 37/PL 93-205) to provide additional protection for fish, wildlife, and plants facing immediate extinction.[17] The authors of this bill, the Endangered Species Conservation Act (ESA) of 1973, wrote that "various species of fish, wildlife and plants in the United States have been rendered extinct as a consequence of economic growth and development untempered by adequate concern and conservation."[18] They also said, "These species of fish, wildlife, and plants are of esthetic, ecological, educational, historical, recreational, and scientific value to the Nation and its people."[19] In the end, the law was one of the most comprehensive pieces of environmental legislation to protect wild animals.

Specifically, the ESA mandated that the Fish and Wildlife Service and the National Marine Fisheries Service list any endangered species as either "threatened" or "endangered."[20] Any federal action that could potentially damage these animals or their habitats was prohibited, as was the "taking" of these species by anyone.[21] The Fish and Wildlife Service and National Marine Fisheries Service were then to develop recovery plans for any species on the list. The laws made it such that endangered wildlife was protected even when this resulted in possible economic losses to businesses.[22] But, according to the law, businesses had to demonstrate that a proposed economic activity would not further endanger a species that appeared on the list.[23] Even federal, state, and private landowners were required to protect certain designated species.

In other provisions, the legislation protected smaller animals, even invertebrates and plants. It also authorized a grant program to assist state programs, as well as coordination among all federal agencies to protect the animals. Moreover, the ESA allowed for both civil and criminal penalties if any part of the act was violated. The law allowed for forfeiture of any item used to aid in the violation of the act.[24]

ESA and the Courts

The ESA contains a variety of protections designed to save designated species from extinction. Section 9 of the act makes it unlawful for any person to "take" any endangered or threatened species. *Take* was defined by the law as "harassing, harming, pursuing, hunting, shooting, wounding, killing, trapping, capturing or collection any of the protected wildlife."[25]

In the early 1990s, the secretary of the interior modified part of the law that defined the statute's prohibition on taking to include "significant habitat modification or degradation where it actually kills or injures wildlife."[26] A group called the Sweet Home Chapter of Communities for a Great Oregon, made up of small landowners, logging companies, and families dependent on the forest products industry of the Pacific Northwest, filed a lawsuit with the court alleging that the secretary of the interior exceeded his authority by making that regulation. The group argued that the legislative history of the ESA demonstrated that Congress had considered, and then rejected, such a broad definition of *take*. Further, they questioned the word *harm*. The group argued the regulation as applied to the habitat of two species of birds, the northern spotted owl and the red-cockaded woodpecker, had injured them economically, as there were now vast areas of land that could not be logged. They argued that if the secretary of the interior chose to protect these endangered species, then the secretary would have to buy their land.

The case was eventually heard in the U.S. Supreme Court in the case *Babbitt v. Sweet Home Chapter of Communities for a Great Oregon* (515 U.S. 687, 1995). The Supreme Court upheld the Department of the Interior's regulation and concluded that the secretary's definition of *harm* was reasonable. Further, the Court concluded that the writing of the technical and science-based regulation was a complex policy choice that Congress had entrusted the secretary with the discretion to complete, and the Court expressed a reluctance to overturn the secretary's decision.[27]

President Ford

During his presidency, Ford considered sanctions against other countries (particularly the Union of Soviet Socialist Republics and Japan) that had exceeded the International Whaling Commission quotas for minke whale catches. He explained that quotas on the number of whales that could be killed each year were set by the International Whaling Commission. He noted that in considering trade sanctions against these countries, serious economic impacts had to be considered. He chose not to impose such sanctions at the time but kept the option open for imposing them in the future.[28] Ford asked the Senate to ratify an International Agreement on the Conservation of Polar Bears to protect them from hunting, killing, and capture.[29] He also sent to the Senate the Convention for the Conservation of Antarctic Seals, which would provide valuable protection for that animal.[30]

1975–76: 94th Congress

The House passed a bill (HR 5512/PL 94-223) that required all areas within the National Wildlife Refuge System to be administered by the Fish and Wildlife Service, found within the Department of the Interior. It also permitted cooperative agreements for the administration of Alaskan refuges to continue under the supervision of the wildlife service.[31] The law was needed because most transfers of land from the system were prohibited unless they had congressional approval. Some environmental groups and many members of Congress objected to the transfer, arguing that the land management agency was more concerned with commercial activities such as grazing and mining than with protecting wildlife.

The 1975 Convention on International Trade in Endangered Species (CITES) was passed this year. It is an important international treaty with the goal of halting the trade in endangered species.[32] The convention was established to prevent international trade from threatening species with extinction. Over 150 countries agreed to be part of CITES. In doing so, they agreed to control the import and export of an agreed-upon list of animals that were endangered, or at risk of becoming endangered, because of a lack of regulation in their products or the animals themselves.

President Carter

Early in his administration, Carter proposed providing $50 million to purchase wetlands to protect waterfowl habitat.[33] He suggested that the federal government refrain from supporting proposed development in wetlands. He also expressed his support for the Federal Water Pollution Control Act to regulate the filling and disposal of dredged materials in all wetlands.[34]

In addition, the president asked Congress to provide $295 million for a program to rehabilitate and improve the wildlife refuge system, as well as improve protection for nongame wildlife. This would be enhanced by an accelerated effort to identify habitat that is critical to the survival and recovery of endangered species. In another effort to protect wildlife, Carter promised to present an executive order to restrict the introduction of potentially harmful foreign plants and animals into the United States. He also wanted legislation that would prohibit commercial whaling in the United States' 200-mile fisheries zone.[35]

When Carter signed the Endangered Species Amendments of 1978, he noted that the new law would authorize a special committee that would have the power to exempt federal agencies from a requirement that their actions not jeopardize the existence of endangered or threatened species. He went on to ask the committee members to be exceedingly cautious in using that power to consider exemptions and use it only when there were clear threats to national security.[36]

1977–78: 95th Congress

A bill to extend the program of grants to states for conservation programs (S 1316/PL 95-212) was passed by Congress. The new law made it easier for states to qualify for the funding. The money would go to help protect endangered species.[37]

Congress passed a law (S 2899/HR 14104/PL 95-632) to enact the Endangered Species Act Amendments. The bill set up a cabinet-level board that would be responsible for considering federal construction projects, even though they might harm or kill off animal species that were protected by the 1973 law. In other words, the bill would allow for exemptions under the original law so that future conflicts with public works projects could be avoided.[38] The law was a response to a Supreme Court decision barring operation of the Tennessee Valley Authority's Tellico Dam, which had already been built. In that case, the Court ruled that opening the floodgates of the dam would destroy the critical habitat of a three-inch fish called the snail darter.[39]

Another law passed was HR 10730 (PL 95-316), which authorized $14.8 million for fiscal year 1979, $15.6 million for fiscal year 1980, and $16.8 million for fiscal year 1981 in funding for matching grants to state marine mammal protection programs for a variety of marine mammal research projects.[40]

A bill that did not pass was S 1140, which would have provided federal matching grants to states to help them develop and carry out nongame projects, such as songbird-viewing areas or long-range nongame conservation plans. The plans could have included acquisition of land for fish and wildlife habitat.[41]

HR 2329 (PL 95-616), or the Interoceanic Canal Study Act, enacted the Fish and Wildlife Improvement Act of 1978. The law mandated that the secretaries of the interior and commerce establish national training programs for state fish and wildlife law enforcement personnel, and it authorized the appropriation of funds necessary for conducting these programs. Further, the new law amended the Migratory Bird Conservation Act to broaden the power of the secretary of the interior to acquire certain lands, and it extended administrative protection to migratory birds listed in the Convention between the United States and Japan for the Protection of Migratory Birds and Birds in Danger of Extinction.

1979–80: 96th Congress

After the decision about the Tellico Dam was made public, Congress amended the Endangered Species Act to set up a special board with the power to grant exemptions to the law (S 1143/PL 96-159). Although the panel refused to exempt Tellico at first, the senators eventually gave the go-ahead to the project. President Carter signed the bill over the objections of environmentalists.[42]

This legislation also provided for a three-year authorization of the ESA, and it strengthened the endangered species protection program by including plant as well as animal species in the emergency listing.[43]

Another bill to protect animals was HR 3292/S 2181 (PL 96-366), the Fish and Wildlife Conservation Act of 1980, which provided funds for states to develop comprehensive plans to protect fish and wildlife species that were not hunted for sport, food, or fur. It was designed to protect animals that were considered nongame species, which made up 83 percent of all animals within the United States.[44]

Members of the House debated HR 6839 (PL 96-246), which increased the amount of financial assistance available to states to develop endangered species management programs. The Congress authorized $6 million per year for fiscal years 1981 and 1982. The increase was necessary because the number of states participating in the program had doubled since 1977.[45]

In 1980 Carter signed a law that would protect vast tracts of land in Alaska. HR 39 (PL 96-487), the Alaska National Interest Lands Conservation Act, was intended to preserve certain lands and waters in Alaska that had nationally significant natural, scenic, historic, archaeological, geological, scientific, wilderness, cultural, recreational, and wildlife values.

A bill requested by the president but not cleared was HR 5604/S 1882. Carter requested it in his second environmental message after federal investigators found a thriving illegal trade in endangered species. The underground trade was found to be threatening domestic agricultural and pet industries because diseases were often imported with the animals. Investigators also realized that since the criminal penalties for violating existing laws were so small, wildlife smugglers were risking ignoring the regulations to obtain the enormous profits available for illegal fish and animals.[46]

President Reagan

In 1985, President Reagan chose not to take action against the Soviet Union for exceeding agreed-upon limits for the taking of minke whales He did this because a punishment could have had repercussions for the United States, including high unemployment of U.S. fishermen and other workers and dissolution of other joint-venture companies.[47]

1981–82: 97th Congress

A bill to strengthen federal controls over illegal imports or trafficking in protected fish and wildlife was signed by Reagan (S 736/HR 1638/PL 97-79).

The Lacey Act Amendments combined two existing wildlife control laws and stiffened the penalties for illegal importation or interstate trafficking of protected wildlife. The bill was designed to crack down on a lucrative illegal trade in fish, wildlife, and wild plants.[48] It made it illegal to import or trade in live fish or wildlife or animal products that are taken, transported, or sold in violation of state or foreign laws.

HR 6133 (S 2309/PL 97-304) reauthorized the 1973 Endangered Species Act. The new act set a one-year deadline for the Department of the Interior to decide whether to list a species as endangered after it received evidence of such. It also encouraged introduction of captive-bred populations of endangered species into the wild populations.[49]

HR 4084 reauthorized the federal marine mammal protection program while modifying existing law to deal with the accidental killing of porpoises by tuna fishermen. Reagan signed the measure (PL 97-58). The new law retained the 1972 act's goal of reducing accidental porpoise deaths to insignificant levels approaching a zero mortality and serious injury rate. However, it specified that yellowfin tuna fishermen could satisfy this objective through the use of the best equipment and techniques practicable. The law allowed other commercial fishermen, deep seabed mining groups, oceanographic researchers, and oil and gas drillers to accidentally kill a small number of marine mammals as long as there was a "negligible" impact on the species.[50]

In HR 5662 (PL 97-347), the Congress extended the appropriations for certain programs under the Fish and Wildlife Act of 1956. Specifically, the law provided money for the fisheries loan fund used to make loans for financing or refinancing commercial fishing vessels and gear. It also extended the authorizations for volunteer services in fish and wildlife programs through 1983.

1983–84: 98th Congress

In HR 2395 (PL 98-200), Congress extended for one year the Wetlands Loan Act of 1961, through which hunters help pay for federal purchases of waterfowl habitat. The original law, passed in 1961, was last renewed in 1976.[51] Another law passed by Congress was HR 4997 (PL 98-364), which reauthorized the Marine Mammal Protection Act for four years and renewed the limits on the accidental killing of porpoises in tuna fishing. This new law was called the Commercial Fishing Industry Vessel Act.[52]

HR 3082, a bill to expand funding for protection of the nation's dwindling wetlands, died. Instead, Congress cleared another, simpler bill (HR 5271/PL 98-548).[53] The second bill extended the Wetlands Loan Act, which helped fund preservation of waterfowl breeding and feeding grounds for 10 years.[54]

1985–86: 99th Congress

A bill to renew the Endangered Species Act (HR 1027) was not passed.[55] The proposal would have amended the Endangered Species Act of 1973 to require the secretary of the interior to implement a monitoring system for species for which a petition to add them to the endangered or threatened lists had been evaluated but not yet acted upon. The proposal would have extended appropriations to carry out the act through fiscal year 1988.

In HR 1404/S 585, land acquisition was approved for a national wildlife refuge at Cape Charles on the eastern shore of Virginia and a wildlife service training center at Cape Charles, Virginia. The area was threatened by development.[56] However, Reagan vetoed the bill, explaining that the bill "does not simply provide protection for this valuable habitat. It would also require the Secretary of the Interior to develop a training center at the refuge for use by the Service, other Federal and State agencies, educational institutions, and private organizations and individuals. In this time of fiscal constraint, the Federal Government must limit its expenditures to matters of significant national concerns."[57]

1987–88: 100th Congress

HR 1467/S 675/PL 100-478 reauthorized the Endangered Species Act at increased funding levels for fiscal years 1988–92: $57 million for fiscal year 1988, $59 million for year 1989, $61 million for 1990, $64 million for 1991, and $66 million for 1992.[58]

The Congress cleared S 1389 (PL 100-240) to increase the annual authorization for the National Fish and Wildlife Foundation from $1 million to $5 million for fiscal years 1988–93. Congress originally created the foundation in 1984 (S 1271/ PL 98-244) as an independent conservation organization to develop private support for the programs and activities of the U.S. Fish and Wildlife Service. Since 1984, foundation funds had been used to support activities such as restoration of the bald eagle population and the tracking of endangered sea turtles.[59] A new law (HR 4209/PL 100-536) authorized appropriations to carry out Title 1 of the Marine Protection, Research and Sanctuaries Act of 1972.

President G.H.W. Bush

President Bush sent a letter to congressional leaders in which he stated that Norway violated an agreement on taking of minke whales. However, he noted that the country had made improvements in their research programs and had reduced their take of whales. Because of that, Bush did not support sanctions against Norwegian fish products.[60] Bush chose to sign the Wild Bird Conservation Act of 1992, a new law to promote the conservation of wild exotic birds.[61]

1989–90: 101st Congress

Congress passed, and the president signed, the North American Wetlands Conservation Act (S 804/ HR 2323/ PL 101-233), a bill to help increase the continent's declining waterfowl population through federal preservation of North American wetlands. The bill authorized up to $15 million a year for fiscal years 1990–93, plus about $11 million annually from interest earned on the unused portion of a federal trust fund made up of excise taxes on hunters' equipment. With the money, the Department of the Interior was to purchase wetlands in the United States, Canada, and Mexico in an effort to protect the habitats of migratory birds and double the waterfowl population by the year 2000. The bill also established a nine-member North American Wetlands Conservation Council to make recommendations about which wetlands to buy.[62]

In HR 89, the Endangered Species Protection Act of 1989, the House proposed increases in the maximum fine for motor vehicle speed-limit violations in the National Forest System, the National Park System, and the National Wildlife Refuge System to $1,000. These areas are regularly inhabited by endangered wildlife species. The proposal did not have enough support to pass.

In HR 132, the International Fish and Wildlife Protection Act, restrictions were imposed on the importation of fish or wildlife products from countries that violate international endangered or threatened species programs. The proposal was to allow the president to prohibit the importation of any products from the offending country into the United States. This did not pass.

HR 544, the State Fish and Wildlife Assistance Act, did not pass. It would have amended the Mineral Leasing Act for Acquired Lands to require half of the lease revenues from lands acquired by the United States under the Weeks Act to be deposited into a state fish and wildlife assistance fund in the Treasury and 25 percent to be paid to the state where the leased lands were located.

1991–92: 102nd Congress

The Striped Bass Protection Bill was passed by Congress during this session. The bill (HR 2387/PL 102-130) renewed the 1984 Atlantic Striped

Bass Conservation Act (PL 98-613) for three years. The law required states along the Atlantic seaboard to follow the recommendations of the federally chartered Atlantic States Marine Fisheries Commission or face a moratorium on commercial bass fishing.[63] Help for salmon was included in the fiscal year 1992 Interior appropriations bill (HR 2686/PL 102-154).

If HR 4045 had been passed, the federal government would have been forced to take steps to protect declining species; it would have required the government to consider habitat preservation in protecting species. It also sought to tighten deadlines for recovery plans and to make it easier for legal action to be taken in emergency cases to stop a species's decline.[64] The bill went nowhere during this session.

The Army's Rocky Mountain Arsenal in Colorado was to be transformed into a de facto wildlife refuge under a bill cleared by Congress (HR 1435/S 2260/PL 102-402), even though parts of the proposed refuge had become so contaminated that they were on the list of the nation's worst hazardous waste sites.[65]

HR 61, the Endangered Species Protection Act of 1991, was not passed by the House and Senate. It would have increased the maximum fine for motor vehicle speed-limit violations in units of the National Forest System, the National Park System, and the National Wildlife Refuge System to $1,000. These are areas regularly inhabited by endangered wildlife species.

The Dolphin Protection and Fair Fishing Act of 1991 (HR 261) would have amended the Marine Mammal Protection Act of 1972 to prohibit, with regard to fishing for yellowfin tuna, promulgation of regulations or issuance of permits allowing the intentional setting of purse seine nets on marine mammals. The law would have required observers on tuna fishing vessels in the eastern tropical Pacific Ocean to ensure that the taking of any marine mammal was reported to officials.

HR 330, the Refuge Wildlife Protection Act of 1991, would have amended the National Wildlife Refuge System Administration Act to require that any wildlife management or other activity that affects wildlife in any area of the system be conducted in the most humane manner possible. It would have allowed the secretary of the interior to authorize the killing of a member of a wildlife species in an area of the system, if there was evidence that such killing is necessary for the health and habitat of the species in the area. If an emergency existed, public hearings could be held. This was not passed. In HR 367, the Florida panther would be considered an endangered species under the Endangered Species Act of 1973. This did not pass.

President Clinton

President Clinton was very concerned about protecting the wildlife and endangered species found both in the United States and globally. Clinton supported biological diversity conservation that would assist in managing our lands and waters and the animals that live in these areas.[66] In 1997, Clinton expressed his support for legislation (HR 1420) to create a strong mission for the National Wildlife Refuge System,[67] and he proclaimed a week in April as National Wildlife Week to educate people about the importance of protecting the wildlife across the country.[68] In 1999, he was pleased to announce that the bald eagle had come back from being endangered and started the process to remove it from the list of endangered species.[69]

To protect endangered specials internationally, Clinton sent to the Senate the protocol to the Caribbean Environmental Convention. This convention asked countries to establish special protected areas to protect the habitats of endangered species.[70] He also sent a message to the Congress reporting that the People's Republic of China (PRC) and Taiwan were trading rhinoceros and tiger parts, both of which were listed as threatened species at risk of extinction.[71] In 1994, he called for trade sanctions against Taiwan for practicing in such trading behaviors.[72] Later, the president signed HR 2807 into law (PL 105-312), a measure to reauthorize the Rhinoceros and Tiger Conservation Act and prohibit the sale, importation and exportation derived from those animals.[73] Before leaving office, Clinton sent the Inter-American Convention for the Protection and Conservation of Sea Turtles to the Senate for their ratification, which they did.[74]

1993–94: 103rd Congress

The Congress gave commercial fishermen a six-month extension of an exemption from the ban on killing marine mammals in a new proposal (HR 3049 /PL 103-86). This law also gave a temporary extension to lawmakers, to give them more time to overhaul the 1972 Marine Mammal Protection Act.[75] Congress voted to reauthorize the Marine Mammal Protection Act for five years in S 1636/HR 2760 (PL 103-238). It was a six-year reauthorization that sought to reduce accidental killing of marine mammals without harming the economic well-being of commercial fishermen. The bill retained the goal of the 1972 act to reduce the accidental killing of porpoises, whales, seals, and other marine mammals to "insignificant levels," meaning as close to zero as possible. The bill required the goal to be reached within seven years, and it authorized $20 million annually in fiscal years 1994–99 to carry out conservation measures. The bill outlawed the intentional killing of marine mammals by commercial fishermen. However, it allowed the intentional killing of "nuisance" seals and sea lions that had been eating endangered and threatened fish stocks at dams. It included regulations for the public display of marine mammals in zoos, aquariums, and theme parks. The bill also contained a controversial provision lifting a ban on importing trophies, typically skins, of polar bears that were legally hunted in Canada.[76]

In HR 1845, the National Biological Survey Act, Congress would have authorized an inventory of all plant and animal species in the country. It would have provided for the creation of an office within the Department of the Interior to undertake the survey and supply information on all species. The measure also authorized the creation of a policy board to advise the survey's director and a science council to work with other governmental agencies and private groups.[77] This did not become law.

S 476 (PL 103-232), the National Fish and Wildlife Foundation Improvement Act of 1994, reauthorized and amended the National Fish and Wildlife Foundation Establishment Act. It created the Brownsville Wetlands Policy Center and the Walter B. Jones Center for the Sounds at the Pocosin Lakes National Wildlife Refuge.

The black bear would have been protected under HR 55/S 181, the Black Bear Protection Act of 1993. If it had passed, the law would have directed the secretary of commerce under the Export Administration Act of 1979 to prohibit the export of the innards of the American black bear.

HR 236 (PL 103-64) was a proposal to establish the Snake River Birds of Prey National Conservation Area in the state of Idaho. The law provided for the conservation, protection, and enhancement of raptor populations, habitats, and associated natural resources and of the scientific, cultural, and educational resources of the public lands. The secretary of the interior would create a comprehensive management plan for the area and review and revise it at least once every five years.

HR 833, the National Wildlife Refuge System Management and Policy Act of 1993, would have prohibited the secretary of the interior from initiating the new use of a refuge, or expanding, renewing, or extending an existing refuge, unless it was determined that this was compatible with the purposes of the system and the refuge. This did not pass.

In HR 1391, wildlife on public lands would have been protected from airborne hunting. If passed, the law would have amended the Fish and Wildlife Act of 1956 to prohibit airborne hunting of a species that is listed as endangered or threatened under the Endangered Species Act. This did not pass.

1995–96: 104th Congress

Congress did not take final action on a bill to overhaul the Endangered Species Act during this session (HR 2275/S 1364).[78] The bill would have prohibited the federal government from using privately owned property or limiting the use of privately owned property when such action would diminish the value of the property, unless the government paid the fair market value of the private property to the owner. The bill would have forced consideration of the potential economic impacts and of property owners' rights while encouraging practices that protect wildlife. Among other things, the law would have directed the secretary of the interior to publish a conservation objective and a conservation plan for each species determined to be an endangered or threatened species. It would have also established a National Biological Diversity Reserve composed of units of federal and state lands to protect wildlife. It authorized states to establish and maintain programs for the conservation of endangered and threatened species and would have provided funds for that purpose.

In HR 321, it was proposed that the Florida panther be made an endangered species under the Endangered Species Act of 1973. This law did not pass. The Endangered Species Recovery and Conservation Incentive Act of 1995 (HR 2364) would have required the secretary of the interior to develop a plan for the conservation or recovery of an endangered species within two years after the animal was listed as endangered. The bill did not pass.

S 768, the Endangered Species Act Reform Act of 1995, did not pass during this session. The proposal was intended to make the law of 1973 more effective and less burdensome. When the secretary of the interior was determining what species to include on the endangered species list, the results should be made public. It would allow the secretary to determine whether a species is endangered or threatened because of the inadequacy of any existing federal, state, local governmental, and international regulatory mechanisms.

Finally, S 1915, the Rhino and Tiger Product Labeling Act, also did not pass. The proposal would have amended the Endangered Species Act of 1973 to make it unlawful for persons to sell or offer for sale in interstate or foreign commerce any product labeled as containing any listed endangered or threatened species of animals, plants, or fish.

1997–98: 105th Congress

In HR 1420 (HR 511/PL 105-57), Congress amended the National Wildlife Refuge System Improvement Act of 1966. It amended the mission of the system to being the administration of a national network of lands and waters for the conservation, management, and restoration of fish, wildlife, and plant resources and their habitats. It added requirements that the secretary of the interior should ensure that the system's mission and policies are carried out and that the status of the fish, wildlife, and plants in each refuge is monitored.

The Rhino and Tiger Product Labeling Act (HR 2807/HR 2863/S 361) became law this session (PL 105-312). This would prohibit the sale, import, or export, or the attempted sale, import, or export, of any product, item, or substance (product) intended for human consumption or application that contained, or was labeled or advertised as containing, any substance derived from any species of rhinoceros or tiger.

The bill also included the Migratory Bird Treaty Reform Act to make it illegal for anyone to hunt migratory birds by the aid of baiting. The law increased penalties for doing so. Also in the bill was the National Wildlife Refuge System Improvement Act, which made changes in the operation of the law. The Wetlands and Wildlife Enhancement Act was included in part of the bill, as was the Chesapeake Bay Initiative Act of 1998. This directed the secretary of the interior to provide technical and financial assistance to identify, conserve, restore, and interpret natural, recreational, historical, and cultural resources within the Chesapeake Bay Watershed.

In S 2094 (PL 105-328), the Fish and Wildlife Revenue Enhancement Act, Congress agreed to amend the Fish and Wildlife Improvement Act of 1978 to enable the secretary of the interior to more effectively use the proceeds from sales of abandoned items derived from fish, wildlife, and plants to cover costs incurred in such items' shipping, storage, appraisal, lien clearance, and disposal, including processing and shipping of eagles and migratory birds or parts of migratory birds for Native American religious purposes. It prohibited the secretaries of commerce and the interior from selling any species of fish, wildlife, plants, or derivatives for which the sale was prohibited by another federal law.

S 1180 was an attempt to rewrite the federal Endangered Species Act. This bill proposed a new process for recovering species and offered greater involvement to ranchers, property owners, and communities.[79] This proposal did not pass.

Congress considered many laws to protect different animals in this session. One of those was HR 39 (PL 105-217), the African Elephant Conservation Reauthorization Act of 1998 through 2002. Another was HR 408 (PL 105-42), the International Dolphin Conservation Program Act. This would allow authorizations for the incidental taking of marine mammals during commercial purse seine yellowfin tuna fishing in the eastern tropical Pacific Ocean. It removed provisions requiring that the goal of reducing incidental kill or serious injury to insignificant levels approaching zero be satisfied by the best safety techniques and equipment that were economically and technologically practicable. Bears were the topic of HR 619/S 263, the Bear Protection Act. If this had passed, it would have been illegal to import bear viscera into, or export it from, the United States. Finally, in HR 1787 (PL 105-96), the Asian Elephant Conservation Act of 1997, the secretary of the interior was required to provide financial assistance for projects for the conservation of Asian elephants for which the secretary had approved final project proposals. The law also established the Asian Elephant Conservation Fund.

HR 4555, the Endangered Species Criminal and Civil Penalties Liability Reform Act, did not pass during this congressional session. The proposal would have amended the Endangered Species Act of 1973 and would have prohibited a person from being liable for any criminal or civil penalty for a violation committed while conducting an otherwise lawful activity.

1997–98: Courts

The Endangered Species Act was taken to court during this session. In *Bennett v. Spears*, the Supreme Court overturned a ruling by the 9th Circuit Court of Appeals and ruled that property owners and other citizens could sue the federal government for actions taken under the Endangered Species Act.[80] The justices ruled that if government officials take actions to protect certain plants and animals and those actions cause economic harm, the landowners could sue those officials.[81] This would be true even if the officials were seeking less, not more, protection for the endangered species.

1999–2000: 106th Congress

Congress chose to amend the Endangered Species Act during this session (S 1744/PL 106-201). Under the new law, certain species conservation reports would have to be submitted. Many proposals did not pass in this session. For example, HR 187 was a proposal to deem the Florida panther to be an endangered species for the purposes of the Endangered Species Act of 1973. Another bill, HR 495, otherwise known as the Endangered Species Land Management Reform Act, also did not pass. This was a proposal to amend the Endangered Species Act of 1973 to prohibit a federal agency from taking an action that might affect privately owned property if it might lower the value of the property. A third proposal, HR 496, the Endangered Species Criminal and Civil Penalties Liability Reform Act, did not pass. This proposal would have amended the Endangered Species Act of 1973 to define *take* to mean to knowingly and intentionally perform any act with the knowledge that the act would constitute harassing, harming, pursuing, hunting, shooting, wounding, killing, trapping, capturing, or collecting an individual member of a species that was present at the time and location of the act, or to attempt to engage in such conduct.

Other bills defeated in Congress included HR 1101, a bill to amend the Endangered Species Act of 1973 to improve the ability of individuals and local, state, and federal agencies to prevent natural flood disasters. Under the proposal, consultations with the secretary of the interior would no longer have been required if a project revolved around repairing a levee, for example. HR 2343, which also failed, would have required the secretary of the interior and the National Academy of Sciences to review the list of endangered species every five years and recommend whether any animals should be removed from the list.

A bill that did not pass was HR 4496, a bill to protect America's wolves. This bill would have directed the secretary of the interior to prepare and publish a recovery plan for the eastern timber wolf in the northeastern United States, including a plan for releasing such wolves in the Catskill Mountains, New York.

Another bill that did not pass was S 1210, the Foreign Endangered Species Conservation Act of 1999, which would have directed the secretary of the interior to use money from the Foreign Endangered and Threatened Species Conservation Account to provide financial assistance for projects for the conservation of endangered or threatened species in foreign countries.

One bill that did have enough support to make it through Congress was the Great Ape Conservation Act of 2000 (HR 4320/S 1007/PL 106-411). The new law directed the secretary of the interior to provide financial assistance for projects designed to help the conservation of great apes (chimpanzees, gorillas, bonobos, orangutans, and gibbons).

President G. W. Bush

President Bush signed legislation to establish the Detroit River International Wildlife Refuge, the first international wildlife refuge. The law was intended to protect this area along the Michigan-Canada border to conserve and restore the wetland habitats found there.[82]

2001–02: 107th Congress

A bill to reauthorize the African Elephant Conservation Act of 1997 (HR 643/PL 107-111) was passed in this session. The new law authorized appropriations to the Multinational Species Conservation Fund. It also directed the secretary to convene an advisory group from both public and private organizations that are actively involved in the conservation of African elephants to assist in carrying out the act.

Similarly, a bill was proposed and passed (HR 645/PL 107-112) to reauthorize the Rhinoceros and Tiger Conservation Act of 1994. It provided for appropriations to the Multinational Species Conservation Fund and authorized the Secretary of the Interior to convene an advisory group of individuals representing public and private organizations actively involved in the conservation of rhinoceros and tiger species to assist in carrying out the act.

The same provisions were found in HR 700 (PL 107-141), a bill to reauthorize the Asian Elephant Conservation Act of 1997. This law amended the Asian Elephant Conservation Act of 1997 to authorize appropriations to the Multinational Species Conservation Fund through fiscal year 2007 to carry out the act.

An effort to protect bears did not pass Congress (HR 397/S 1125). The Bear Protection Act of 2001 would have prohibited any person from (1) importing bear viscera into, or exporting it from, the United States; or (2) selling bear viscera, bartering for it, offering it for sale or barter, or purchasing, possessing, transporting, delivering, or receiving it in interstate or foreign commerce. Any person who violated the law would have faced criminal penalties.

HR 1404 was a proposal called the Endangered Species Criminal and Civil Penalties Liability Reform Act, but it did not pass. The proposal would have barred criminal or civil punishments for a violation of the law if the violation was committed while conducting an otherwise lawful activity and not for the purpose of a prohibited taking.

2003–04: 108th Congress

A new law (HR 2619/PL 108-481) provided for the expansion of Kilauea Point National Wildlife Refuge. HR 1194, the Species Rescue Act, did not pass during this session. The law would have provided for expedited consultation and conferencing with respect to the impacts on endangered or threatened species of federal agencies' flood control actions.

S 2009/HR 1662 also did not pass. The proposal would require the secretary of the interior, when determining that a species is endangered or threatened, to ensure that the data used included timely field survey data to the extent possible.

HR 3545, the Southern Sea Otter Recovery and Research Act, did not pass. The proposal would have required the secretary of the interior and the Fish and Wildlife Service to carry out a recovery program for southern sea otter populations along the coast of California. The Preble's meadow jumping mouse was also the focus of legislation (HR 4466). The proposal would have excluded the mouse from the list of endangered species, but it did not pass.

HR 4475, the Endangered Species Recovery Act of 2004, did not pass. The proposal would have amended the Endangered Species Act of 1973 to redefine *endangered species* as any species that is in danger of extinction throughout all of its global range. The Secretary of the Interior would be asked to identify, from among endangered or threatened species, the 109 species that were in greatest danger of extinction. That list would be published. The listing period would be five years.

2005–06: 109th Congress

There were many bills concerning wildlife that did not pass this session. One of those, HR 1707, was a bill called the Great Cats and Rare Canids Act of 2005. The proposal would ask the secretary of the interior to provide assistance for projects for the conservation of rare felids and rare canids. Additionally, the Secretary of the Interior would be required to convene an advisory group of individuals representing public and private organizations actively involved in the conservation of felids and canids.

The Species Rescue Act (HR 2779) also did not pass. It was a proposal to change the consultation process with respect to the impacts on endangered or threatened species of federal agencies' flood control actions. The Endangered Species Improvement Act (HR 3300) did not have enough support to pass. Under the proposal, the secretary of the interior would be asked to establish a model form for entering into a species recovery agreement with a person for habitat protection and restoration services.

And, once again, a bill (S 1250) to reauthorize the Great Ape Conservation Act of 2000 did not pass. This time, the proposal was to amend the Great Ape Conservation Act of 2000 to authorize the secretary of the interior to approve projects to address the causes of the threats faced by great apes. The bill would have asked the secretary to convene an expert panel to identify the possible conservation needs for the great ape. The bill would also have increased funding for administrative expenses.

2007–08: 110th Congress

Two laws were passed during the 110th congressional session to reauthorize laws regarding two groups of animals. One was HR 50 (PL 110-132), which reauthorized the African Elephant Conservation Act and the Rhinoceros and Tiger Conservation Act of 1995. The second was HR 465 (PL 110-133) to reauthorize the Asian Elephant Conservation Act of 1997.

But many others did not become law. The Endangered Species Recovery Act (HR 1422/S 700/S 2223/S 2242) would have amended the Internal Revenue Code to allow certain landowners to get a tax credit for costs relating to the restoration of their property if they had entered into a habitat protection agreement. This would occur when the property contained the habitat of an endangered or threatened species. However, limits were placed on the amount of credits that could be granted. A bill to designate part of the Arctic National Wildlife Refuge as a wilderness area (S 2316) was referred to committee but went no further.

A bill proposed in the previous session, the Great Cats and Rare Canids Act of 2008 (HR 1464), was again proposed but did not pass. This bill would have directed the secretary of the interior to provide assistance for projects for the conservation of rare felids and rare canids. In addition, the secretary would be asked to convene an advisory group of experts from both public and private organizations actively involved in the conservation of felids and canids.

Again, the Species Rescue Act (HR 1917), also proposed earlier, did not pass. It would provide for quicker consultation with respect to possible impacts on endangered or threatened species of federal agencies' flood control actions. Congress also refused to pass the Bear Protection Act of 2007 (HR 3029) in this session. The bill would have prohibited any person from importing bear viscera into, or exporting it from, the United States; selling it or offering it for sale; or purchasing, possessing, transporting, delivering, or receiving it in interstate or foreign commerce. The bill would have subjected violators to criminal penalties.

A bill to protect otters, the Southern Sea Otter Recovery and Research Act (HR 3639), did not pass. This bill would have required the secretary of the interior to carry out a recovery program for southern sea otter populations along the coast of California. Another bill denied passage by Congress during this session was the Protect America's Wildlife Act of 2007

(HR 3663). The bill would have amended the Fish and Wildlife Act of 1956 to add airborne hunting offenses by prohibiting any person from shooting or attempting to shoot any bird, fish, or other animal before 3:00 A.M. on the day following after the person has traveled by aircraft, other than on a regularly scheduled commercial aircraft. It would also increase the fine for such offenses to not more than $50,000.

President Obama

President Obama supported wildlife conservation efforts. He signed a memorandum to help restore the Endangered Species Act, increasing the role of the scientific process in determining policy. Claiming that the previous administrations had damaged the act, he wanted to improve it and protect our nation's most endangered animals.[83]

Oddly, a fund for state wildlife conservation that helps pay for restoring depleted populations of many types of game animals, such as turkeys, pronghorn antelopes, wood ducks, beavers, black bears, elk, bighorn sheep, bobcats, and mountain lion, actually increased, in some states by 40 percent, since Obama became president. Some say the influx of money is largely due to Obama's policies toward weapons. Many hunters were afraid that Obama was going to reinstitute a ban on assault weapons that was originally passed during the Clinton administration. As a result, many people have been buying more firearms and ammunition, which are then taxed to create the fund for state wildlife conservation.[84]

2009–10: 111th Congress

In January, the Supreme Court considered a request from several Great Lakes states that Illinois should be forced to close two locks that connect Chicago-area waterways to Lake Michigan in order to keep out Asian carp, a foreign species of fish that could disrupt the lakes' ecosystem. The fish is an invasive fish.[85]

The Crane Conservation Act of 2009 (S 197/HR 388) did not pass. This would have required the secretary of the interior to provide financial assistance for approved projects relating to the conservation of cranes, using amounts in the Crane Conservation Fund established by this act.

Many laws that were proposed earlier were reintroduced in this session. One was the Great Cats and Rare Canids Act of 2009 (S 529/HR 411). It would direct the secretary of the interior to provide assistance for projects for the conservation of rare felids and rare canids. Additionally, an advisory group of experts would be convened to assist in carrying out this act. Another bill that was reintroduced was the Southern Sea Otter Recovery and Research Act (S 1748/HR 556), but it did not pass. It would have required the secretary of the interior, acting through the Fish and Wildlife Service and the United States Geological Survey, to carry out a recovery and research program for southern sea otter populations along the coast of California.

HR 1914 would suspend provisions of the Endangered Species Act of 1973 during periods of drought. This did not pass. Another bill, S 1535/HR 3381, a bill called Protect America's Wildlife Act of 2009, also did not pass. It was a proposal to amend the Fish and Wildlife Conservation Act by imposing criminal penalties on anyone who knowingly violated regulations prohibiting the shooting or harassing of birds, fish, or other animals from aircraft (airborne hunting). It would have increased the monetary penalty for airborne hunting from $5,000 to $50,000.

Other proposals also did not receive enough support to become law. One was the Bear Protection Act of 2009 (HR 3480). This law would have prohibited someone from importing bear viscera into, or exporting it from, the United States. Another was the Great Ape Conservation Reauthorization Amendments Act of 2010 (HR 4416). The bill would have amended the Great Ape Conservation Act and authorized the secretary of the interior to award grants to people to implement a great ape conservation project that would be a long-term conservation strategy for great apes and their habitats.

Conclusion

This chapter indicates the importance that Congress places on protecting wildlife both in the United States and worldwide. The members of Congress have proposed many different laws, and have passed some, to identify endangered species and to make efforts to increase their populations. Their efforts have protected many species from extinction and will continue to do so in the coming years.

Chapter 10

Conclusion

When the government first began creating legislation to control pollution and protect natural resources, many issues of concern to today's population were then unknown. Things like the depletion of the ozone layer or the health effects of breathing polluted air are commonly recognized concerns today.[1] It is now widely accepted that a clean and healthy environment is essential for people's health. Since most pollution is invisible, people don't know how it will affect them or when. Nonetheless, it is widely accepted that an unhealthy environment accounts for many preventable illnesses each year. A clean environment is also needed for a strong economy. A clean environment can be a major drawing factor for economic development as more and more companies follow the trend toward "greening" their industries. A clean environment can attract people and businesses to a community. Today's customers are seeking cleaner products and support companies that provide them. Banks are more willing to lend to companies that prevent pollution rather than paying for cleanup, and insurance companies are more likely to cover clean companies. Many employees would choose to work for environmentally responsible corporations.[2]

The goal of this book was to show how lawmakers are working to help solve the problems of environmental pollution. It is obvious that the topics surrounding the environment have been the subject of numerous proposals in Congress to either clean up past damage or prevent future damage to the air and water. The elected representatives and presidents have all spent numerous hours trying to design new laws to prevent environmental pollution. In recent years, the government has recognized the international aspects of pollution and has worked with other nations to help protect against further damage. The United States has agreed to be part of many regional and international treaties because of the understanding that the quality of the environment is indeed a global concern.

As a result of this legislation, there have been many improvements in the environment. The levels of dangerous toxics in the air have been reduced. Major waterways have been cleaned up, becoming viable for aquatic life again. The number of abandoned hazardous waste sites has been reduced. Limits have been placed on manufacturers and distributors of hazardous or toxic chemicals, and new rules have been made for the approval of new chemicals. The emissions from the automobiles we drive each day have been reduced so that less air pollution is created. The EPA has reported that there is less pollution in the air, and the levels of pesticides in animals are down. Many plant and animal species that were once close to extinction, including the American bald eagle and the American panther, have been restored. Many of these changes stem from a general agreement in Congress that protecting the environment needs to be a subject of governmental action.[3]

However, the changes in environmental laws have been slow and incremental. The basic structure of the laws has varied only slightly over time. When change has occurred, it has been marginal and often simply a revision of past legislation. Real changes have occurred only in the aftermath of a crisis, such as a massive oil spill.[4]

Even though environmental legislation has been effective, environmental problems still exist today. As President G. W. Bush put it in 2003, "Over the past three decades we've reduced the nation's air pollution by half. But there is more to do."[5] Water pollution remains a serious problem, and wetlands continue to disappear. The air quality remains dangerous and even unhealthy for millions of Americans in some parts of the country. Ozone and smog remain as major public health concerns. Waste disposal is expensive and becoming a more difficult problem for many cities. Many Americans (including children) are exposed to thousands of chemicals in their lives each day, posing serious potential health risks.[6] More needs to be done to improve and protect the environment because of its impact on the public that lives within it.

Many people have argued that state and local governments should have the responsibility for the creation and enforcement of environmental statutes and regulatory policies.[7] They argue that local governmental units have a better understanding of the specific problems facing their communities and the needs of their particular constituents when it comes to the natural resources in their own jurisdiction. Because of this, they are more capable of making policies to address those circumstances than the federal government would be.[8] Others have called for privatization of environmental policies. They point out that those public bureaucracies that are responsible for implementing environmental policies are inefficient and thus are not able to protect the environment. They claim that privatizing environmental policies could enable the use of market-determined prices to provide a more efficient implementation of environmental laws.[9]

On the other hand, others have presented arguments that environmental policy is best created and implemented by the federal government because states vary as to their resources and willingness to comply. Additionally, many companies may simply refuse to voluntarily pass environmentally friendly policies on their own, forcing federal government involvement. Since most pollution does not respect state boundaries, and cleanup can be very expensive, there is little doubt that the federal government will continue to be involved in the environmental arena.

It is virtually impossible to make any precise prediction as to exactly what laws will be made in the future by Congress or even what specific problems the Congress will address.[10] But no matter what new laws are created and how they are implemented in the next few years, it is clear that politics will continue to play a crucial role in shaping those laws. Future policies will depend on many factors such as the economy, the influence of interest groups, political leaders, and even public opinion and media coverage of events. It is certain that there will continue to be disagreement and debate about the need for environmental law and the specific content of any future environmental legislation. In fact, the political conflict surrounding environmental law may even increase as members of Congress consider different options and methods to regulate environmental laws as they apply not only to American businesses but also to the global markets and international governments as well.[11]

It is also clear that many policies considered by Congress in the future will necessitate some form of international cooperation. It has become obvious that environmental problems are global and can no longer be based on national boundaries. Traditional responses to environmental issues will no longer be sufficient. Many problems are much more complex than once thought, and solutions to air and water pollution will require action by several nations. This means that the success of any future environmental policy will partly depend on the actions of other nations as well as those of the United States.[12] Future environmental policy will need to go in a new direction and have possible worldwide implications.

Additionally, future environmental policy, to be successful, will require social changes. The government will need to work closely with the private sector so that they can develop new and innovative methods to cope with the problem of environmental pollution. Education will be needed to help teach people of all ages about the need to protect the environment. Business will need to be encouraged to create more environmentally friendly practices, including manufacturing and production of goods. This may include incentives such as tax breaks, grants, loans, and other forms of technical assistance that will encourage companies to support environmentally safe methods.

Protecting the environment is a daunting and never-ending task. To date, Congress and the president have acted in many ways to address these issues. In the future they will continue to act to protect the public from environmental harm in new and innovative ways.

Appendix A

Environmental Interest Groups

American Coalition for Clean Coal Electricity (**ACCE**)—advocates the use of coal as an affordable source of energy such as electricity. ACCE also advocates for safe and clean technology to acquire coal. The ACCE was formed in 2008, as a combination of the Center for Energy and Economic Development (CEED) and Americans for Balanced Energy Choices (ABEC). Their website is www.cleancoalusa.org.

American Rivers—advocates the protection and restoration of rivers in the United States so the rivers can be beneficial to both people and wildlife. American Rivers was formed in 1973. They have five programs to protect and restore rivers: Rivers and Global Warming, River Restoration, River Protection, Clean Water, and Water Supply. Their website is www.americanrivers.org.

Apollo Alliance—works to create new, high-quality, green jobs through the use of clean energy. They wish to create the domestic job growth in America that is necessary for the 21st century. Their interests include energy efficiency, clean power, mass transit, next-generation vehicles, and emerging technology. Founded after September 11, 2001, Apollo Alliance was named after the Apollo missions to inspire creations in new technology. Their website is www.apolloalliance.org.

Bat Conservation International (**BCI**)—seeks to protect bats and their habitats through education, conservation, and research. BCI is based in Austin, Texas, and was founded in 1982. BCI has produced a television documentary, *The Secret World of Bats,* to educate people on bat ecosystems. BCI has also achieved protection for bat caves and has saved bats from being buried when mines have closed down. BCI's website is www.batcon.org.

Council on Environmental Quality (**CEQ**)—develops environmental policies through working with federal agencies, including the White House. The CEQ was established by Congress as part of the National

Environmental Policy Act of 1969 (NEPA). The CEQ also makes sure federal agencies meet their environmental obligations, as stated in NEPA. Its website is www.whitehouse.gov/administration/eop/ceq.

Cousteau Society—is dedicated to protecting water ecosystems on the entire planet. The Cousteau Society provides people with educational material to understand water systems. The Cousteau Society offers programs to teach communities how to live in harmony with nature and the water system without causing damage for future generations. The Cousteau Society was founded in 1973 by Captain Jacques-Yves Cousteau, and its headquarters is in Hampton, Virginia. The society's website is www.Cousteau.org.

Defenders of Wildlife—protects and restores all wildlife in its natural communities in the United States. The Defenders of Wildlife uses the Endangered Species Act as a means to protect and restore the endangered habitats and species. The group also seeks to educate people about the problems with building new developments and the loss of habitat experienced by the wildlife. The group was founded in 1947. Their website is www.defenders.org.

Earth First!—is considered a movement, not an organization, and is made up of small, regionally based movements. The group encourages people to form their own movements. The main goal is biocentrism, or putting the life of the Earth first. Earth First! seeks to educate the public through grassroots organization and legal actions regarding environmental issues, and it even advocates civil disobedience to get its views heard. The movement began in 1979. The group's website is www.earthfirst.org.

Earth Island Institute—seeks to preserve, restore, and conserve ecosystems on the planet. Through its Project Support program, the institute provides assistance to other groups to start up and finance other ecological preservation groups. Earth Island Institute publishes *Earth Island Journal,* a magazine dedicated to providing information about environmental activism. The Earth Island Institute was founded in 1982 by David Brower. Their website is www.earthisland.org.

Environmental Action—was formed from the organization that began Earth Day in 1970. It lobbies against pollution and advocates environmental protection, focusing on the government and special interest groups. Environmental Action lobbied President Nixon to sign the Clean Air Act, the Clean Water Act, and the Endangered Species Act. Its website is www.environmental-action.org.

Environmental Defense Fund (EDF)—uses four core strategies to deal with environmental issues: sound science, economic incentives, corporate partnerships, and getting the law right. The EDF was created in 1967 and seeks to protect environmental rights, such as clean air and water, safe and healthy food, and conserving and maintaining ecosystems. The group's website is www.edf.org.

Environmental Law Institute (ELI)—focuses on environmental law policy in order to protect and sustain the environment. The main aspect of their work focuses on protecting water resources, land, and biodiversity as well as improving environmental law. The ELI was founded in 1969. Their website is www.eli.org.

Friends of the Earth—has been in existence for 40 years and is currently led by Erich Pica. They support a healthy environment free from pollution, and current initiatives include promoting clean energy and other solutions to climate change, promoting safe foods free of toxins, and protecting marine ecosystems. Friends of the Earth is also part of Friends of the Earth International. Their website is www.foe.org.

Greenpeace—was created in 1971 to protest nuclear testing near Alaska. They are dedicated to protecting the environment by protesting in nonviolent ways. Greenpeace is focused on preservation of the ocean and forests, as well as bans on nuclear testing and toxic chemicals. Some successes include banning commercial whaling, protecting Antarctica, and stopping nuclear testing in some countries. Their website is www.greenpeace.org.

Heritage Foundation—is a conservative research institute founded on free enterprise and traditional American values. The Heritage Foundation was created in 1973. One of the many issues that the Heritage Foundation is involved in is energy and the environment: to balance demand; create free, fair markets; and allow for responsible use of energy. The institute's website is www.heritage.org.

Izaak Walton League of America—was formed in 1922 and named after the 17th-century author Izaak Walton, who wrote *The Compleat Angler,* a famous book on fishing. The league seeks to protect and conserve soil, forests, water, and other natural resources and to educate the public on these issues. Some success has been achieved by their efforts to promote sustainable agriculture and clean energy. Their website is www.iwla.org.

League of Conservation Voters (LCV)—provides scorecards of congressional voting records on environmental issues to the public, as well as for other public officials. The LCV also endorses candidates who are pro-environment in order to further their agenda on environmental issues and get environmental legislation passed. The league's website is www.lcv.org.

National Audubon Society—attempts to conserve and restore birds' natural ecosystems and habitats in the United States. The Audubon Society also works with other wildlife, but their main focus is birds. Some of their successes include the protection of the Arctic National Wildlife Refuge and the preservation efforts for the endangered California condor and brown pelican. Their website is www.audobon.org.

National Parks and Conservation Association (NPCA)—seeks to protect and maintain national parks. They educate the public and also lobby Congress to create new laws or uphold existing laws to protect national parks. The NPCA also assess the health of public parks by examining conditions in and threats to certain parks. The association's website is www.npca.org.

National Wildlife Federation (NWF)—aims to protect wildlife and habitats from global warming and loss of habitat due to technology and the increase in human population. NWF sees global warming as the biggest threat to wildlife and is actively involved in attempting to get legislation passed. The NWF also seeks to educate people and has a program for kids and families called "Get Out There," so people can learn to enjoy the outdoors. The NWF also publishes the *Ranger Rick* magazine for kids. Their website is www.nwf.org.

Natural Resources Defense Council (NRDC)—was founded in 1970 by a group of environmentally conscious law students and attorneys. The NRDC seeks to protect the environment and life on Earth, which depends on natural resources. Some of their main concerns include global warming, toxic chemicals, production of clean energy, oceans and wildlife, and assistance to help China go green. Their website is www.nrdc.org.

Nature Conservancy—aims to protect ecologically important land and water. They work with federal, state, and local agencies, as well as other international agencies, to obtain protection for endangered ecosystems. Founded in 1951, the Nature Conservancy has protected 119 million acres of land and 5,000 miles of river. The group's website is www.nature.org.

North American Bluebird Society (NABS)—preserves and recovers bluebirds and other native cavity-nesting birds in North America. The NABS was founded in 1978 by Dr. Lawrence Zeleny, a bluebird activist. The NABS does not seek tax money or political influence, and it is funded by donations, dues, profits from sales, and grants. The society's website is www.nabluebirdsociety.org.

Pew Center on Global Climate Change—provides information on climate issues while sustaining economic growth. Pew researches climate issues and distributes the findings to Congress, major corporations, academics, students, and other relevant audiences in order to improve the understanding climate issues. One of the key issues is to lower the carbon footprint. Its website is www.pewclimate.org.

Population Connection (formerly Zero Population Growth)—advocates the stabilization of population growth so resources will not be depleted. Population Connection was founded in 1968. One of its main campaigns is to make sure women around the world have access to affordable birth control, in order to prevent overpopulation. Its website is www.populationconnection.org.

Rainforest Action Network (RAN)—seeks to protect forests from destruction by logging, oil drilling, and other harmful activities. RAN uses education and lobbying to influence corporations to engage in nondestructive behaviors and to make advances in sustainability. RAN was founded in 1985. Its website is www.ran.org.

Republicans for Environmental Protection (REP)—is a group of Republicans who advocate for environmental conservation. REP endorses Republican candidates who are actively involved in environmental legislation. REP also developed an annual scorecard for Republican members of Congress that tracks their environmental voting records. The group's website is www.rep.org.

Scenic Hudson Club—was created in 1963 to protect the Hudson Valley, especially to save Storm King Mountain. Since then, Scenic Hudson has promoted conservation and community planning to protect the Hudson River Valley. Scenic Hudson also promotes clean and safe water and a healthy environment. Their website is www.scenichudson.org.

Sea Shepherd Conservation Society—was founded in 1977 by Captain Paul Watson. The goal is to protect marine wildlife from destruction of their habitat. Sea Shepherd's mandate is to act as a law enforcement agency to protect and conserve marine wildlife based on the UN World Charter for Nature. The society's website is www.seashepherd.org.

Sierra Club—was founded by John Muir in 1892 and is currently the oldest and largest environmental organization in the United States. The goal of the Sierra Club is to protect and conserve the environment and wildlife. Some of the club's current projects include clean energy, green transportation, safeguarding of communities, and protection of habitats. The club's website is www.sierraclub.org.

Student Conservation Association (SCA)—was founded in 1955 by Elizabeth Cushman, a college student at Vassar. It allows high school and college students to experience conservation through hands-on experiences. Some of the programs that the SCA offers allow students to build trails in national parks that will allow others to experience nature. Others are community programs that allow urban students the opportunity to experience wildlife. A Conservation Corps program is also offered to allow college students the opportunity to address conservation problems. The group's website is www.thesca.org.

The Wilderness Society—was created in 1935. The Wilderness Society has helped protect over 110 million acres in 44 states. Their goal is to conserve public land in the United States. They use education and lobbying in order to achieve their goals. Their website is www.wilderness.org.

Worldwatch Institute—was created in 1974. Worldwatch uses research to inform and promote environmental protection. They have three main program areas: Climate and Energy, Food and Agriculture, and the Green Economy. Through these programs, Worldwatch seeks to find solutions for climate change, resource depletion, and population growth and poverty. Their website is www.worldwatch.org.

World Wildlife Fund (WWF)—was founded in 1961. The WWF currently is active in 100 countries with more than 1 million members in the United States. The purpose of the WWF is to protect and conserve wildlife. They seek to protect endangered species, promote sustainable uses of renewable natural resources, and promote clean energy solutions. Their website is www.worldwildlife.org.

Appendix B

List of Environmental Laws

Chapter 2: The Politics of Air Pollution

Bill Number	Public Law	Title of Law
S 928	84–159	Air Pollution Control Act
HR 7464	86–365	Extension of Air Pollution Control Act
S 455	87–761	Surgeon general to investigate air pollution
HR 6518	88–206	Clean Air Act
S 306	89–272	Motor Vehicle Air Pollution Control Act
S 3112	89–675	Expansion of Clean Air Act of 1963
S 780	90–148	Air Quality Act of 1967
HR 17255	91–604	Clean Air Act of 1970
HR 5445	93–15	Funding for 1970 Clean Air Act
HR 14368	93–319	Energy Supply and Environment Coordination Act
HR 6161	95–95	Clean Air Amendments
SJ Res 188	96–300	National Commission on Air Quality
HJ Res 395	100–202	Extension of Clean Air standards
S 1630/HR 3030	101–549	Clean Air Act Amendments of 1990
SJ Res 187	102–187	Clean Air Act Technical Amendments
HR 1627	104–170	Food Quality Protection Act
S 880	106–40	Chemical Safety Information, Site Security and Fuels Regulatory Relief Act
S 551	108–336	Southern Ute and Colorado Intergovernmental Agreement Implementation Act of 2004
HR 6	109–58	Energy Policy Act of 2005
S 2146	110–255	Diesel emissions reduction projects
HR 3435	111–47	Consumer Assistance to Recycle and Save Program
S 1660	111–199	Formaldehyde Standards for Composite Wood Products Act

Chapter 3: The Politics of Water Pollution

Bill Number	Public Law	Title of Law
S 936	226	Water Powers Act
HR 8143	204	Alaskan Fisheries Act
HR 8750	84–1018	Watershed Protection and Flood Prevention Act
S 890/HR 9540	84–660	Water Pollution Control Act Amendments
HR 6441/S 120	87–88	Water Pollution Control Act Amendments
S 2	88–379	Water Resources Research Act
S 4	89–234	Water Quality Act
S 2947	89–753	Clean Water Restoration Act
S 1766	89–240	Grants to states
S 20	90–515	Creation of the National Water Committee Commission
S 119	90–542	Wild and Scenic Rivers Act
HR 25	90–454	Estuary Preservation Act
HR 4148	91–224	Water Quality Improvement Act
S 2770	92–500	Clean Water Act Amendments
S 433	93–523	Safe Drinking Water Act
S 2812	93–243	Grants to states
HR 3199	95–217	Federal Water Pollution Control Act Amendments
S 1528/HR 6827	95–190	Safe Drinking Water Amendments
S 2701/HR 11655	95–404	Water Resources Planning Act
S 2704/HR 11126	95–467	Water Research and Development Act
S 1146	96–63	Safe Drinking Water Act Reauthorization
S 1640	96–457	Extension of water resources research
S 901/HR 4023	96–148	To amend the Clean Water Act of 1977
HR 8117	96–502	Extension of deadlines in Safe Drinking Water Act
HR 1650/S 124	99–339	To amend the Safe Drinking Water Act
HR 1	100–4	Water Quality Act of 1987
HR 4939	100–572	Lead Contamination Control Act of 1998
HR 1465	101–380	Oil Pollution Act of 1990
HR 4223	101–596	Great Lakes Critical Programs Act of 1990
HR 1101	101–397	Water Resources Research Act Reauthorization
S 1316	104–182	Safe Drinking Water Act Amendment
HR 2024	104–142	Mercury-containing and Rechargeable Battery Management Act
S 1004	104–324	Coast Guard Authorization Act of 1996
HR 999	106–284	Beaches Environment Assessment and Coastal Health Act of 2000
HR 1495	110–114	Water Resource Development Act
S 2766/HR 5949	110–288	Clean Boating Act of 2008
HR 6460	110–365	Great Lakes Legacy Reauthorization Act

Chapter 4: The Politics of Ocean Pollution

Bill Number	Public Law	Title of Law
HR 8181	87–758	National Fisheries Center/Aquarium
HR 6845	87–396	Coast Guard research
S 944	89–454	Oceanography
HR 16559	89–688	National Sea Grant College
HR 8794	91–15	National Council on Marine Resources
HR 10420	92–522	Marine mammals
HR 9727	92–532	Ocean dumping
HR 8140	92–340	Oil spills
HR 15540	93–472	Marine Protection, Research, and Sanctuaries Act
HR 5451	93–119	International Convention for Prevention of Pollution of the Sea by Oil Amendments
S 1070	93–248	Intervention on the High Seas Act
HR 5710	94–62	Marine Protection, Research, and Sanctuaries Act Amendments
S 682	95–474	Port and Tanker Safety Act
S 1522/HR 10730	95–136	Extension of the Marine Mammal Protection Act
HR 4297	95–153	Amendments to Marine Protection Act
S 1617	95–273	Ocean Pollution Research Program Act
S 1148	96–572	Marine Protection, Research, and Sanctuaries Act reauthorization
S 1123	96–381	Research on marine life
S 1140	96–332	Marine sanctuaries
S 1213	97–16	Marine Protection, Research, and Sanctuaries Act reauthorization
S 1003	97–109	Designation of land as marine sanctuaries
HR 4997	98–364	Commercial Fishing Industry Vessel Act
HR 3515	100–582	Medical Waste Tracking Act
HR 3674	100–220	United States-Japan Fishery Agreement Approval Act
S 2030	100–688	Ocean Dumping Ban Act
S 1986	100–556	Plastic Pollution Control Act
HR 1465/S 686	101–380	Oil Pollution Prevention, Response, Liability, and Compensation Act
HR 1668	101–224	National Oceanic and Atmospheric Administration Ocean and Coastal Programs Authorization Act
HR 5909	101–605	Florida Keys National Marine Sanctuary and Protection Act
HR 2130	102–567	National Oceanic and Atmospheric Administration Authorization Act
S 1636	103–238	Marine Mammal Protection Act Reauthorization
HR 5176	103–431	Ocean Pollution Reduction Act
HR 1965	104–150	Coastal Zone Protection Act

HR 3908	107–308	North American Wetlands Conservation Reauthorization Act Amendments
HR 5782	109–468	Pipeline Inspection, Protection, Enforcement, and Safety Act
HR 3552	109–226	Coastal Barrier Resources Reauthorization Act
HR 260	109–294	Partners for Fish and Wildlife Act
HR 2430	109–326	Great Lakes Fish and Wildlife Restoration Act
S 3692	109–449	Marine Debris Research, Prevention and Reduction Act

Chapter 5: The Politics of Pesticides

Bill Number	Public Law	Title of Law
S 1605	88–305	Amendments to Federal Insecticide, Fungicide, and Rodenticide Act (FIFRA)
HR 10729	92–516	Federal Environmental Pesticide Control Act
HR 8841	94–140	FIFRA extension
/S 1678	95–396	Federal Pesticide Act
HR 7018	96–539	FIFRA funding
HR 2785	98–201	FIFRA reauthorization
S 659	100–532	FIFRA Amendments of 1988
S 1913	103–231	FIFRA reauthorization
HR 1627	104–170	FIFRA reauthorization
S 1983	110–94	Pesticide Registration Improvement Renewal Act
S 2571	110–193	FIFRA corrections

Chapter 6: The Politics of Solid Waste

Bill Number	Public Law	Title of Law
HR 8248	89–272	Solid Waste Disposal Act
S 2005/HR 11833	90–574	Resource Recovery Act
HR 11833	91–512	Resource Recovery Act of 1969
HR 5446	93–14	Solid Waste Disposal Act Reauthorization
HR 16045	93–611	Solid Waste Disposal Act Reauthorization
S 2150	94–580	Resource Conservation and Recovery Act of 1976
S 2412	96–463	Used Oil Recycling Act
S 1156/HR 3994	96–482	Solid Waste Disposal Act Amendments
HR 5288	97–278	New Hampshire/Vermont compact
S 1430	101–610	National and Community Service Act of 1990
HR 2194	102–386	Resource Conservation and Recovery Act Amendments
HR 2194	102–386	Federal Facilities Compliance Act
HR 2024	104–142	Mercury-Containing and Rechargeable Battery Management Act

Chapter 7: The Politics of Toxic and Hazardous Waste

Bill Number	Public Law	Title of Law
S 1717	79–585	Atomic Energy Act
S 3149	94–469	Toxic Substances Control Act
S 2150	94–580	Resource Conservation and Recovery Act
S 2498	94–305	Small Business Export Development Act
HR 13650	95–604	Uranium Mill Tailings Radiation Control Act
HR 7020/S 1480	96–510	Comprehensive Environmental Response, Compensation, and Liability Act
S 1156	96–482	Solid Waste Disposal Act Amendments
S 2189	96–573	Nuclear Waste Policy Act
S 2412	96–463	Used Oil Recycling Act
S 1211/HR 3495	97–129	Extension of Toxic Substances Control Act
HR 3809	97–425	Nuclear Waste Policy Act of 1982
HR 2867/ S 757	98–616	Hazardous and Solid Waste Amendments of 1984
S 2706/HR 5530	98–559	Hazardous Materials Transportation Act Amendments of 1984
HR 2005	99–499	Superfund Amendments and Reauthorization Act
HR 1083	99–240	Low-Level Radioactive Waste Policy Amendments
HR 3545	100–203	Omnibus Budget Reconciliation Act
HR 776	102–486	Energy Policy Act
HR2637	102–579	Waste Isolation Pilot Plant Land Withdrawal Act
HR 2194	102–386	Federal Facilities Compliance Act
HR 629/S 270	105–236	Texas Low-Level Radioactive Waste Disposal Compact Consent Act
HJ Res 87	107–200	Yucca Mountain
HR 2869	107–118	Small Business Liability and Brownfields Revitalization Act

Chapter 8: The Politics of Land

Bill Number	Public Law	Title of Law
S 979	81–128	Survey of forest research
HR 7391	87–383	Purchase of wetlands
S 4/HR 9070	88–577	National Wilderness Preservation System
HR 3846	88–578	Land and Water Conservation Fund Act
HR 8070	88–606	Public Land Law Review Commission
S 889	90–271	National wilderness land
S 2531	90–318	National wilderness land
S 2751	90–548	National wilderness land
S 1321	90–544	National wilderness land
S 2515	90–545	National wilderness land

HR 12121	90–213	Public Land Law Review Commission
S 3014	91–504	National wilderness land
S 719	91–631	Mining Act of 1872 Amendments
S 713	91–82	National wilderness land
S 714	91–58	National wilderness land
S 2564	91–88	National wilderness land
HR 11069	91–42	National wilderness land
S 3014	91–504	National wilderness land
HR 736	92–364	National wilderness land
HR 10655	92–510	National wilderness land
S 141	92–537	National wilderness land
S 1852	92–592	National wilderness land
HR 5838	92–493	National wilderness land
S 3129	92–475	National wilderness land
HR 15597	92–533	National wilderness land
S 1497	92–501	National wilderness land
HR 6957	92–400	National wilderness land
S 960	92–241	National wilderness land
S 3153	92–275	National wilderness land
S 493	92–521	National wilderness land
S 27	92–593	National wilderness land
S 1198	92–528	National wilderness land
HR 16444	92–589	National wilderness land
S 484	92–395	National wilderness land
S 3433	93–622	Eastern Wilderness Areas Act
HR 6395	93–429	National wilderness land
HR 8352	93–535	Cascade Head Scenic-Research Area
S 507	94–579	National Resource Lands Management Act
HR 7792	94–357	National wilderness land
S 3091/HR 12503/HR 15069	94–588	National Forest Management Act
HR 3130	94–83	Amendments to National Environmental Policy Act
HR 2	95–87	Surface Mining Control and Reclamation Act
HR 3453	95–237	Endangered American Wilderness Act
HR 2676	96–229	Environmental Research, Development, and Demonstration Authorization Act
S 2009	96–312	Central Idaho Wilderness Act
HR 39	96–487	Alaska National Interest Lands Conservation Act
HR 2043	96–182	Funding for wetland preservation
HR 5487	96–560	Colorado National Forest Wilderness Act
HR 8298	96–550	New Mexico Wilderness Act of 1980
S 1119	97–250	Crater Lake National Park

S 2125	98–508	Arkansas Wilderness Act
HR 4707	98–406	Arizona Wilderness Act
HR 1473	98–425	California Wilderness Act
HR 9	98–430	Florida Wilderness Act
S 2773	98–514	Georgia Wilderness Act
S 2808	98–515	Mississippi National Forest Wilderness Act
S 64	98–289	Irish Wilderness Act
HR 3921	98–323	New Hampshire Wilderness Act
HR 6296	98–603	San Juan Basin Wilderness Protection Act
HR 3960	98–324	North Carolina Wilderness Act
HR 1149	98–328	Oregon Wilderness Act
HR 5076	98–585	Pennsylvania Wilderness Act
HR 4263	98–578	Tennessee Wilderness Act
HR 3788	98–574	Texas Wilderness Act
S 2155	98–428	Utah Wilderness Act
HR 4198	98–322	Vermont Wilderness Act
HR 5121	98–586	Virginia Wilderness Act
S 837	98–339	Washington State Wilderness Act
HR 3578	98–321	Wisconsin Wilderness Act
S 543	98–550	Wyoming Wilderness Act
HR 5496	99–555	Georgia Wilderness Act
S 816	99–504	Nebraska Wilderness Act
HR 5166	99–490	Tennessee Wilderness Act
S 2506	99–565	Great Basin National Park Act
HR 4362	100–648	Recreation and Public Purposes Amendment Act
HR 1963	100–34	Surface Mining Control Act Amendments
HR 14	100–33	National wilderness land
HR 148/S 1036	100–184	Michigan Wilderness Act of 1987
S 974	101–195	Nevada Wilderness Protection Act
HR 2570	101–628	National wilderness land
HR 987	101–626	Tongass Timber Reform Act
S 21	103–433	California Desert Protection Act
HR 631	103–77	Colorado Wilderness Act
HR 698	103–169	Lechuguilla Cave Protection Act
HR 38	103–104	Jemez National Recreation Area
HR 236	103–64	Snake River Birds of Prey National Conservation Area
HR 1745/S 884	104–396	National wilderness land
HR 449/S 94	105–263	Southern Nevada Public Land Management Act
HR 15	106–145	Otay Mountain Wilderness Act
HR 451/S1205/ S1894	107–334	Mount Nebo Wilderness Area

HR 1576/ S 1711	107–216	James Peak Wilderness Protection Area Act
HR 1904	108–148	Healthy Forests Initiative
HR 233/S 128	109–362	Northern California Coastal Wild Heritage Wilderness Act
HR 539/ S 272	109–118	Caribbean National Forest Act
S 22/HR 146	111–11	Omnibus Public Land Management Act

Chapter 9: The Politics of Endangered Species and Wildlife

Bill Number	Public Law	Title of Law
HR 9424	89–669	Endangered Species Conservation Act
HR 11363	91–135	Endangered Species Conservation Act
HR 12475	91–503	Federal aid for Fish and Wildlife Restoration Act
HR 10420	92–522	National Marine Mammal Protection Act
HR 11091	92–558	Funds for wildlife restoration projects
HR 12143	92–330	San Francisco Bay National Wildlife Refuge
HR 5060	92–159	Protection of wildlife
HR 12186	92–535	Bald Eagle Protection Act
HR 16074	92–604	Controlling jellyfish and other pests
S 1983/HR 37	93–205	Endangered Species Conservation Act
HR 5512	94–223	National Wildlife Refuge System
S 1316	95–212	Funding for endangered species
S 2899/HR 14104	95–632	Endangered Species Act Amendments
HR 10730	95–316	Funding for Marine Mammal Protection
HR 2329	95–616	Interoceanic Canal Study Act
S 1143	96–159	Endangered Species Act Amendments
HR 3292/S 2181	96–366	Fish and Wildlife Conservation Act of 1980
HR 6839	96–246	Financial assistance for endangered species management
HR 39	96–487	Alaska National Interest Lands Conservation Act
S 736/HR 1638	97–79	Lacey Act Amendments of 1981
HR 6133/ S 2309	97–304	Endangered Species Act Reauthorization
HR 4084	97–58	Marine Mammal Protection Program
HR 5662	97–347	Fish and Wildlife Act Amendments
HR 2395	98–200	Wetlands Loan Act extension
S 1271	98–244	National Fish and Wildlife Foundation
HR 4997	98–364	Commercial Fishing Industry Vessel Act
HR 5271	98–548	Wetlands Funding
HR 1467/S 675	100–478	Reauthorization of Endangered Species Act

S 1389	100–240	National Fish and Wildlife Foundation funding
HR 4209	100–536	Appropriations for Marine Protection, Research and Sanctuaries Act
S 804/HR 2322	101–233	North American wetlands Conservation Act
HR 2387	102–130	Atlantic Striped Bass Conservation Bill
HR 2686	102–154	To help salmon
HR 1435/S 2260	102–402	Rocky Mountain Arsenal Wildlife Refuge Act
HR 5013	102–440	Wild Bird Conservation Act of 1992
HR 3049	103–86	Extension of ban on killing marine mammals
S 1636/HR 2760	103–238	Marine Mammal Protection Act reauthorization
S 476	103–232	National Fish and Wildlife Foundation Improvement Act
HR 236	103–64	Snake River Birds of Prey National Conservation Area
HR 1420/HR 511	105–57	National Wildlife Refuge System Improvement Act
HR 2807/HR 2863/S 361	105–312	Rhino and Tiger Product Labeling Act
S 2094	105–328	Fish and Wildlife Revenue Enhancement Act
HR 39	105–217	African Elephant Conservation Reauthorization Act
HR 408	105–42	International Dolphin Conservation Program Act
HR 1787	105–96	Asian Elephant Conservation Act
S 1744	106–201	Amendments to Endangered Species Act
HR 4320/S 1007	106–411	Great Ape Conservation Act
HR 643	107–111	Reauthorization of African Elephant Conservation Act
HR 645	107–112	Multinational Species Conservation Fund
HR 700	107–141	Asian Elephant Conservation Act
HR 2619	108–481	National Wildlife Refuge expansion
HR 50	110–132	African Elephant Conservation Act and Rhinoceros and Tiger Conservation Act reauthorization
HR 465	110–133	Asian Elephant Conservation Act reauthorization

Notes

Chapter 1 Introduction

1. CNN, "Oil Estimate Raised to 35,000–60,000 Barrels a Day," June 15, 2010, http://edition.cnn.com/2010/US/06/15/oil.spill.disaster/inex.html (accessed October 11, 2010); Ray Henry, "Scientists Up Estimate of Leaking Gulf Oil," MSNBC, June 15, 2010, http://www.msnbc.com/id/3771735/#slice-2 (accessed June 15, 2010); Campbell Robertson, "Estimates Suggest Spill Is Biggest in U.S. History," *New York Times,* June 27, 2010, http://www.nytimes.com/2010/05/28/us/28flow.html (accessed June 15, 2010).

2. Ed Crooks, "BP Oil Spill Well Effectively Dead, Says U.S.," *Financial Times,* September 19, 2010, http://www.ft.com/cms/s/o/a708b568-c40f-11df-b827-00144 (accessed November 3, 2010); Allen G. Breed, "After Pressure Test, Government Pronounces BP Oil Well Dead," *Boston Globe,* September 20, 2010, http://www.boston.com/news/science/articles/2010/09/20/after-pressure-test-government-pronouncing-bp-oil-well-doner (accessed November 3, 2010).

3. "U.S. Oil Spill in Gulf 'Making Dolphins Act Drunk,'" BBC News, June 18, 2010, http://www.bbc.co.uk/news/10346092 (accessed October 11, 2010); Harte Research Institute for Gulf of Mexico Studies, Texas A&M University, "Biodiversity of the Gulf of Mexico: Applications to the Deep Horizon Oil Spill," May 2010, http://www.harteresearchinstitute.org/images/press_releases/biodiversity.pdf (accessed October 11, 2010).

4. Geof Koss, Emma Dumain, and Niels Lesniewski, "Legislative Response to Spill Takes Shape," *CQ Weekly Online,* July 5, 2010: 1634–1635, http://library.cqpress.com/cqweekly/weeklyreport111-000003697177 (accessed September 2, 2010).

5. Joanna Anderson and Jennifer Scholtes, "House and Senate Advance Oil Spill Bills," *CQ Weekly Online,* June 28, 2010: 1580, http://library.cqpress.com/cqweekly/weeklyreport111-000003691799 (accessed September 2, 2010).

6. Jennifer Scholtes, "Lawmakers Clear Legislation to Allow Quicker Release of Oil Spill Funds," *CQ Weekly Online,* June 14, 2010: 1460, http://library.cqpress.com/cqweekly/weeklyreport111-000003681825 (accessed September 2, 2010); Anderson and Scholtes, "House and Senate"; Koss, Dumain, and Lesniewski, "Legislative Response."

7. Joseph J. Schatz and Coral Davenport, "Spill's Impact Spreads to Washington," *CQ Weekly Online,* May 10, 2010: 1132–1135, http://library.cqpress.com/cqweekly/weeklyreport111-000003658684 (accessed September 2, 2010).

8. Coral Davenport, "Drilling Halt Heats Up Green Rhetoric," *CQ Weekly Online,* May 31, 2010: 1334–1335, http://library.cqpress.com/cqweekly/weeklyreport111-000003675763 (accessed September 2, 2010).

9. Schatz and Davenport, "Spill's Impact Spreads."

10. Brandon Payne, "Under Many Microscopes," *CQ Weekly Online,* June 7, 2010: 1395, http://library.cqpress.com/cqweekly/weeklyreport111-000003677635 (accessed September 2, 2010).

11. Alan K. Ota, "Hill Now a Regular Stop for CEOs," *CQ Weekly Online,* May 10, 2010: 1134–1135, http://library.cqpress.com/cqweekly/weeklyreport111-000003658685 (accessed September 2, 2010).

12. Michael E. Kraft, *Environmental Policy and Politics* (New York: Longman, 2001): 12.

13. Ibid.

14. Richard N. L. Andrews, *Managing the Environment, Managing Ourselves* (New Haven, CT: Yale University Press, 1999): x.

15. Freda Adler, "Offender-Specific vs. Offense-Specific Approaches to the Study of Environmental Crime," in Sally M. Edwards, Terry D. Edwards, and Charles B. Fields, eds., *Environmental Crime and Criminality* (New York: Garland, 1996), 35–54: 35.

16. Mary Clifford and Terry D. Edwards, "Defining Environmental Crime," in Mary Clifford, ed., *Environmental Crime* (Gaithersburg, MD: Aspen, 1998), 5–30: 6.

17. Scott Hays, Michael Esler, and Carol Hays, "Radical Environmentalism and Crime," in Edwards, Edwards, and Fields, *Environmental Crime and Criminality,* 163–181: 163.

18. Jeffrey J. Pompe and James R. Rinehart, *Environmental Conflict* (Albany: State University of New York, 2002): 96–99.

19. Samuel P. Hays, *Environmental Politics since 1945* (Pittsburgh: University of Pittsburgh Press, 2000): 16.

20. Richard N. L. Andrews, *Managing the Environment, Managing Ourselves,* 2nd ed. (New Haven, CT: Yale University Press, 2006): 4.

21. Walter A. Rosenbaum, *Environmental Politics and Policy,* 6th ed. (Washington, DC: CQ Press, 2005): 72; Kraft, *Environmental Policy:* 12.

22. Rosemary O'Leary, "Environmental Policy in the Courts," in Norman J. Vig and Michael E. Kraft, eds., *Environmental Policy: New Directions for the Twenty-First Century* (Washington, DC: CQ Press, 2006), 148–168: 151.

23. Daniel J. Fiorino, *Making Environmental Policy* (Berkeley: University of California Press, 1995).

24. Gerhard O. Mueller, "An Essay on Environmental Criminality," in Edwards, Edwards, and Fields, eds., *Environmental Crime and Criminality,* 3–33: 5; Andrews, *Managing the Environment* (1999): xi.

25. Rosenbaum, *Environmental Politics:* 10–11.

26. Kraft, *Environmental Policy:* 12.

27. Ibid., 13.

28. Hays, *Environmental Politics:* 186.

29. Andrews, *Managing the Environment* (1999): 2–4.

30. Earl R. Kruschke and Byron M. Jackson, *The Public Policy Dictionary* (Santa Barbara, CA: ABC-CLIO, 1987): 108.

31. Fiorino, *Making Environmental Policy:* xiv.

32. Jerry W. Calvert, "Party Politics and Environmental Policy," in James P. Lester, ed., *Environmental Politics and Policy* (Durham, NC: Duke University Press, 1989), 158–211: 161; Henry C. Kenski and Margaret Corgan Kenski, "Partisanship, Ideology and Constituency Differences on Environmental Issues in the U.S. House of Representatives and Senate: 1973–1978," in Dean E. Mann, ed., *Environmental Policy Formation* (Lexington, MA: Lexington Books, 1981): 87–102.

33. Hays, *Environmental Politics:* 187.

34. Ibid., 118.

35. Michael E. Kraft, "Environmental Policy in Congress," in Vig and Kraft, *Environmental Policy: New Directions,* 124–147: 126.

36. Kraft, *Environmental Policy:* 126.

37. Calvert, "Party Politics": 161.

38. Hays, *Environmental Politics:* 110.

39. Andrews, *Managing the Environment* (1999): x.

40. Hays, *Environmental Politics.*

41. Janet S. Conary, "An Emerging Nation: The Early Presidents and the Development of Environmental Policy," in Dennis L. Soden, ed., *The Environmental Presidency* (Albany: State University of New York Press, 1999), 15–39: 18.

42. Hays, *Environmental Politics.*

43. Norman J. Vig and Michael E. Kraft, "Introduction," in Normal J. Vig and Michael E. Kraft, eds., *Environmental Policy in the 1980s: Reagan's New Agenda* (Washington, DC: CQ Press, 1984), 3–26: 8.

44. Ibid., 8, 11.

45. Andrews, *Managing the Environment* (1999): 94.

46. Ibid., 95.

47. Fiorino, *Making Environmental Policy:* 202.

48. Andrews, *Managing the Environment* (1999): 98.

49. George Alderson, "The Public Lands," in James Rathlesberger, ed., *Nixon and the Environment: The Politics of Devastation* (New York: Village Voice, 1972), 110–128: 112.

50. Andrews, *Managing the Environment:* 99.

51. Ibid., 144.

52. Ibid., 150.

53. Ibid., 150.

54. David H. Rosenbloom and Rosemary O'Leary, *Public Administration and the Law* (New York: Marcel Dekker, 1997): 88; Mary Clifford, "A Review of Federal Environmental Legislation," in Clifford, *Environmental Crime,* 95–109: 100.

55. David Zwick, "Water Pollution," in Rathlesberger, *Nixon and the Environment,* 30–58: 37.

56. Andrews, *Managing the Environment* (1999): 204.

57. Clarence J. Davies III and Barbara S. Davies, *The Politics of Pollution* (Indianapolis: Pegasus, 1975): 27.

58. Ibid., 27.

59. Vig and Kraft, "Environmental Policy from the 1970s to 2000: An Overview," in Norman J. Vig and Micheal E. Kraft, eds, *Environmental Policy: New Directions:* 1–31: 1, 11.

60. Andrews, *Managing the Environment* (1999): 205.

61. Rosenbloom and O'Leary, *Public Administration:* 88.

62. Matthew J. Linstrom and Zachary A. Smith, *The National Environmental Policy Act* (College Station: Texas A&M University Press, 2001): 21, 4.

63. Ibid., 21.

64. Rosenbloom and O'Leary, *Public Administration:* 88; Vig and Kraft, "Introduction," in Normal J. Vig and Michael E. Kraft, eds., *Environmental Policy in the 1980s: Reagan's New Agenda* (Washington, DC: CQ Press, 1984), 3–26: 11.

65. Mueller, "Essay on Environmental Criminality": 6.

66. Phillip F. Cramer, *Deep Environmental Politics* (Westport, CT: Praeger, 1998): 85, citing National Environment Policy Act, U.S. Code, title 42 (the Public Health and Welfare), chapter 55, section 4321.

67. Cramer, *Deep Environmental Politics:* 86, citing National Environment Policy Act, U.S. Code, title 42 (the Public Health and Welfare), chapter 55, sec. 4331.

68. Cramer, *Deep Environmental Politics:* 85, citing National Environment Policy Act, U.S. Code, title 42 (the Public Health and Welfare), chapter 55, sec. 4321.

69. "What Congress Did," *Congressional Quarterly Almanac* 25 (1969): 79; "Environment: Action Completed," *Congressional Quarterly Almanac* 25 (1969): 86.

70. "Environment: Action Completed," *Congressional Quarterly Almanac* 26 (1970): 79.

71. Vig and Kraft, "Environmental Policy" in *Environmental Policy in the 1980s:* 12–13; Linstrom and Smith, *National Environmental Policy Act:* 4.

72. Donald J. Rebovich, Dangerous Ground (New Brunswick: Transaction Publishers, 1992): 4–5.

73. Linstrom and Smith, *National Environmental Policy Act:* 19.

74. Rosenbaum, *Environmental Politics:* 7.

75. Vig and Kraft, *Environmental Policy from the 1970s:* 1; Andrews, *Managing the Environment* (1999): 227.

76. Ibid., ix.

77. Mueller, "Essay on Environmental Criminality": 5–6.

78. Kraft, *Environmental Policy:* 4.

79. Hays, *Environmental Politics.*

80. Vig and Kraft, *Environmental Policy from the 1970s:* 14.

81. Ibid., 15.

82. "Quayle and Bentsen, Running Mates under Fire," CQ Press Electronic Library, CQ Almanac Online Edition, cqal88-852-25653-1140204, http://library.cqpress.com/cqalmanac/cqal88-852-25653-1140204 (accessed August 30, 2010), originally published in *CQ Almanac 1988* (Washington, DC: Congressional Quarterly, 1989).

83. Vig and Kraft, *Environmental Policy from the 1970s:* 15.

84. Fiorino, *Making Environmental Policy:* 33.

85. Hays, *Environmental Politics:* 118–119.

86. Kraft, *Environmental Policy:* 4.

87. Vig and Kraft, *Environmental Policy from the 1970s:* 6.

88. Ibid., 6–7.

89. Ibid., 10.

90. Christopher J. Bosso and Deborah Lynn Guber, "Maintaining Presence: Environmental Advocacy and the Permanent Campaign," in Vig and Kraft, *Environmental Policy: New Directions,* 78–99: 80.

91. Vig and Kraft, *Environmental Policy from the 1970s:* 4.

92. Thomas R. Dye, *Understanding Public Policy,* 8th ed. (Englewood Cliffs, NJ: Prentice Hall, 1995): 2–3.

93. D. Easton, *A Systems Analysis of Political Life* (New York: John Wiley and Sons, 1965): 212.

94. J. E. Anderson, *Public Policy Making: An Introduction* (Boston: Houghton Mifflin, 1990): 8.

95. R. W. Cobb and C. D. Elder, *Participation in American Politics: The Dynamics of Agenda Building* (Baltimore: Johns Hopkins University Press, 1975).

96. Lester, ed. *Environmental Politics and Policy:* 1; Lynton K. Caldwell, *Environment: Challenge to Modern Society* (Garden City, NY: Natural History Press, 1970): 23.

97. Kraft, *Environmental Policy:* 126.

98. Rosenbaum, *Environmental Politics:* 100–101.

99. Davies and Davies, *Politics of Pollution:* 61.

100. National Academy of Public Administration (NAPA), *Setting Priorities Getting Results: A New Direction for EPA* (Washington, DC: NAPA, 1995): 124–125.

101. Henry P. Caulfield, "The Conservation and Environmental Movements: An Historical Analysis," in Lester, *Environmental Politics and Policy,* 13–56: 18.

102. Robert A. Shanley, *Presidential Influence and Environmental Policy* (Westport, CT: Greenwood Press, 1992): 7.

103. Rosenbaum, *Environmental Politics:* 66; Shanley, *Presidential Influence:* 9.

104. Vig and Kraft, *Environmental Policy from the 1970s:* 4.

105. Sally M. Edwards, "A History of U.S. Environmental Movement," in Clifford, *Environmental Crime,* 31–56.

106. Andrea K. Gerlak, and Patrick J. McGovern, "The Twentieth Century: Progressivism, Prosperity, and Crisis," in Soden, *Environmental Presidency,* 41–75: 49.

107. Edwards, "History of U.S. Environmental Movement."

108. Shanley, *Presidential Influence:* 14.

109. Gerlak and McGovern, "Twentieth Century": 49.

110. Andrews, *Managing the Environment* (1999): 153.

111. Ibid., 227; Dennis L. Soden and Brent S. Steel, "Evaluating the Environmental Presidency," in Soden, *Environmental Presidency,* 313–354: 323.

112. Soden and Steel, "Evaluating the Environmental Presidency": 323.

113. Caulfield, "Conservation and Environmental Movements": 18.

114. Ibid.

115. Ibid.

116. Soden and Steel, "Evaluating the Environmental Presidency."

117. "President Seeks Conservation Program," CQ Press Electronic Library, CQ Almanac Online Edition, cqal62-878-28126-1324166, http://library.cqpress.com/cqal62-878-28126-1324166 (accessed August 23, 2010), originally published in *CQ Almanac 1962* (Washington, DC: Congressional Quarterly, 1963).

118. Caulfield, "Conservation and Environmental Movements": 34.

119. "Message to Congress: Johnson's Conservation Message," CQ Press Electronic Library, CQ Almanac Online Edition, cqal68-1124522, http://library.cqpress.com/cqal68-1124522 (accessed August 23, 2010), originally published in *CQ Almanac 1968* (Washington, DC: Congressional Quarterly, 1969).

120. Richard M. Nixon, "Annual Budget Message to the Congress: Fiscal Year 1974," January 29, 1973, in *Public Papers of the President of the United States* (Washington, DC: U.S. Government Printing Office, 1974), 32–48: 45.

121. Richard M. Nixon, "Annual Budget Message to the Congress: Fiscal Year 1973," January 24, 1972, in *Public Papers of the President of the United States* (Washington, DC: U.S. Government Printing Office, 1973), 78–99: 93.

122. Richard M. Nixon, "Annual Budget Message to the Congress: Fiscal Year 1973," January 24, 1972, in *Public Papers of the President of the United States* (Washington, DC: U.S. Government Printing Office, 1973), 78–99: 96.

123. Ibid., 45.

124. Andrews, *Managing the Environment* (1999): 229.

125. Richard M. Nixon, "Annual Budget Message to the Congress, FY 1971," February 1, 1970, in *Public Papers of the President of the United States* (Washington, DC: U.S. Government Printing Office, 1971), 46–68: 61.

126. Caulfield, "Conservation and Environmental Movements": 39.

127. Shanley, *Presidential Influence:* 52.

128. "Major Anti-Pollution Measures Clear 91st Congress," *Congressional Quarterly Almanac* 26 (1970): 127; "Statement on Signing an Executive Order for the Control of Air and Water Pollution at Federal Facilities," *Congressional Quarterly Almanac* 26 (1970): 78.

129. "Presidential Statement to Congress: Nixon's 1970 Environmental Message," CQ Press Electronic Library, CQ Almanac Online Edition, cqal71-869-26707-1254975, http://library.cqpress.com/cqal71-869-26707-1254975 (accessed August 23, 2010), originally published in *CQ Almanac 1971* (Washington, DC: Congressional Quarterly, 1972).

130. "Nixon Presents 'Six Great Goals' to Congress: Stresses Welfare Reform and Stabilization of the Economy," *Congressional Quarterly Almanac* 27 (1971): 2A.

131. "Environment," *Congressional Quarterly Almanac* 27 (1971): 48; "Text of Environment Message," *Congressional Quarterly Almanac* 27 (1971): 27A–33A.

132. "Special Message to the Congress Proposing the 1971 Environmental Program," *Congressional Quarterly Almanac* 27 (1971): 125–142.

133. Richard M. Nixon, "Special Message to the Congress Outlining the 1972 Environmental Program," February 8, 1972, in *Public Papers of the President*

of the United States (Washington, DC: U.S. Government Printing Office, 1973): 173–189.

134. Richard M. Nixon, "Fourth Annual Report to the Congress on U.S. Foreign Policy," May 3, 1973, in *Public Papers of the President of the United States* (Washington, DC: U.S. Government Printing Office, 974), 348–520: 514.

135. "Second Annual Report to the Congress on U.S. Foreign Policy," *Congressional Quarterly Almanac* 27 (1971): 331–332.

136. Richard M. Nixon, "Third Annual Report to the Congress on U.S. Foreign Policy," February 9, 1972, in *Public Papers of the President of the United States* (Washington, DC: U.S. Government Printing Office, 1973), 294–346: 340.

137. Richard M. Nixon, "Special Message to the Congress on National Legislative Goals," September 10, 1973, in *Public Papers of the President of the United States* (Washington, DC:U.S. Government Printing Office, 1974), 761–786: 772.

138. Richard M. Nixon, "Annual Message to the Congress on the State of the Union," January 30, 1974, in *Public Papers of the President of the United States* (Washington, DC: U.S. Government Printing Office, 1975), 56–100: 81.

139. Gerald R. Ford, "Annual Message to the Congress on the State of the Union," January 30, 1974, in *Public Papers of the President of the United States* (Washington, DC: U.S. Government Printing Office, 1975), 56–100: 81–83.

140. Soden and Steel, "Evaluating the Environmental Presidency."

141. Gerald R. Ford, "Special Message to the Congress Proposing Oil Pollution Control Legislation," July 9, 1975, in *Public Papers of the President of the United States* (Washington, DC: U.S. Government Printing Office, 1976), 944–945.

142. "Energy and Environment," CQ Press Electronic Library, CQ Almanac Online Edition, cqal77-1203738, http://library.cqpress.com/cqal77-1203738 (accessed August 23, 2010), originally published in *CQ Almanac 1977* (Washington, DC: Congressional Quarterly, 1978).

143. Caulfield, "Conservation and Environmental Movements": 44; "Protection of the Environment," *Congressional Quarterly Almanac* 33 (1977): 30E–37E.

144. Jimmy Carter, "Message to the Congress on the Environment," May 23, 1977, in *Public Papers of the President of the United States* (Washington, DC: U.S. Government Printing Office, 1978), 967–986.

145. Jimmy Carter, "Annual Message to the Congress on the State of the Union," January 19, 1978, in *Public Papers of the President of the United States* (Washington, DC: U.S. Government Printing Office, 1979), 98–123: 115.

146. Rosenbaum, *Environmental Politics:* 68.

147. Carter, "State of the Union," January 19, 1978: 115.

148. Jimmy Carter, "Federal Water Policy," June 6, 1978, in *Public Papers of the President of the United States* (Washington, DC: U.S. Government Printing Office, 1979), 1043–1044.

149. Shanley, *Presidential Influence:* 54.

150. Ibid., 57.

151. Ibid., 57.

152. Caulfield, "Conservation and Environmental Movements": 46.

153. Norman J. Vig, "Presidential Leadership and the Environment: From Reagan to Clinton," in Norman J. Vig and Michael E. Kraft eds., *Environmental*

Policy:New Directions for the Twenty-First Century (Washington, D.C.: CQ Press), 98–130: 104.

154. V. Kerry Smith, *Environmental Policy under Reagan's Executive Order* (Chapel Hill: University of North Carolina Press, 1984): vii.

155. Ronald W. Reagan, "Question-and-Answer Session with High School Students on Domestic and Foreign Policy Issues," February 25, 1983, in *Public Papers of the President of the United States* (Washington, DC: U.S. Government Printing Office, 1984), 301–308: 305.

156. John Buell and Tom DeLuca, *Sustainable Democracy* (Thousand Oaks, CA: Sage, 1996): ix.

157. Soden and Steel, "Evaluating the Environmental Presidency": 334.

158. "Environment 1981," CQ Press Electronic Library, CQ Almanac Online Edition, cqal81-1171517, http://library.cqpress.com/cqal81-1171517 (accessed August 23, 2010), originally published in *CQ Almanac 1981* (Washington, DC: Congressional Quarterly, 1982).

159. Calvert, "Party Politics": 161.

160. Soden and Steel, "Evaluating the Environmental Presidency": 334.

161. Vig and Kraft, "Introduction" in *Environmental Policy in the 1980s:* 21.

162. "Presidential Veto Message: Reagan Vetoes Environmental Research Act," CQ Press Electronic Library, CQ Almanac Online Edition, cqal82-858-25803-1162805, http://library.cqpress.com/cqal82-858-24803-1162805 (accessed August 31, 2010), originally published in *CQ Almanac 1982* (Washington, DC: Congressional Quarterly, 1983); "Reagan Vetoes EPA Research Authorization," CQ Press Electronic Library, CQ Almanac Online Edition, cqal82-1163069, http://library.cqpress.com/cqal82-1163069 (accessed August 31, 2010), originally published in *CQ Almanac 1982* (Washington, DC: Congressional Quarterly, 1983); "Message to the Senate Returning without Approval an Environmental Research, Development and Demonstration Bill," October 22, 1982, in *Public Papers of the President of the United States* (Washington, DC: U.S. Government Printing Office,): 1374.

163. Vig and Kraft, "Introduction" in *Environmental Policy in the 1980s:* 104–105; Rosenbaum, *Environmental Politics:* 69.

164. Kraft, *Environmental Policy:* 30.

165. Rosenbaum, *Environmental Politics:* 98.

166. Robert Cameron Mitchell, "Public Opinion and Environmental Politics in the 1970s and 1980s," in Vig and Kraft, *Environmental Policy in the 1980s*, 51–74: 53–54.

167. Ronald W. Reagan, "Address before a Joint Session of the Congress on the State of Union," January 25, 1984, in *Public Papers of the President of the United States* (Washington, DC: U.S. Government Printing Office, 1985), 87–94; Ronald W. Reagan, "Radio Address to the Nation on Environmental Issues," July 14, 1984, in *Public Papers of the President of the United States* (Washington, DC: U.S. Government Printing Office, 1985), 1045–1215; Ronald W. Reagan, "Statement on Proposed Superfund Reauthorization Legislation," February 2, 1985, in *Public Papers of the President of the United States* (Washington, DC: U.S. Government Printing Office, 1986), 205.

168. Ronald W. Reagan, "Remarks of the President and Prime Minister Brian Mulroney in Quebec City, Canada, Announcing the Appointment of Special Envoys on Acid Rain," March 17, 1985, in *Public Papers of the President of the United States* (Washington, DC: U.S. Government Printing Office, 1986), 298–299; Ronald W. Reagan, "Statement on the Canada-United States Report on Acid Rain," January 8, 1986, in *Public Papers of the President of the United States* (Washington, DC: U.S. Government Printing Office, 1987), 25; Larry M. Speakes, "Statement by Principal Deputy Press Secretary Speakes on the Canada-United States Report on Acid Rain," March 19, 1986, in *Public Papers of the President of the United States* (Washington, DC: U.S. Government Printing Office, 1987), 373.

169. Ronald W. Reagan, "Message to the Senate Transmitting an Amendment to the Convention on International Trade in Endangered Species of Wild Fauna and Flora," October 4, 1983, in *Public Papers of the President of the United States* (Washington, DC: U.S. Government Printing Office, 1984), 1408.

170. Ronald W. Reagan, "Message to the Senate Transmitting the Vienna Convention for the Protection of the Ozone Layer," September 4, 1985, in *Public Papers of the President of the United States* (Washington, DC: U.S. Government Printing Office, 1986), 1038–1039.

171. Rosenbaum, *Environmental Politics:* 69.

172. Mitchell, "Public Opinion": 53–54.

173. Rosenbaum, *Environmental Politics:* 7.

174. "Environment," *Congressional Quarterly Almanac* 37 (1981): 503–504.

175. Daniel D. Chiras, *Beyond the Fray* (Boulder, CO: Johnson Books, 1990): 83–84.

176. Vig and Kraft, *Environmental Policy from the 1970s:* 15.

177. Glen Sussman and Mark Andrew Kelso, "Environmental Priorities and the President as Legislative Leader," in Soden, *Environmental Presidency,* 113–146: 117.

178. Cited in Rosenbaum, *Environmental Politics:* 8.

179. George H. W. Bush, "Remarks on Transmitting to the Congress Proposed Legislation to Amend the Clean Air Act," July 21, 1989, in *Public Papers of the President of the United States* (Washington, DC: U.S. Government Printing Office, 1990), 997–999.

180. George H. W. Bush, "White House Fact Sheet on Environmental Initiatives," September 18, 1989, in *Public Papers of the President of the United States* (Washington, DC: U.S. Government Printing Office, 1990), 1212–1215; George H. W. Bush, "Letter to Congressional Leaders to Amend the Clean Air Act," September 26, 1990, in *Public Papers of the President of the United States* (Washington, DC: U.S. Government Printing Office, 1991), 1303–1305.

181. George H. W. Bush, "Joint Statement Announcing Canada-United States Air Quality Negotiations," July 8, 1990, in *Public Papers of the President of the United States* (Washington, DC: U.S. Government Printing Office, 1991), 977.

182. Byron Daynes, "Bill Clinton: Environmental President," in Soden, *Environmental Presidency,* 259–312: 259.

183. Vig and Kraft, *Environmental Policy from the 1970s:* 15.

184. Sussman and Kelso, "Environmental Priorities": 117.

185. Rosenbaum, *Environmental Politics:* 70.

186. Ibid., 8.

187. Vig and Kraft, *Environmental Policy from the 1970s:* 15; Sussman and Kelso, "Environmental Priorities": 118.

188. Rosenbaum, *Environmental Politics:* 8.

189. Daynes, "Bill Clinton": 262.

190. Vig and Kraft, *Environmental Policy from the 1970s:* 16.

191. Andrews, *Managing* 1999: 315.

192. William J. Clinton, "Remarks on the Observance of Earth Day," April 21, 1994, in *Public Papers of the President of the United States* (Washington, DC: U.S. Government Printing Office, 1995), 740–745.

193. Ibid.

194. William J. Clinton, "Remarks on Earth Day," April 21, 1993, in *Public Papers of the President of the United States* (Washington, DC: U.S. Government Printing Office, 1994), 468–472.

195. William J. Clinton, "Remarks at the Children's Town Meeting," February 20, 1993, in *Public Papers of the President of the United States* (Washington, DC: U.S. Government Printing Office, 1994), 146–165; William J. Clinton, "Remarks to the U.S. Chamber of Commerce National Business Action Rally," February 23, 1993, in *Public Papers of the President of the United States* (Washington, DC: U.S. Government Printing Office, 1994), 185–195; Clinton, "Observance of Earth Day," April 21, 1994: 740–745.

196. Clinton, "Observance of Earth Day": 740–745.

197. William J. Clinton, "Remarks on Regulatory Reform in Arlington, Virginia," March 16, 1995, in *Public Papers of the President of the United States* (Washington, DC: U.S. Government Printing Office, 1996), 358–362: 359.

198. William J. Clinton, "Remarks to Participants in Project XL," November 3, 1995, in *Public Papers of the President of the United States* (Washington, DC: U.S. Government Printing Office, 1996), 1713–1716; Clinton, "Regulatory Reform in Arlington, Virginia," March 16, 1995: 361.

199. William J. Clinton, "Remarks on Environmental Protection in Baltimore, MD," August 8, 1995, in *Public Papers of the President of the United States* (Washington, DC: U.S. Government Printing Office, 1996), 1215–1218.

200. William J. Clinton, "Remarks Announcing the Creation of the White House Office on Environmental Policy," February 8, 1993, in *Public Papers of the President of the United States* (Washington, DC: U.S. Government Printing Office, 1994), 62–63.

201. William J. Clinton, "Remarks to the United Nations Special Session on Environment and Development in New York City," June 26, 1997, in *Public Papers of the President of the United States* (Washington, DC: U.S. Government Printing Office, 1998), 826–828: 827.

202. William J. Clinton, "Remarks at the National Geographic Society," October 22, 1997, in *Public Papers of the President of the United States* (Washington, DC: U.S. Government Printing Office, 1998), 1408–1412: 1410.

203. Vig and Kraft, *Environmental Policy from the 1970s:* 16.

204. Ibid., 16.

205. Sussman and Kelso, "Environmental Priorities": 141.

206. Andrews, *Managing the Environment* (1999): 363.

207. Daynes, "Bill Clinton": 262.

208. Vig and Kraft, *Environmental Policy from the 1970s:* 15.

209. Ibid., 100.

210. Ibid., 16.

211. Ibid., 100.

212. Rosenbaum, *Environmental Politics:* 9.

213. Vig and Kraft, *Environmental Policy from the 1970s:* 16.

214. George W. Bush, "Message to the Congress on Reporting on Environmental Goals," March 24, 1992, in *Public Papers of the President of the United States* (Washington, DC: U.S. Government Printing Office, 1993), 498–502; George W. Bush, "Message to the Congress Reporting on Environmental Quality," April 18, 1991, in *Public Papers of the President of the United States* (Washington, DC: U.S. Government Printing Office, 1992), 404–406; George W. Bush, "Statement on Earth Day," April 22, 1992, in *Public Papers of the President of the United States* (Washington, DC: U.S. Government Printing Office, 1993), 624.

215. Bush, "Message on Environmental Goals"; Bush, "Statement on Earth Day"; George W. Bush, "Remarks on Environmental Policy," January 23, 1992, in *Public Papers of the President of the United States* (Washington, DC: U.S. Government Printing Office, 1993), 138–139.

216. George W. Bush, "Statement on Attending the United Nations Conference on Environment and Development in Rio de Janeiro, Brazil," May 12, 1992, in *Public Papers of the President of the United States* (Washington, DC: U.S. Government Printing Office, 1993), 753.

217. Bush, "Message on Environmental Goals."

218. George W. Bush, "Remarks by the President and Prime Minister Brian Mulroney of Canada at the Air Quality Agreement Signing Ceremony in Ottawa," March 13, 1991, in *Public Papers of the President of the United States* (Washington, DC: U.S. Government Printing Office, 1992), 254–257; George W. Bush, "Remarks at a White House Briefing for the Associated General Contractors of America," April 15, 1991, in *Public Papers of the President of the United States* (Washington, DC: U.S. Government Printing Office, 1992), 369–371; George W. Bush, "Message to the Congress on Japanese Importation of Sea Turtles," May 17, 1991, in *Public Papers of the President of the United States* (Washington, DC: U.S. Government Printing Office, 1992), 521.

219. Marlin Fitzwater, "Statement by Press Secretary Fitzwater on the Phaseout of Ozone-Depleting Substances," February 11, 1992, in *Public Papers of the President of the United States* (Washington, DC: U.S. Government Printing Office, 1993), 232.

220. George W. Bush, "Message to the Senate Transmitting the Antarctic Treaty Protocol on Environmental Protection," February 14, 1992, in *Public Papers of the President of the United States* (Washington, DC: U.S. Government Printing Office, 1993), 244–245.

221. Vig, "Presidential Leadership,": 100.

222. Robert Gottlieb, *Forcing the Spring* (Washington, DC: Island Press, 2005): 5.

223. Barry G. Rabe, "Power to the States: The Promise and Pitfalls of Decentralization," in Vig and Kraft, *Environmental Policy: New Directions*, 34–56: 48.

224. Adriel Bettelheim, "Empty-Handed before the World on Cap-and-Trade," *CQ Weekly Online*, September 28, 2009: 2146–2153, http://library.cqpress.com/cqweekly/weeklyreport111000003211645 (accessed September 2, 2010).

225. Coral Davenport, "Climate Change Pledge Is No Easy Sell on Hill," *CQ Weekly Online*, January 25, 2010: 211–212, http://library.cqpress.com/cqweekly/weeklyreport111-000003283402 (accessed September 2, 2010).

226. Barack Obama, "Proclamation 8392—National Oceans Month, 2009," June 12, 2009 *The Federal Register*, June 17, 2009.

227. Obama, "Proclamation 8392"; Barack Obama, "Proclamation 8531—National Oceans Month, 2010," May 28, 2010, C: *The Federal Register*, June 7, 2010.

228. Barack Obama, "Memorandum on National Policy for the Oceans, Our Coasts, and the Great Lakes," June 12, 2009 *The Federal Register*, June 17, 2009.

229. Barack Obama, "Remarks on the 160th Anniversary of the Department of the Interior," March 3, 2009 (Washington, DC: *The Federal Register*, March 6, 2009); Barack Obama, "Memorandum on the Endangered Species Act," March 3, 2009 *The Federal Register*, March 6, 2009.

230. Barack Obama, "Remarks on Signing the Omnibus Public Land Management Act of 2009," March 30, 2009 *The Federal Register*, April 3, 2009.

231. Barack Obama, "Proclamation 8409—National Wilderness Month, 2009," September 3, 2009 *The Federal Register*, September 8, 2009; Barack Obama, "Proclamation 8553—National Wilderness Month, 2010," August 31, 2010 *The Federal Register*, September 7, 2010.

232. O'Leary, "Environmental Policy": 151.

233. "Environmental Law," *Congressional Quarterly Almanac* 35 (1979): 35-A.

234. O'Leary, "Environmental Policy": 150.

235. Vig and Kraft, *Environmental Policy: New Directions:* 4.

236. Cramer, *Deep Environmental Politics:* 149.

237. Davies and Davies, *Politics of Pollution:* 123.

238. Lettie M. Wenner, "Judicial Oversight of Environmental Deregulation," in Vig and Kraft, *Environmental Policy in the 1980s*, 181–199: 182.

239. David H. Rosenbloom and Rosemary O'Leary, *Public Administration:* 90.

240. Rosenbaum, *Environmental Politics:* 96.

241. Davies and Davies, *Politics of Pollution:* 124.

242. Ibid., 125.

243. Rosenbloom and O'Leary, *Public Administration:* 95; see *United States v. Conservation Chemical Co.*, 1985; *United States v. Ottati & Goss*, 1985; *United States v. Waste Ind*, 1984; *United States v. Reilly Tar*, 1982; *United States v. NEPACCO*, 1984; *United States v. Seymour Recycling Corp*, 1984; *United States v. Vertack Chemical Corp*, 1980.

244. Bosso and Guber, "Maintaining Presence": 94.

245. John Rehfuss, *Public Administration as Political Process* (New York: Charles Scribner's Sons, 1973): 158.

246. Carl E. VanHorn, Donald C. Baumer, and William T. Gormley Jr., *Politics and Public Policy* (Washington, DC: CQ Press, 1992).

247. Richard M. Nixon, "Special Message to the Congress about Reorganization Plans to Establish the Environmental Protection Agency and the National Oceanic and Atmospheric Administration," July 9, 1970, in *Public Papers of the President of the United States* (Washington, DC: U.S. Government Printing Office, 1971), 578–586.

248. Richard M. Nixon, "Statement Announcing the Creation of the Environmental Quality Council and the Citizens' Advisory Committee on Environmental Quality," May 29, 1969, in *Public Papers of the President of the United States* (Washington, DC: U.S. Government Printing Office, 1970), 422–423; "Environmental Problems," *Congressional Quarterly Almanac* 25 (1969): 513; Richard M. Nixon, "Statement about the Council on Environmental Quality," January 29, 1970, in *Public Papers of the President of the United States* (Washington, DC: U.S. Government Printing Office, 1971), 34–35.

249. Richard M. Nixon, "Statement on Establishing the National Industrial Pollution Control Council," April 9, 1970, in *Public Papers of the President of the United States* (Washington, DC: U.S. Government Printing Office, 1971), 344–345.

250. Jeffrey M. Berry, *The Interest Group Society* (Boston: Little, Brown, 1984).

251. Peter Bachrach and Morton S. Baratz, *Power and Poverty: Theory and Practice* (New York: Oxford University Press, 1970); James Houston and William W. Parsons, *Criminal Justice and the Policy Process* (Chicago: Nelson Hall, 1998).

252. Hays, *Environmental Politics:* 189–190.

253. Ibid., 189–190.

254. William R. Lowry, "A Return to Traditional Priorities in Natural Resource Policies," in Vig and Kraft, *Environmental Policy: New Directions,* 311–332: 315.

255. Hays, *Environmental Politics:* 27–28.

256. Ibid., 29–32.

257. John Buell and Tom DeLuca, *Sustainable Democracy* (Thousand Oaks, CA: Sage, 1996): x.

258. Ibid., xi.

259. Rosenbaum, *Environmental Politics:* 52.

260. Rosenbaum, *Environmental Politics:* 51.

261. Riley E. Dunlap, "Public Opinion and Environmental Policy," in Lester, *Environmental Politics and Policy,* 87–134: 96.

262. Nixon Poll, methodology: interviewing conducted by Opinion Research Corporation, May 7–25, 1971, and based on 1,506 telephone interviews, sample: national adult, http://webapps.ropercenter.uconn.edu/CFIDE/cf/action/ipoll/questionDetail.cfm? (accessed July 25, 2010).

263. Harris Survey, November 1976, conducted by Louis Harris and Associates during November 1976 and based on 1,532 telephone interviews, sample: national adult, July 25, 2010.

264. Ibid.

265. Survey by the League of Conservation Voters, methodology: interviewing conducted by the Tarrance Group and Mellman, Lazarus & Lake, January 10–12, 1994, and based on 1,000 telephone interviews, sample: national registered voters, http://webapps.ropercenter.uconn.edu/CFIDE/cf/action/ipoll/questionDetail.cfm? (accessed July 25, 2010).

266. ABC News/*Washington Post* Poll, March 2001, methodology: conducted by ABC News/*Washington Post,* March 22–25, 2001, and based on 903 telephone interviews, sample: national adult, http://webapps.ropercenter.uconn.edu/CFIDE/cf/action/ipoll/questionDetail.cfm? (accessed July 25, 2010).

267. Gallup/CNN/*USA Today* Poll, January 2003, survey by Cable News Network, *USA Today,* methodology: interviewing conducted by Gallup Organization, January 3–4, 2003, and based on 1,000 telephone interviews, sample: national adult, http://webapps.ropercenter.uconn.edu/CFIDE/cf/action/ipoll/questionDetail.cfm? (accessed July 25, 2010).

268. CBS News/*New York Times* Poll, April 2007, conducted by CBS News/*New York Times,* April 20–24, 2007, and based on 1,052 telephone interviews, sample: national adult, http://webapps.ropercenter.uconn.edu/CFIDE/cf/action/ipoll/questionDetail.cfm? (accessed July 25, 2010).

269. Associated Press/Stanford University/Ipsos-Public Affairs Poll, June 2007, survey by Associated Press, Stanford University, methodology: interviewing conducted by Ipsos-Public Affairs, June 12–14, 2007, and based on 1,001 telephone interviews, sample: national adult, http://webapps.ropercenter.uconn.edu/CFIDE/cf/action/ipoll/questionDetail.cfm? (accessed July 25, 2010).

270. Pew Research Center for the People & the Press Political Survey, January 2009, survey by Pew Research Center for the People & the Press, methodology: interviewing conducted by Princeton Survey Research Associates International, January 7–11, 2009, and based on 1,503 telephone interviews, sample: national adult, http://webapps.ropercenter.uconn.edu/CFIDE/cf/action/ipoll/questionDetail.cfm? (accessed July 25, 2010).

271. CNN/Opinion Research Corporation Poll, February 2009, survey by Cable News Network, methodology: interviewing conducted by Opinion Research Corporation, February 18–19, 2009, and based on 1,046 telephone interviews, sample: national adult, http://webapps.ropercenter.uconn.edu/CFIDE/cf/action/ipoll/questionDetail.cfm? (accessed July 25, 2010).

272. Kraft, "Environmental Policy in Congress": 126.

273. National Survey for Republican National Committee/National Republican Congressional Committee, October 1979, survey by Republican National Committee, methodology: interviewing conducted by Market Opinion Research, October 21–November 15, 1979, and based on 1,506 telephone interviews, sample: national adult, http://webapps.ropercenter.uconn.edu/CFIDE/cf/action/ipoll/questionDetail.cfm? (accessed July 25, 2010).

274. ABC News/*Washington Post* Poll, April 1995; methodology: conducted by ABC News/*Washington Post,* April 4–5, 1995, and based on 1,026 telephone

interviews, sample: national adult, http://webapps.ropercenter.uconn.edu/CFIDE/cf/action/ipoll/questionDetail.cfm? (accessed July 25, 2010).

275. *New York Times* Poll, June 1996, methodology: conducted by *New York Times,* June 20–23, 1996, and based on 1,121 telephone interviews, sample: national adult, http://webapps.ropercenter.uconn.edu/CFIDE/cf/action/ipoll/questionDetail.cfm? (accessed July 25, 2010).

276. *Los Angeles Times* Poll, January 2002; methodology: conducted by *Los Angeles Times,* January 31–February 3, 2002, and based on 1,545 telephone interviews, sample: national adult, http://webapps.ropercenter.uconn.edu/CFIDE/cf/action/ipoll/questionDetail.cfm? (accessed July 25, 2010).

277. ABC News/*Washington Post* Poll, April 2003, methodology: conducted by ABC News/*Washington Post,* April 27–30, 2003, and based on 1,105 telephone interviews, sample: national adult including an oversample of blacks, http://webapps.ropercenter.uconn.edu/CFIDE/cf/action/ipoll/questionDetail.cfm? (accessed July 25, 2010).

278. Pew News Interest Index Poll, January 2004, survey by Pew Research Center for the People & the Press, methodology: interviewing conducted by Princeton Survey Research Associates, January 6–11, 2004, and based on 1,503 telephone interviews, sample: national adult, http://webapps.ropercenter.uconn.edu/CFIDE/cf/action/ipoll/questionDetail.cfm? (accessed July 25, 2010).

279. ABC News/*Time*/Stanford University Poll, March 2006, survey by *Time,* Stanford University, methodology: interviewing conducted by ABC News, March 9–14, 2006, and based on 1,002 telephone interviews, sample: national adult, interviews were conducted by TNS Intersearch, http://webapps.ropercenter.uconn.edu/CFIDE/cf/action/ipoll/questionDetail.cfm? (accessed July 25, 2010).

280. CBS News/*New York Times* Poll, April 2007, conducted by CBS News/*New York Times,* April 20–24, 2007, and based on 1,052 telephone interviews, sample: national adult, http://webapps.ropercenter.uconn.edu/CFIDE/cf/action/ipoll/questionDetail.cfm? (accessed July 25, 2010).

Chapter 2 The Politics of Air Pollution

1. J. S. Kidd and Renee A. Kidd, *Air Pollution: Problems and Solutions* (New York: Chelsea House, 2006), x.

2. Ibid., 58.

3. Michael E. Kraft, *Environmental Policy and Politics* (New York: Longman, 2001), 32.

4. Kidd and Kidd, *Air Pollution:* xi.

5. Daniel D. Chiras, *Beyond the Fray* (Boulder, CO: Johnson Books, 1990), 4.

6. Jeffrey J. Pompe and James R. Rinehart, *Environmental Conflict* (Albany: State University of New York, 2002), 128.

7. Walter A. Rosenbaum, *Environmental Politics and Policy,* 6th ed. (Washington, DC: CQ Press, 2005), 2.

8. Kidd and Kidd, *Air Pollution:* x.

9. Ibid., 121.

10. Kraft, *Environmental Policy:* 27.

11. Harry S. Truman, "Message to the United States Technical Conference on Air Pollution," May 3, 1950, in *Public Papers of the President of the United States* (Washington, DC: U.S. Government Printing Office, 1951), 281–282.

12. Dwight D. Eisenhower, "Annual Budget Message to the Congress for Fiscal Year 1957," January 16, 1956 in *Public Papers of the President of the United States,* (Washington, DC: U.S. Government Printing Office, 1957), 75–156: 119.

13. Dwight D. Eisenhower, "Special Message to the Congress Recommending a Health Program," January 31, 1955, in *Public Papers of the President of the United States* (Washington, DC: U.S. Government Printing Office, 1956), 216–223: 221.

14. "Air Pollution," *Congressional Quarterly Almanac* 17 (1961): 266; "Air Pollution Control," CQ Press Electronic Library, CQ Almanac Online Edition, cqal62-1326129, http://library.cqpress.com/cqal62-1326129 (accessed August 23, 2010), originally published in *CQ Almanac 1962* (Washington, DC: Congressional Quarterly, 1963).

15. "Pollution Control," *Congressional Quarterly Almanac* 15 (1959): 68.

16. "Air Pollution," *Congressional Quarterly Almanac* 15 (1959): 267.

17. "Kennedy Health Program," *Congressional Quarterly Almanac* 18 (1962): 211.

18. John F. Kennedy, "Special Message to the Congress on National Health Needs," February 27, 1962, in *Public Papers of the President of the United States* (Washington, DC: U.S. Government Printing Office, 1963), 165–173.

19. John F. Kennedy, "Special Message to the Congress on Improving the Nation's Health," February 7, 1963, in *Public Papers of the President of the United States* (Washington, DC: U.S. Government Printing Office, 1964), 140–147.

20. "Air Pollution" (1961): 266; "Air Pollution Control," cqal62-1326129.

21. Lyndon B. Johnson, "Remarks upon Signing the Clean Air Act," December 17, 1963, in *Public Papers of the President of the United States* (Washington, DC: U.S. Government Printing Office, 1964), 60–61.

22. Lyndon B. Johnson, "Special Message to the Congress on Conservation and Restoration of Natural Beauty," February 8, 1965, in *Public Papers of the President of the United States* (Washington, DC: U.S. Government Printing Office, 1966), 155–165.

23. Lyndon B. Johnson, "Annual Message to the Congress: The Economic Report of the President," January 27, 1966, in *Public Papers of the President of the United States* (Washington, DC: U.S. Government Printing Office, 1967), 96–109.

24. "Natural Beauty Sought as U.S. Goal," *Congressional Quarterly Almanac* 21 (1965): 722–723.

25. Johnson, "Message on Conservation and Restoration."

26. "Text of Message on Pollution, Resources, Road Safety," *Congressional Quarterly Almanac* 23 (1967): 36-A–40-A.

27. Ibid.; Lyndon B. Johnson, "Special Message to the Congress: Protecting Our Natural Heritage," January 30, 1967, in *Public Papers of the President of the United States* (Washington, DC: U.S. Government Printing Office, 1968), 93–103.

28. "Message to Congress: Johnson's Conservation Message," CQ Press Electronic Library, CQ Almanac Online Edition, cqal68-1124522, http://library.cqpress.com/cqal68-1124522 (accessed August 23, 2010), originally published in *CQ Almanac 1968* (Washington, DC: Congressional Quarterly, 1969).

29. Phillip F. Cramer, *Deep Environmental Politics* (Westport, CT: Praeger, 1998), 71.

30. Cramer, *Deep Environmental Politics:* 71, citing Air Pollution, Prevention and Control, U.S. Code, title 42 (Public Health and Welfare), Chapter 85, Section 7401.

31. Ibid., 72.

32. Ibid., citing "Air Pollution, Prevention and Control," U.S. Code, title 42 (Public Health and Welfare), Chapter 85, Section 7401.

33. Ibid., 71.

34. "Review of the Session," *Congressional Quarterly Almanac* 19 (1963): 75.

35. Paul R. Portney, "Air Pollution Policy," in Paul R. Portney and Robert N. Stavins, eds., *Public Policies for Environmental Protection,* 2nd ed. (Washington, DC: Resources for the Future, 2000), 77–123: 78.

36. Cramer, *Deep Environmental Politics:* 72, citing Air Pollution, Prevention and Control, U.S. Code, title 42 (Public Health and Welfare), Chapter 85, Section 7521.

37. Ibid., 78, citing Air Pollution, Prevention and Control, U.S. Code, title 42 (Public Health and Welfare), Chapter 85, Section 7554.

38. "Congress Enacts New Air Pollution Law," CQ Press Electronic Library, CQ Almanac Online Edition, cqal63-1316972, http://library.cqpress.com/cqal63-1316972 (accessed August 23, 2010), originally published in *CQ Almanac 1963* (Washington, DC: Congressional Quarterly, 1964).

39. "Auto Pollution, Waste Disposal Act Passed," *Congressional Quarterly Almanac* 21 (1965): 780–786.

40. Portney, "Air Pollution Policy": 79.

41. "Clean Air," *Congressional Quarterly Almanac* 22 (1966), 77; "Air Pollution Control," *Congressional Quarterly Almanac* 22 (1966): 685–687.

42. "Congress Acts on Pollution but Clears No Major Bills," *Congressional Quarterly Almanac* 24 (1968): 569–575.

43. "Congress Strengthens Air Pollution Control Powers," *Congressional Quarterly Almanac* 23 (1967), 875–887: 887.

44. Portney, "Air Pollution Policy": 80.

45. "What Congress Did," *Congressional Quarterly Almanac* 23 (1967): 76–87: 77.

46. "Congress Strengthens Air Pollution Control Powers."

47. Kraft, *Environmental Policy:* 33.

48. "Special Message to the Congress on Environmental Quality," *Congressional Quarterly Almanac* 26 (1970): 96–109.

49. "Annual Message to the Congress on the State of the Union," *Congressional Quarterly Almanac* 26 (1970): 8–16; "Remarks on Transmitting Special Message to the Congress on Environmental Quality," *Congressional Quarterly Almanac* 26 (1970): 95–96.

50. "Special Message on Environmental Quality"; "Presidential Statement to Congress: Nixon on the Environment 1970," CQ Press Electronic Library, CQ

Almanac Online Edition, cqal70-1290732, http://library.cqpress.com/cqal70-1290732 (accessed August 23, 2010), originally published in *CQ Almanac 1970* (Washington, DC: Congressional Quarterly, 1971).

51. "Presidential Statement to Congress: Nixon's 1970 Environmental Message," CQ Press Electronic Library, CQ Almanac Online Edition, cqal71-869-26707-1254975, http://library.cqpress.com/cqal71-869-26707-1254975 (accessed August 23, 2010), originally published in *CQ Almanac* 1971 (Washington, DC: Congressional Quarterly, 1972).

52. Ibid.

53. "Environment: Action Completed," *Congressional Quarterly Almanac* 26 (1970): 79.

54. "Major Anti-Pollution Measures Clear 91st Congress," *Congressional Quarterly Almanac* 26 (1970): 128.

55. Daniel J. Fiorino, *Making Environmental Policy* (Berkeley: University of California Press, 1995), 25.

56. "Major Anti-Pollution Measures Clear 91st Congress," *Congressional Quarterly Almanac* 26 (1970): 128; "Clean Air Bill Cleared with Auto Emission Deadline," *Congressional Quarterly Almanac* 26 (1970): 472–486.

57. "Air Quality Act Amendments," CQ Press Electronic Library, CQ Almanac Online Edition, cqal69-1248478, http://library.cqpress.com/cqal69-1248478 (accessed August 23, 2010), originally published in *CQ Almanac 1969* (Washington, DC: Congressional Quarterly, 1970).

58. Richard N. L. Andrews, *Managing the Environment, Managing Ourselves* (New Haven, CT: Yale University Press, 1999), 233–234; Fiorino, *Making Environmental Policy:* 25.

59. Fiorino, *Making Environmental Policy:* 25.

60. Samuel P. Hays, *Environmental Politics since 1945* (Pittsburgh: University of Pittsburgh Press, 2000), 160.

61. J. Clarence Davies III and Barbara S. Davies, *The Politics of Pollution* (Indianapolis: Pegasus, 1975), 139.

62. Ibid.

63. Ibid.

64. "Inner City Environment," CQ Press Electronic Library, CQ Almanac Online Edition, cqal72-1250880, http://library.cqpress.com/cqalmanac/cqal72-1250880 (accessed August 30, 2010), originally published in *CQ Almanac 1972* (Washington, DC: Congressional Quarterly, 1973).

65. Ibid., 139–141.

66. "Low Emission Vehicles," CQ Press Electronic Library, CQ Almanac Online Edition, cqal70-1293756, http://library.cqpress.com/cqalmanac/cqal70-1293756 (accessed August 23, 2010), originally published in *CQ Almanac 1970* (Washington, DC: Congressional Quarterly, 1971).

67. "Clean Air, Waste Disposal," CQ Press Electronic Library, CQ Almanac Online Edition, cqal73-1227314, http://library.cqpress.com/cqal73-1227314 (accessed August 23, 2010), originally published in *CQ Almanac* (Washington, DC: Congressional Quarterly, 1972).

68. Kidd and Kidd, *Air Pollution:* 103.

69. "Congress Votes to Delay Clean Air Standards," *Congressional Quarterly Almanac* 30 (1974): 738–744.

70. "Auto Pollution Control Deadline Postponed," CQ Press Electronic Library, CQ Almanac Online Edition, cqal73-1227355, http://library.cqpress.com/cqalmanac/cqal73-1227355 (accessed August 23, 2010), originally published in *CQ Almanac 1973* (Washington, DC: Congressional Quarterly, 1974).

71. "Clean Air Standards Relaxed," *Congressional Quarterly Almanac* (29) 1973: 656–657.

72. Kidd and Kidd, *Air Pollution:* 104.

73. Ibid.

74. Gerald R. Ford, "Special Message to the Congress Urging Action on Pending Legislation," July 22, 1975, in *Public Papers of the President of the United States* (Washington, DC: U.S. Government Printing Office, 1976), 2058–2080: 2069.

75. Gerald R. Ford, "The President's News Conference of February 4, 1975," in *Public Papers of the President of the United States* (Washington, DC: U.S. Government Printing Office, 1976), 184–195: 192.

76. Gerald R. Ford, "Remarks and a Question-and-Answer Session at the Vail Symposium in Vail, Colorado," August 15, 1975, in *Public Papers of the President of the United States* (Washington, DC: U.S. Government Printing Office), 1150–1699.

77. Kidd and Kidd, *Air Pollution:* 102.

78. Ibid., x.

79. "Ozone-Aerosol Issue," CQ Press Electronic Library, CQ Almanac Online Edition, cqal75-1213807, http://library.cqpress.com/cqalmanac/cqal75-1213807 (accessed August 23, 2010), originally published in *CQ Almanac 1975* (Washington, DC: Congressional Quarterly, 1976).

80. "Energy and Environment: Action Not Taken," *Congressional Quarterly Almanac* 31 (1975): 15; "Congress Faces Hard Choices on Clean Air Act," *Congressional Quarterly Almanac* 31 (1975): 245–250.

81. "Energy and Environment: Action Not Completed," *Congressional Quarterly Almanac* 32 (1976): 11; "Clean Air Amendments Die at Session's Close," *Congressional Quarterly Almanac* 32 (1976): 128–143; "Energy and Environment," *Congressional Quarterly Almanac* 32 (1976): 91–94.

82. Jimmy Carter, "The Environment," May 23, 1977, in *Public Papers of the President of the United States* (Washington, DC: U.S. Government Printing Office), 967–986.

83. Ibid.

84. Jimmy Carter, "Remarks at a Meeting with Environmental, Community, and Governmental Leaders," May 4, 1978, in *Public Papers of the President of the United States* (Washington, DC: U.S. Government Printing Office), 832–834.

85. "Energy and Environment: Clean Air Amendments," *Congressional Quarterly Almanac* 33 (1977): 20.

86. "Major Clean Air Amendments Enacted," *Congressional Quarterly Almanac* 33 (1977): 627–637; Jimmy Carter, "Statement on Signing HR 6161 into Law,"

August 8, 1977, in *Public Papers of the President of the United States* (Washington, DC: U.S. Government Printing Office, 1978), 1460–1461; Andrews, *Managing the Environment:* 234.

87. Pompe and Rinehart, *Environmental Conflict:* 126.

88. Ronald W. Reagan, "Message to the Senate Transmitting the Vienna Convention for the Protection of the Ozone Layer," September 4, 1985, in *Public Papers of the President of the United States* (Washington, DC: U.S. Government Printing Office, 1986), 1038–1039.

89. Ronald W. Reagan, "Address to the Nation on the Economy," October 13, 1982, in *Public Papers of the President of the United States* (Washington, DC: U.S. Government Printing Office, 1983), 1307–1312: 1312.

90. Ronald W. Reagan, "Remarks at the Swearing in Ceremony of William D. Ruckelshaus as Administrator of the Environmental Protection Agency," May 18, 1983, in *Public Papers of the President of the United States* (Washington, DC: U.S. Government Printing Office, 1984), 733–735.

91. Ronald W. Reagan, "Message to the Congress on 'A Quest for Excellence,'" January 27, 1987, in *Public Papers of the President of the United States* (Washington, DC: U.S. Government Printing Office, 1988), 61–79; Ronald W. Reagan, "Statement on Acid Rain," March 18, 1987, in *Public Papers of the President of the United States* (Washington, DC: U.S. Government Printing Office, 1988), 254–255.

92. "Commission on Air Quality," *Congressional Quarterly Almanac* 37 (1981): 513–514.

93. Ibid.

94. "Congress Begins Rewrite of Clean Air Act," *Congressional Quarterly Almanac* 37 (1981): 505–513.

95. "Congress Fails to Act on Clean Air Rewrite," *Congressional Quarterly Almanac* 38 (1982): 425–434.

96. "No Action on Clean Air Bill," *Congressional Quarterly Almanac* 39 (1983): 339; "Environment: Clean Air/Acid Rain," *Congressional Quarterly Almanac* 40 (1984): 21.

97. "Clean Air Bill Stalled by Acid Rain Dispute," *Congressional Quarterly Almanac* 40 (1984): 339–342; "Clean Water Act Rewrite," *Congressional Quarterly Almanac* 39 (1983): 360–361.

98. "Acid Rain Bill Stalls," *Congressional Quarterly Almanac* 42 (1986): 137.

99. "House Sets Stage for Clean-Air Debate in 1988," *Congressional Almanac Quarterly* 43 (1987): 299–301.

100. "Energy/Environment: Clean Air," *Congressional Quarterly Almanac* 44 (1988): 18; "Clean-Air Bill Fails to Move," *Congressional Quarterly Almanac* 44 (1988); 142–48.

101. "Environment: Clean Air," *Congressional Quarterly Almanac* 43 (1987): 16.

102. George H. W. Bush, "Remarks on Transmitting to the Congress Proposed Legislation to Amend the Clean Air Act," July 21, 1989, in *Public Papers of the President of the United States* (Washington, DC: U.S. Government Printing Office, 1990), 997–999.

103. George H. W. Bush, "Statement on Air Pollution Regulatory Relief," March 18, 1992, in *Public Papers of the President of the United States* (Washington, DC: U.S. Government Printing Office, 1993), 470; George H. W. Bush, "Remarks to Giddings and Lewis Employees and Local Chambers of Commerce in Feraser, Michigan," April 14, 1992, in *Public Papers of the President of the United States* (Washington, DC: U.S. Government Printing Office, 1993), 595–599.

104. George H. W. Bush, "White House Fact Sheet on Environmental Initiatives," September 18, 1989, in *Public Papers of the President of the United States* (Washington, DC: U.S. Government Printing Office, 1990), 1212–1215.

105. George H. W. Bush, "Joint Statement Announcing Canada-United States Air Quality Negotiations," July 8, 1990, in *Public Papers of the President of the United States* (Washington, DC: U.S. Government Printing Office, 1991), 977.

106. Kraft, *Environmental Policy:* 133.

107. George H. W. Bush, "Statement on Signing the Bill Amending the Clean Air Act," November 15, 1990, in *Public Papers of the President of the United States* (Washington, DC: U.S. Government Printing Office, 1991), 1602–1604; George H. W. Bush, "Remarks on Signing the Bill Amending the Clean Air Act," November 15, 1990, in *Public Papers of the President of the United States* (Washington, DC: U.S. Government Printing Office, 1991), 1600–1602; "Clean Air Act Rewritten, Tightened," *Congressional Quarterly Almanac* 46 (1990): 229–276.

108. Glen Sussman and Mark Andrew Kelso, "Environmental Priorities and the President as Legislative Leader," in Dennis L. Soden, ed., *The Environmental Presidency* (Albany: State University of New York Press, 1999), 113–146: 125.

109. Ibid.

110. Fiorino, *Making Environmental Policy:* 27.

111. "Energy/Environment: Clean Air," *Congressional Quarterly Almanac* 45 (1989): 22; "Clean-Air Bill Moves in Both Chambers," *Congressional Quarterly Almanac* 44 (1988): 665–667.

112. Fiorino, *Making Environmental Policy:* 28.

113. Mary Clifford, "A Review of Federal Environmental Legislation," in Mary Clifford, ed., *Environmental Crime* (Gaithersburg, MD: Aspen, 1998), 95–109: 102.

114. "Bush Proposes Changes to Air Pollution Rules," *Congressional Quarterly Almanac* 56 (2003): 9–14; George W. Bush, "Remarks on Earth Day in Wilmington, New York," April 22, 2002, in *Public Papers of the President of the United States* (Washington, DC: U.S. Government Printing Office, 2003), 645–648; George W. Bush, "Remarks Announcing the Clear Skies and Global Climate Change Initiatives in Silver Spring, Maryland," February 14, 2002, in *Public Papers of the President of the United States* (Washington, DC: U.S. Government Printing Office, 2003), 226–231; George W. Bush, "Statement on the Clear Skies Initiative," July 1, 2002, in *Public Papers of the President of the United States* (Washington, DC: U.S. Government Printing Office, 2003), 1160; George W. Bush, "Remarks on the Proposed Clear Skies Legislation," September 16, 2003, in *Public Papers of the President of the United States* (Washington, DC: U.S. Government Printing Office, 2004), 1170–1173.

115. George W. Bush, "Message to the Senate Transmitting the Protocol of 1997 to Amend the International Convention for the Prevention of Pollution from Ships," May 15, 2003, in *Public Papers of the President of the United States* (Washington, DC: U.S. Government Printing Office, 2004), 496–497.

116. George W. Bush, "Remarks on Energy Policy in Columbus, Ohio," March 9, 2005, in *Public Papers of the President of the United States* (Washington, DC: U.S. Government Printing Office, 2006), 389–396; George W. Bush, "Remarks to the 16th Annual Energy Efficiency Forum," June 15, 2005, in *Public Papers of the President of the United States* (Washington, DC: U.S. Government Printing Office, 2006), 993–999.

117. Ray Michalowski, "International Environmental Issues," in Clifford, *Environmental Crime,* 315–340: 315.

118. William J. Clinton, "The President's Radio Address," January 25, 1997, in *Public Papers of the Presidents of the United States;* John T. Woolley and Gerhard Peters, *The American Presidency Project,* http://www.presidency.ucsb.edu//ws/?pid=54416 (accessed May 12, 2011).

119. William J. Clinton, "The President's Radio Address," November 4, 1995, in *Public Papers of the Presidents of the United States;* John T. Woolley and Gerhard Peters, *The American Presidency Project,* http://www.presidency.ucsb.edu//ws/?pid=50736 (accessed May 12, 2011).

120. William J. Clinton, "Remarks to the Community in Louisville," January 24, 1996, in *Public Papers of the Presidents of the United States;* John T. Woolley and Gerhard Peters, *The American Presidency Project,* http://www.presidency.ucsb.edu//ws/?pid=53125 (accessed May 12, 2011).

121. William J. Clinton, "Remarks to the Community in Hackensack, New Jersey," March 11, 1996, in *Public Papers of the Presidents of the United States;* John T. Woolley and Gerhard Peters, *The American Presidency Project,* (http://www.presidency.ucsb.edu//ws/?pid=52526 (accessed May 12, 2011); William J. Clinton, "Remarks in Kalamazoo, Michigan," August 28, 1996, in *Public Papers of the Presidents of the United States;* John T. Woolley and Gerhard Peters, *The American Presidency Project,* http://www.presidency.ucsb.edu//ws/?pid=53248 (accessed May 12, 2011).

122. William J. Clinton, "Remarks to the Community in Los Alamos, New Mexico," May 17, 1993, in *Public Papers of the Presidents of the United States;* John T. Woolley and Gerhard Peters, *The American Presidency Project,* Available at http://www.presidency.ucsb.edu//ws/?pid=46468 (accessed May 12, 2011).

123. William J. Clinton, "Remarks on the Federal Fleet Conversion to Alternative Fuel Vehicles," December 8, 1993, in *Public Papers of the Presidents of the United States;* John T. Woolley and Gerhard Peters, *The American Presidency Project,* http://www.presidency.ucsb.edu//ws/?pid=46221 (accessed May 12, 2011).

124. William J. Clinton, "The President's Radio Address," May 1, 1999, in *Public Papers of the Presidents of the United States;* John T. Woolley and Gerhard Peters, *The American Presidency Project,* http://www.presidency.ucsb.edu//ws/?pid=57503 (accessed May 12, 2011).

125. William J. Clinton, "Statement on the Environmental Protection Agency Proposal to Reduce Emissions from Trucks and Busses," May 17, 2000, in *Public*

Papers of the Presidents of the United States; John T. Woolley and Gerhard Peters, *The American Presidency Project,* http://www.presidency.ucsb.edu//ws/?pid=51630 (accessed May 12, 2011).

126. William J. Clinton, "Remarks at the National Geographic Society," June 11, 1998, in *Public Papers of the Presidents of the United States;* John T. Woolley and Gerhard Peters, *The American Presidency Project,* http://www.presidency.ucsb.edu//ws/?pid=56121 (accessed May 12, 2011); William J. Clinton, "Remarks at 'Strengthening Democracy in the Global Economy: An Opening Dialogue' in New York City," September 21, 1998, in *Public Papers of the Presidents of the United States;* John T. Woolley and Gerhard Peters, *The American Presidency Project,* http://www.presidency.ucsb.edu//ws/?pid=51630 (accessed May 12, 2011).

127. William J. Clinton, "Proclamation 6664—Cancer Control Month, 1994," April 7, 1994, in *Public Papers of the Presidents of the United States;* John T. Woolley and Gerhard Peters, *The American Presidency Project,* http://www.presidency.ucsb.edu//ws/?pid=49924 (accessed May 12, 2011).

128. William J. Clinton, "Statement on the Departments of Veterans Affairs and Housing and Urban Development Appropriations Legislation," July 18, 1995, in *Public Papers of the Presidents of the United States;* John T. Woolley and Gerhard Peters, *The American Presidency Project,* http://www.presidency.ucsb.edu//ws/?pid=51630 (accessed May 12, 2011).

129. William J. Clinton, "Address Before a Joint Session of the Congress on the State of the Union," January 23, 1996, in *Public Papers of the Presidents of the United States;* John T. Woolley and Gerhard Peters, *The American Presidency Project,* http://www.presidency.ucsb.edu//ws/?pid=53091 (accessed May 12, 2011).

130. William J. Clinton, "Remarks to the Florida State Democratic Convention in Orlando, Florida," December 11, 1999, in *Public Papers of the Presidents of the United States;* John T. Woolley and Gerhard Peters, *The American Presidency Project,* http://www.presidency.ucsb.edu//ws/?pid=57059 (accessed May 12, 2011); William J. Clinton, "Remarks at a Democratic National Committee Luncheon in Los Angeles, California," January 22, 2000, in *Public Papers of the Presidents of the United States;* John T. Woolley and Gerhard Peters, *The American Presidency Project,* http://www.presidency.ucsb.edu//ws/?pid=58664 (accessed May 12, 2011).

131. "Senate Sought to Reduce Indoor Air Pollution," CQ Press Electronic Library, CQ Almanac Online Edition, cqal91-1110464, http://library.cqpress.com/cqalmanac/cqal91-1110464 (accessed August 23, 2010), originally published in *CQ Almanac 1991* (Washington, DC: Congressional Quarterly, 1992).

132. Kidd and Kidd, *Air Pollution:* 125.

133. George H. W. Bush, "Remarks on Signing the Giant Sequoia in National Forests Proclamation in Sequoia National Forest," July 14, 1992, in *Public Papers of the President of the United States* (Washington, DC: U.S. Government Printing Office, 1993), 1113–1115.

134. "Bush Proposes Changes to Air Pollution Rules," *Congressional Quarterly Almanac* 56 (2003): 9–14; George W. Bush, "Remarks on Earth Day in Wilmington, New York," April 22, 2002, in *Public Papers of the President of the United States* (Washington, DC: U.S. Government Printing Office, 2003), 645–648; George W.

Bush, "Remarks Announcing the Clear Skies and Global Climate Change Initiatives in Silver Spring, Maryland," February 14, 2002, in *Public Papers of the President of the United States* (Washington, DC: U.S. Government Printing Office, 2003), 226–231; George W. Bush, "Statement on the Clear Skies Initiative," July 1, 2002, in *Public Papers of the President of the United States* (Washington, DC: U.S. Government Printing Office, 2003), 1160; George W. Bush, "Remarks on the Proposed Clear Skies Legislation," September 16, 2003, in *Public Papers of the President of the United States* (Washington, DC: U.S. Government Printing Office, 2004), 1170–1173.

135. George W. Bush, "Message to the Senate Transmitting the Protocol of 1997 to Amend the International Convention for the Prevention of Pollution from Ships," May 15, 2003, in *Public Papers of the President of the United States* (Washington, DC: U.S. Government Printing Office, 2004), 496–497.

136. George W. Bush, "Remarks on Energy Policy in Columbus, Ohio," March 9, 2005, in *Public Papers of the President of the United States* (Washington, DC: U.S. Government Printing Office, 2006), 389–396; George W. Bush, "Remarks to the 16th Annual Energy Efficiency Forum," June 15, 2005, in *Public Papers of the President of the United States* (Washington, DC: U.S. Government Printing Office, 2006), 993–999.

137. Kidd and Kidd, *Air Pollution:* 116.

138. "Hill Cool to Global Warming Pact," *Congressional Quarterly Almanac* 53 (1997): 4-13–4-15.

139. "Environmental Agency Tightens Clean Air Rules," CQ Press Electronic Library, CQ Almanac Online Edition, cqal97-0000181000, http://library.cqpress.com/cqalmanac/cqal97-0000181000 (accessed August 23, 2010, originally published in *CQ Almanac 1997* (Washington, DC: Congressional Quarterly, 1998).

140. Rosemary O'Leary, "Environmental Policy in the Courts," in Vig and Kraft, *Environmental Policy: New Directions,* 148–168: 156–157.

141. Pompe and Rinehart, *Environmental Conflict:* 124.

142. Michael E. Kraft, "Environmental Policy in Congress," in Vig and Kraft, *Environmental Policy: New Directions,* 124–147: 140–141.

143. "Energy Overhaul Includes Many Bush Priorities—but Not ANWR," *Congressional Quarterly Almanac* 61 (2005): 8-3–8-20.

144. Shawn Zeller, "EPA Might Allow More Radiation," *CQ Weekly Online,* January 11, 2010: 95, http://library.cqpress.com/cqweekly/weeklyreport111-000003276726 (accessed September 2, 2010).

145. Joseph J. Schatz and Geof Koss, "BP Pressed, Energy Bill Pushed," *CQ Weekly Online,* June 21, 2010: 1506–1507, http://library.cqpress.com/cqweekly/weeklyreport111-000003687060 (accessed September 2, 2010).

146. Coral Davenport, "Its Windy, but What about the View," *CQ Weekly Online,* September 7, 2009: 1938, http://library.cqpress.com/cqweekly/weeklyreport111-000003197308 (accessed September 2, 2010).

147. Adriel Bettelheim, "Empty-Handed before the World on Cap-and-Trade," *CQ Weekly Online,* September 28, 2009: 2146–2153, http://library.cqpress.com/cqweekly/weeklyreport111-000003211645 (accessed September 2, 2010).

148. Coral Davenport, "2009 Key House Votes: 477: Climate Change Mitigation," *CQ Weekly Online,* January 4, 2010: 64, http://library.cqpress.com/cqweekly/weeklyreport111-000003274936 (accessed September 2, 2010).

149. Coral Davenport, "Selling a Climate Change Bill to Coal Country," *CQ Weekly Online,* February 15, 2010: 385–386, http://library.cqpress.com/cqweekly/weeklyreport111-000003293756 (accessed September 2, 2010).

150. Coral Davenport, "Climate Change Pledge Is No Easy Sell on Hill," *CQ Weekly Online,* January 25, 2010: 211–212, http://library.cqpress.com/cqweekly/weeklyreport111-000003283402 (accessed September 2, 2010).

151. Joseph J. Schatz, "House Votes to Top Off Auto Program After Funds Run Out in Only a Week," *CQ Weekly Online,* August 3, 2009: 1853, http://library.cqpress.com/cqweekly/weeklyreport111-000003184523 (accessed September 2, 2010).

152. Shawn Zeller, "Keeping Clunkers on the Road, or at Least in the Junk Yard," *CQ Weekly Online,* April 27, 2009: 957, http://library.cqpress.com/cqweekly/weeklyreport111-000003103287 (accessed September 2, 2010).

153. Colby Itkowitz, "A Very Little Sum to Go a Long Way," *CQ Weekly Online,* June 29, 2009: 1501, http://library.cqpress.com/cqweekly/weeklyreport111-000003155530 (accessed September 2, 2010); Davenport, "2009 Key House Votes": 64; Coral Davenport, "2009 Legislative Summary: Climate Change Mitigation," *CQ Weekly Online,* January 4, 2010: 42, http://library.cqpress.com/cqweekly/weeklyreport111-000003274891 (accessed September 2, 2010); Avery Palmer, "Fall 2009 Outlook: Climate Change Mitigation," *CQ Weekly Online,* September 7, 2009: 1970, http://library.cqpress.com/cqweekly/weeklyreport111-000003197345 (accessed September 2, 2010); Coral Davenport and Avery Palmer, "A Landmark Climate Bill Passes," *CQ Weekly Online,* June 29, 2009: 1516–1517, http://library.cqpress.com/cqweekly/weeklyreport111-000003155539 (accessed September 2, 2010).

154. Coral Davenport, "The Graham Effect on Climate Change," *CQ Weekly Online,* May 3, 2010: 1094, http://library.cqpress.com/cqweekly/weeklyreport111-000003653776 (accessed September 2, 2010); "Key Parts of Senate Climate Change Draft," *CQ Weekly Online,* May 17, 2010: 1222, http://library.cqpress.com/cqweekly/weeklyreport111-000003664056 (accessed September 2, 2010).

155. Coral Davenport, "Emissions Bill Unveiled as Rough Draft," *CQ Weekly Online,* October 5, 2009: 2237, http://library.cqpress.com/cqweekly/weeklyreport111-000003216217 (accessed September 2, 2010); Coral Davenport, "Climate Bill Sponsors Commence Public Hearings, Private Negotiations," *CQ Weekly Online,* November 2, 2009: 2520–2521, http://library.cqpress.com/cqweekly/weeklyreport111-000003236318 (accessed September 2, 2010).

156. Jennifer Scholtes, "Bill to Fund Cleanup of Highly Polluted Areas Advances," *CQ Weekly Online,* May 24, 2010: 64, http://library.cqpress.com/cqweekly/weeklyreport111-000003669541 (accessed September 2, 2010).

157. Geof Koss, "Senate Turns Back GOP Attempt to Reject EPA Emissions Regulation," *CQ Weekly Online,* June 14, 2010: 1460, http://library.cqpress.com/cqweekly/weeklyreport111-000003681826 (accessed September 2, 2010).

158. Jennifer Scholtes, "Formaldehyde Bill Sets New Emissions Standards," *CQ Weekly Online,* June 21, 2010: 1512, http://library.cqpress.com/cqweekly/weekly

report111-000003687064 (accessed September 2, 2010). The House passed the measure a few weeks later. Jennifer Scholtes, "Formaldehyde Emission Limits Cleared for President," *CQ Weekly Online,* June 28, 2010: 1581, http://library.cqpress.com/cqweekly/weeklyreport111-000003691767 (accessed September 2, 2010).

159. Kraft, *Environmental Policy:* 27.

160. Kidd and Kidd, *Air Pollution:* xi.

161. Ibid., xi.

162. Ibid., xii.

Chapter 3 The Politics of Water Pollution

1. Daniel D. Chiras, *Beyond the Fray* (Boulder, CO: Johnson Books, 1990), 1–2.

2. Joanna Burger, *Oil Spills* (Piscataway, NJ: Rutgers University Press, 1997).

3. Ibid.

4. Michael E. Kraft, *Environmental Policy and Politics* (New York: Longman, 2001), 34.

5. Robert A. Shanley, *Presidential Influence and Environmental Policy* (Westport CT: Greenwood Press, 1992), 16.

6. Harry S. Truman, "Special Message to the Congress Recommending a Comprehensive Health Program," November 19, 1945, in *Public Papers of the President of the United States* (Washington, DC: U.S. Government Printing Office, 1946), 479.

7. Harry S. Truman, "Message to the Congress on the State of the Union and on the Budget for 1947," January 21, 1946, in *Public Papers of the President of the United States* (Washington, DC: U.S. Government Printing Office, 1947), 83.

8. Harry S. Truman, "The President's News Conference of April 24," April 24, 1947, in *Public Papers of the President of the United States* (Washington, DC: U.S. Government Printing Office, 1948), 218.

9. Harry S. Truman, "Memorandum of Disapproval of Bill Relating to Garbage Originating outside the Continental United States," August 7, 1947, in *Public Papers of the President of the United States* (Washington, DC: U.S. Government Printing Office, 1948), 372–373.

10. Harry S. Truman, "Special Message to the Congress Summarizing the New Reorganization Plans," March 13, 1950, in *Public Papers of the President of the United States* (Washington, DC: U.S. Government Printing Office, 1951), 195–199; Harry S. Truman, "Special Message to the Congress Transmitting Reorganization Plans 15, 16 and 17 of 1950," March 13, 1950, in *Public Papers of the President of the United States* (Washington, DC: U.S. Government Printing Office), 211–215; Harry S. Truman, "Special Message to the Congress Transmitting Reorganization Plan 16 of 1950," March 13, 1950, in *Public Papers of the President of the United States* (Washington, DC: U.S. Government Printing Office), 216.

11. "Reorganization Plans," *Congressional Quarterly Almanac* 6 (1950), 362–374: 370.

12. "Water Pollution," *Congressional Quarterly Almanac* 4 (1948): 152; Mary Clifford, "A Review of Federal Environmental Legislation," in Mary Clifford, ed., *Environmental Crime* (Gaithersburg, MD: Aspen, 1998), 95–109: 99.

13. Clarence J. Davies and Barbara S. Davies, *The Politics of Pollution* (Indianapolis: Pegasus, 1975), 29.

14. David Zwick, "Water Pollution," in James Rathlesberger, ed., *Nixon and the Environment: The Politics of Devastation* (New York: Village Voice, 1972), 30–58: 37–38.

15. Davies and Davies, *Politics of Pollution:* 29.

16. Dwight D. Eisenhower, "Special Message to the Congress Recommending a Health Program," January 31, 1955, in *Public Papers of the President of the United States* (Washington, DC: U.S. Government Printing Office, 1956), 216–223: 221; Dwight D. Eisenhower, "Special Message to the Congress on the Nation's Health Program," January 26, 1956, in *Public Papers of the President of the United States* (Washington, DC: U.S. Government Printing Office), 196–204: 202.

17. Dwight D. Eisenhower, "Statement by the President Upon Signing the Water Pollution Act Amendments of 1956," July 9, 1956, in *Public Papers of the President of the United States* (Washington, DC: U.S. Government Printing Office, 1957), 592–593.

18. Dwight D. Eisenhower, "Letter to the Speaker of the House of Representatives Urging Legislation to Carry Out Recommendations of the Joint Federal-State Action Committee," May 14, 1958, in *Public Papers of the President of the United States* (Washington, DC: U.S. Government Printing Office, 1959), 408–410: 409.

19. Dwight D. Eisenhower, "Annual Budget Message to the Congress," January 18, 1960, in *Public Papers of the President of the United States* (Washington, DC: U.S. Government Printing Office), 37–110: 94.

20. Dwight D. Eisenhower, "Veto of Bill to Amend the Federal Water Pollution Control Act," February 23, 1960, in *Public Papers of the President of the United States* (Washington, DC: U.S. Government Printing Office, 1961), 208–210.

21. Dwight D. Eisenhower, "The President's News Conference of October 3, 1957," in *Public Papers of the President of the United States* (Washington, DC: U.S. Government Printing Office, 1958), 704–716: 714.

22. "Water Pollution," *Congressional Quarterly Almanac* 12 (1956): 570–573

23. Davies and Davies, *Politics of Pollution:* 29.

24. Dwight Eisenhower, "Annual Budget Message to the Congress—Fiscal Year 1959," January 13, 1958, in *Public Papers of the President of the United States* (Washington, DC: U.S. Government Printing Office, 1959); John T. Woolley and Gerhard Peters, *The American Presidency Project,* http://www.presidency.ucsb.edu/ws/?pid=11323 (accessed May 16, 2011).

25. "'Socialism' Repulsed in Medical Aid Battle," *Congressional Quarterly Almanac* 16 (1960): 77, 250–251; "Water Pollution," CQ Press Electronic Library, CQ Almanac Online Edition, cqal59-1335396, http://library.cqpress.com/cqal59-1335396 (accessed August 23, 2010), originally published in *CQ Almanac* 1959 (Washington, DC: Congressional Quarterly, 1960).

26. John F. Kennedy, "Special Message to the Congress: Program for Economic Recovery and Growth," February 2, 1961, in *Public Papers of the President*

of the United States (Washington, DC: U.S. Government Printing Office, 1962), 41–53: 51.

27. "Message on Natural Resources Policy," *Congressional Quarterly Almanac* 17 (1961): 876–878; John F. Kennedy, "Special Message to the Congress on Natural Resources," February 23, 1961, in *Public Papers of the President of the United States* (Washington, DC: U.S. Government Printing Office, 1962), 114–121.

28. Kennedy, "Special Message on Natural Resources."

29. John F. Kennedy, "Remarks upon Signing the Federal Water Pollution Control Act Amendments," July 20, 1961, in *Public Papers of the President of the United States* (Washington, DC: U.S. Government Printing Office, 1962), 524–525.

30. John F. Kennedy, "President Seeks Conservation Program," CQ Press Electronic Library, CQ Almanac Online Edition, cqal62-878-28126-1324166, http://library.cqpress.com/cqal62-878-28126-1324166 (accessed August 23, 2010), originally published in *CQ Almanac 1962* (Washington, DC: Congressional Quarterly, 1963).

31. "Water Pollution Program Increased," CQ Press Electronic Library, CQ Almanac Online Edition, cqal61-1373115, http://library.cqpress.com/cqal61-1373115 (accessed August 23, 2010), originally published in *CQ Almanac 1961* (Washington, DC: Congressional Quarterly, 1961).

32. "Water Pollution Program Increased."

33. Lyndon B. Johnson, "Remarks at the University of Michigan," May 22, 1964, in *Public Papers of the President of the United States* (Washington, DC: U.S. Government Printing Office, 1965), 704–707: 705.

34. Ibid., 706.

35. Lyndon B. Johnson, "Remarks at a Meeting of the Water Emergency Conference," August 11, 1965, in *Public Papers of the President of the United States* (Washington, DC: U.S. Government Printing Office), 867–871.

36. "Review of the Session," *Congressional Quarterly Almanac* 20 (1964): 78.

37. "Water Pollution: Committees Investigate Many Aspects," CQ Press Electronic Library, CQ Almanac Online Edition, cqal70-1293767, http://library.cqpress.com/cqal70-1293767 (accessed August 23, 2010), originally published in *CQ Almanac 1970* (Washington, DC: Congressional Quarterly, 1971).

38. Lyndon B. Johnson, "Annual Message to the Congress: The Economic Report of the President," January 27, 1966, in *Public Papers of the President of the United States* (Washington, DC: U.S. Government Printing Office, 1967), 115.

39. "Natural Beauty Sought as U.S. Goal," *Congressional Quarterly Almanac* 25 (1965): 722–723.

40. Lyndon B. Johnson, "Special Message to the Congress on Conservation and Restoration of Natural Beauty," February 8, 1965, in *Public Papers of the President of the United States* (Washington, DC: U.S. Government Printing Office, 1966), 155–165.

41. Lyndon B. Johnson, "State of Union," January 12, 1966, in *Public Papers of the President of the United States* (Washington, DC: U.S. Government Printing Office, 1967), 3–12; "Text of President Johnson's January 27 Economic Report,"

Congressional Quarterly Almanac 22 (1966): 1226; "Text of President Johnson's Natural Beauty Message," *Congressional Quarterly Almanac* 22 (1966): 1265–1268; Johnson, "Economic Report of the President," 195–203.

42. Lyndon Johnson, "Special Message to the Congress on Conservation: 'To Renew a Nation,'" March 8, 1968, in *Public Papers of the President of the United States* (Washington, DC: U.S. Government Printing Office, 1969), 355–370: 358

43. Lyndon Johnson, "Special Message to the Congress Proposing Measures to Preserve America's Natural Heritage," February 23, 1966, in *Public Papers of the Presidents of the United States* (Washington, DC: U.S. Government Printing Office, 1967), 195–203: 196.

44. Lyndon Johnson, "Special Message to the Congress on Urban Problems," February 23, 1968, in *Public Papers of the Presidents of the United States* (Washington, DC: U.S. Government Printing Office, 1969), 248–258.

45. "Message to Congress: Johnson's Conservation Message," CQ Press Electronic Library, CQ Almanac Online Edition, cqal68-1124522, http://library.cqpress.com/cqal68-1124522 (accessed August 23, 2010), originally published in *CQ Almanac 1968* (Washington, DC: Congressional Quarterly, 1969).

46. "What Congress Did Not Do," *Congressional Quarterly Almanac* 19 (1963): 81.

47. "Water Pollution," *Congressional Quarterly Almanac* 19 (1963): 240.

48. "1964 Action on Public Lands, Water and Reclamation," *Congressional Quarterly Almanac* 20 (1964): 501; "Water Pollution," CQ Press Electronic Library, CQ Almanac Online Edition, cqal63-1316985, http://library.cqpress.com/cqal63-1316985 (accessed August 23, 2010), originally published in *CQ Almanac 1963* (Washington, DC: Congressional Quarterly, 1964).

49. "1964 Action": 501.

50. Henry P. Caulfield, "The Conservation and Environmental Movements: An Historical Analysis," in James P. Lester, ed., *Environmental Politics and Policy* (Durham, NC: Duke University Press, 1989), 13–56: 35.

51. Zwick, "Water Pollution": 38.

52. "Water Quality Improvement Act," CQ Press Electronic Library, CQ Almanac Online Edition, cqal69-1248469, http://library.cqpress.com/cqal69-1248469 (accessed August 23, 2010), originally published in *CQ Almanac 1969* (Washington, DC: Congressional Quarterly, 1970).

53. "Review of the Session," *Congressional Quarterly Almanac* 21 (1965): 76; "Anti-Water Pollution Law Strengthened," *Congressional Quarterly Almanac* 21 (1965): 743–750.

54. "Review of the Session," *Congressional Quarterly Almanac* 22 (1966): 77; "Water Pollution Control Funds Expanded," *Congressional Quarterly Almanac* 22 (1966): 632–645.

55. "Water Quality Improvement Act," CQ Press Electronic Library, CQ Almanac Online Edition, cqal69-1248469, http://library.cqpress.com/cqal69-1248469 (accessed August 23, 2010), originally published in *CQ Almanac 1969* (Washington, DC: Congressional Quarterly, 1970).

56. "Review of the Session" (1965): 76.

57. "Congress Strengthens Air Pollution Control Measures," *Congressional Quarterly Almanac* 23 (1967): 877.

58. "Congress Acts on Pollution but Clears No Major Bills," *Congressional Quarterly Almanac* 24 (1968): 569–575.

59. "Senate Passes Bill to Combat Lake, Mine, Oil Pollution," *Congressional Quarterly Almanac* 23 (1967): 1006–1009; "Senate Passes Bill to Combat Lake, Mine, Oil Pollution," CQ Press Electronic Library, CQ Almanac Online Edition, cqal67-1313307, http://library.cqpress.com/cqal67-1313307 (accessed August 23, 2010).

60. "Resources and Public Works: Action Not Completed," *Congressional Quarterly Almanac* 24 (1968): 83.

61. "Water Commission," *Congressional Quarterly Almanac* 23 (1967): 1022–1023; "Review of the Session," *Congressional Quarterly Almanac* 24 (1968): 82.

62. "Review of the Session" (1968): 82.

63. "Pollution Problems," CQ Press Electronic Library, CQ Almanac Online Edition, cqal69-1248491, http://library.cqpress.com/cqal69-1248491 (accessed August 23, 2010), originally published in *CQ Almanac 1969* (Washington, DC: Congressional Quarterly, 1970).

64. "Our Environment Year 1971," *Congressional Quarterly Almanac* 26 (1970): 46–68.

65. Richard M. Nixon, "Remarks Following Inspection of Oil Damage at Santa Barbara Beach," March 21, 1969, in *Public Papers of the President of the United States* (Washington, DC: U.S. Government Printing Office, 1970), 233–234.

66. Richard M. Nixon, "Annual Budget Message to the Congress: Fiscal Year 1971," February 2, 1970, in *Public Papers of the President of the United States* (Washington, DC: U.S. Government Printing Office, 1971), 46–68.

67. Richard M. Nixon, "Remarks to the North Atlantic Council in Brussels," February 24, 1969, in *Public Papers of the President of the United States* (Washington, DC: U.S. Government Printing Office 1970), 134–136.

68. Richard M. Nixon, "Statement about United States Oceans Policy," May 23, 1970, in *Public Papers of the President of the United States* (Washington, DC: U.S. Government Printing Office, 1971), 454–456.

69. "Presidential Statement to Congress: Nixon's 1970 Environmental Message," CQ Press Electronic Library, CQ Almanac Online Edition, cqal71-869-26707-1254975, http://library.cqpress.com/cqal71-869-26707-1254975 (accessed August 23, 2010), originally published in *CQ Almanac 1971* (Washington, DC: Congressional Quarterly, 1972).

70. Ibid.

71. Richard Nixon, "Special Message to the Congress Proposing the 1971 Environment Program," February 8, 1971, in *Public Papers of the President of the United States* (Washington, DC: U.S. Government Printing Office, 1972), 125–142: 129.

72. "Presidential Statement to Congress: Nixon's 1970 Environmental Message."

73. Richard Nixon, "Special Message to the Congress on Marine Pollution From Oil Spills," May 20, 1970, in *Public Papers of the President of the United States* (Washington, DC: U.S. Government Printing Office, 1971), 443–447: 443.

74. "Oil Pollution Text," CQ Press Electronic Library, CQ Almanac Online Edition, cqal70-1290868, http://library.cqpress.com/cqal70-1290868 (accessed August 23, 2010), originally published in *CQ Almanac 1970* (Washington, DC: Congressional Quarterly, 1971).

75. "President Nixon's Message on the Environment," *Congressional Quarterly Almanac* 26 (1970): 22A–27A.

76. Richard M. Nixon, "Annual Message to the Congress on the State of the Union," January 22, 1970, in *Public Papers of the President of the United States* (Washington, DC: U.S. Government Printing Office 1971), 8–16.

77. Richard M. Nixon, "Special Message to the Congress on Environmental Quality," February 10, 1970, in *Public Papers of the President of the United States* (Washington, DC: U.S. Government Printing Office, 1971), 96–109; Nixon, "Annual Budget Message: 1971": 46–68.

78. Richard M. Nixon, "Special Message on Environmental Quality."

79. Richard M. Nixon, "Special Message to the Congress about Waste Disposal," April 15, 1970, in *Public Papers of the President of the United States* (Washington, DC: U.S. Government Printing Office, 1971), 357–359.

80. "Presidential Statement to Congress: Nixon on the Environment 1970," CQ Press Electronic Library, CQ Almanac Online Edition, cqal70-1290732, http://library.cqpress.com/cqal70-1290732 (accessed August 23, 2010), originally published in *CQ Almanac 1970* (Washington, DC: Congressional Quarterly, 1971).

81. "Nixon's First Year," *Congressional Quarterly Almanac* 25 (1969): 107–114: 111.

82. "1969—The Year in Review," *Congressional Quarterly Almanac* 25 (1969): 77–97: 79.

83. "Water Pollution: Committees Investigate Many Aspects," *Congressional Quarterly Almanac* 26 (1970): 489–494.

84. "Legislative Review," *Congressional Quarterly Almanac* 26 (1970): 74; "Environment: Action Completed 1970," CQ Almanac Online, cqal69-1248469, http://library.cqpress.com/cqal69-1248469 (accessed August 23, 2010), originally published in *CQ Almanac 1969* (Washington, DC: Congressional Quarterly, 1970); "Environment: Action Not Completed," *Congressional Quarterly Almanac* 25 (1969): 86; "Environmental Problems," *Congressional Quarterly Almanac* 25 (1969): 513; "Water Quality Improvement Act," *Congressional Quarterly Almanac* 25 (1969): 513–521; "Comprehensive Water Pollution Control Act Cleared," CQ Press Electronic Library, CQ Almanac Online Edition, cqal70-1292833, http://library.cqpress.com/cqal70-1292833 (accessed August 23, 2010), originally published in *CQ Almanac 1970* (Washington, DC: Congressional Quarterly, 1971).

85. "Safe Drinking Water," CQ Press Electronic Library, CQ Almanac Online Edition, cqal72-1251151, http://library.cqpress.com/cqal72-1251151 (accessed August 23, 2010), originally published in *CQ Almanac 1972* (Washington, DC: Congressional Quarterly, 1973).

86. "Water Pollution," CQ Press Electronic Library, CQ Almanac Online Edition, cqal76-1189238, http://library.cqpress.com/cqal76-1189238 (accessed August 23, 2010), originally published in *CQ Almanac 1976* (Washington, DC: Congressional Quarterly, 1977).

87. Phillip F. Cramer, *Deep Environmental Politics* (Westport, CT: Praeger, 1998), citing Water Pollution Prevention and Control, U.S. Code, title 33 (Navigation and Navigable Waters), chapter 26, section 1251.

88. Cramer, *Deep Environmental Politics:* 91–92, citing Water Pollution Prevention and Control, U.S. Code, title 33 (Navigation and Navigable Waters), chapter 26, section 1252.

89. Walter A. Rosenbaum, *Environmental Politics and Policy,* 6th ed. (Washington, DC: CQ Press, 2005), 195–196.

90. Daniel J. Fiorino, *Making Environmental Policy* (Berkeley: University of California Press, 1995), 28–29.

91. Rosenbaum, *Environmental Politics:* 195–196.

92. Cramer, *Deep Environmental Politics:* 91, citing Water Pollution Prevention and Control, U.S. Code, title 33 (Navigation and Navigable Waters), chapter 26, section 1251.

93. "Environment and Resources: Action Completed," *Congressional Quarterly Almanac* 28 (1972): 17; "Clean Water: Congress Overrides Presidential Veto," *Congressional Quarterly Almanac* 28 (1972): 708–722; John H. Baldwin, *Environmental Planning and Management* (Boulder, CO: Westview Press, 1985), 169; Rosenbaum, *Environmental Politics:* 195–196.

94. Rosenbaum, *Environmental Politics:* 195–196.

95. "Water Pollution: Senate Votes $16.8 Billion," CQ Press Electronic Library, CQ Almanac Online Edition, cqal71-1254318, http://library.cqpress.com/cqal71-1254318 (accessed August 23, 2010), originally published in *CQ Almanac 1971* (Washington, DC: Congressional Quarterly, 1972); "Water Pollution Control," CQ Press Electronic Library, CQ Almanac Online Edition, cqal72-1250338, http://library.cqpress.com/cqal71–1250338 (accessed August 23, 2010), originally published in *CQ Almanac 1972* (Washington, DC: Congressional Quarterly, 1973).

96. Cramer, *Deep Environmental Politics:* 90; Water Pollution Prevention and Control, U.S. Code, title 33 (Navigation and Navigable Waters), chapter 26, section 1251.

97. "Environment: Action Not Completed," *Congressional Quarterly Almanac* 27 (1971): 29; "Environment," *Congressional Quarterly Almanac* 27 (1971): 48; "Water Pollution: Senate Votes $16.8 Billion."

98. Richard M. Nixon, "Veto of the Federal Water Pollution Control Act Amendments of 1972," October 17, 1972, in *Public Papers of the President of the United States* (Washington, DC: U.S. Government Printing Office, 1973), 990–993.

99. Gerald R. Ford, "Remarks and a Question-and-Answer Session at the University of New Hampshire in Durham," February 8, 1976, in *Public Papers of the President of the United States* (Washington, DC: U.S. Government Printing Office, 1977), 215–233: 227.

100. Gerald R. Ford, "Statement on Signing the Bill Providing for Loan Guarantees for Construction of Municipal Waste Water Treatment Plants," October 20, 1976, in *Public Papers of the President of the United States* (Washington, DC: U.S. Government Printing Office, 1977), 2584.

101. Richard M. Nixon, "Veto of the Rural Water and Sewer Grant Program Bill," April 5, 1973, in *Public Papers of the President of the United States* (Washington, DC: U.S. Government Printing Office, 1974), 254–256.

102. Andrews, *Managing the Environment:* 242.

103. "Safe Drinking Water," *Congressional Quarterly Almanac* 29 (1973): 516.

104. Gerald R. Ford, "Statement on Signing the Safe Drinking Water Act," December 17, 1974, in *Public Papers of the President of the United States* (Washington, DC: U.S. Government Printing Office), 759.

105. "Water Pollution: New Funding Formula Cleared," *Congressional Quarterly Almanac* 29 (1973): 658–670.

106. Ford, "Safe Drinking Water Act."

107. "Water Quality Report," CQ Press Electronic Library, CQ Almanac Online Edition, cqal76-1189251, http://library.cqpress.com/cqal76-1189251 (accessed August 23, 2010), originally published in *CQ Almanac 1977* (Washington, DC: Congressional Quarterly, 1977).

108. Jimmy Carter, "The Environment," May 23, 1977, in *Public Papers of the President of the United States* (Washington, DC: U.S. Government Printing Office, 1978), 967–986.

109. Jimmy Carter, "State of the Union Address," January 21, 1980, in *Public Papers of the President of the United States* (Washington, DC: U.S. Government Printing Office, 1981), 114–180: 159.

110. "Carter's Second Session Agenda," January 21, 1980, CQ Press Electronic Library, CQ Almanac Online Edition, cqal80-860-25882-1174115, http://library.cqpress.com/cqalmanac/cqal80-860-25882-1174115 (accessed August 30, 2010), originally published in *CQ Almanac 1980* (Washington, DC: Congressional Quarterly, 1981).

111. "Energy and Environment: Water Pollution," *Congressional Quarterly Almanac* 33 (1977): 21; Jimmy Carter, "Statement on Signing HR 3199 into Law," December 28, 1977, in *Public Papers of the President of the United States* (Washington, DC: U.S. Government Printing Office, 1978), 2179–2180.

112. "Water Pollution Compromise Enacted," *Congressional Quarterly Almanac* 33 (1977): 697–703.

113. Fiorino, *Making Environmental Policy:* 30.

114. Baldwin, *Environmental Planning and Management:* 169.

115. "Safe Drinking Water," CQ Press Electronic Library, CQ Almanac Online Edition, cqal77-1204122, http://library.cqpress.com/cqal77-1204122 (accessed August 23, 2010), originally published in *CQ Almanac 1977* (Washington, DC: Congressional Quarterly, 1978).

116. "Senate Judiciary Reports Criminal Code Bill," *Congressional Quarterly Almanac* 35 (1979), 363–369: 368.

117. "Energy and Environment: Action Not Completed," *Congressional Quarterly Almanac* 32 (1976): 11.

118. "Safe Drinking Water," *Congressional Quarterly Almanac* 35 (1979): 680.

119. "Waste Treatment Plant Costs," CQ Press Electronic Library, CQ Almanac Online Edition, cqal79-1184260, http://library.cqpress.com/cqal79-1184260

(accessed August 23, 2010), originally published in *CQ Almanac 1979* (Washington, DC: Congressional Quarterly, 1980).

120. "Water Bank Funds," CQ Press Electronic Library, CQ Almanac Online Edition, cqal79-1184264, http://library.cqpress.com/cqal79-1184264 (accessed August 30, 2010), originally published in *CQ Almanac 1979* (Washington, DC: Congressional Quarterly, 1980).

121. Ronald W. Reagan, "Message to the Senate Returning without Approval the Water Resources Research Bill," February 21, 1984, in *Public Papers of the President of the United States* (Washington, DC: U.S. Government Printing Office, 1985), 243–244.

122. "No Clean Water Act Rewrite," *Congressional Quarterly Almanac* 38 (1982): 459.

123. "Water Board Authorization," *Congressional Quarterly Almanac* 37 (1981): 526–528.

124. "No Clean Water Act Rewrite": 459.

125. "Energy and Environment," *Congressional Quarterly Almanac* 37 (1981): 12A.

126. "House Approves Revision of Clean Water Act," CQ Press Electronic Library, CQ Almanac Online Edition, cqal84-1153057, http://library.cqpress.com/cqal84-1153057 (accessed August 23, 2010), originally published in *CQ Almanac 1984* (Washington, DC: Congressional Quarterly, 1985).

127. "Clean Water Act Rewrite," CQ Press Electronic Library, CQ Almanac Online Edition, cqal83-119317, http://library.cqpress.com/cqal83-119317 (accessed August 23, 2010), originally published in *CQ Almanac 1983* (Washington, DC: Congressional Quarterly, 1984).

128. "Drinking Water Bill Dies," CQ Press Electronic Library, CQ Almanac Online Edition, cqal84-1153065, http://library.cqpress.com/cqal84-1153065 (accessed August 23, 2010), originally published in *CQ Almanac 1984* (Washington, DC: Congressional Quarterly, 1985).

129. Ronald W. Reagan, "Statement on Signing the Safe Drinking Water Act Amendments of 1986," June 19, 1986, in *Public Papers of the President of the United States* (Washington, DC: U.S. Government Printing Office, 1987), 802; "Safe Drinking Water Protections Mandate Regulation of Chemical and Bacteriological Pollutants," CQ Press Electronic Library, CQ Almanac Online Edition, cqal85-1147693, http://library.cqpress.com/cqal85-1147693 (accessed August 23, 2010), originally published in *CQ Almanac 1985* (Washington, DC: Congressional Quarterly, 1986).

130. "Congress Strengthens Safe Drinking Water Act," CQ Press Electronic Library, CQ Almanac Online Edition, cqal86-1149878, http://library.cqpress.com/cqal86-1149878 (accessed August 23, 2010), originally published in *CQ Almanac 1986* (Washington, DC: Congressional Quarterly, 1987).

131. "Final Action Stalls on Clean Water Act Revision," *Congressional Quarterly Almanac* 41 (1985): 204–206.

132. "Reagan Vetoes $20 Billion Clean Water Bill," *Congressional Quarterly Almanac* 42 (1986): 136.

133. Ronald W. Reagan, "Memorandum of Disapproval of the Bill Amending the Clean Water Act," November 6, 1986, in *Public Papers of the President of the United States* (Washington, DC: U.S. Government Printing Office, 1987), 1529.

134. "Presidential Veto Message: Reagan Vetoes Clean Water Act," CQ Press Electronic Library, CQ Almanac Online Edition, cqal87-853-25674-1142889, http://library.cqpress.com/cqal87-853-25674-1142889 (accessed August 23, 2010), originally published in *CQ Almanac 1987* (Washington, DC: Congressional Quarterly, 1988).

135. "Environment: Clean Water," *Congressional Quarterly Almanac* 43 (1987): 16; "Congress Overrides Clean-Water Bill Veto," *Congressional Quarterly Almanac* 43 (1987): 291–296; Ronald W. Reagan, "Remarks on Signing the Message to the House of Representatives Returning without Approval the Water Quality Act of 1987," January 30, 1987, in *Public Papers of the President of the United States* (Washington, DC: U.S. Government Printing Office, 1988), 93–94.

136. Rosenbaum, *Environmental Politics:* 195–196.

137. Fiorino, *Making Environmental Policy:* 30.

138. "Groundwater Research," CQ Press Electronic Library, CQ Almanac Online Edition, cqal87-1145015, http://library.cqpress.com/cqal87-1145015 (accessed August 23, 2010), originally published in *CQ Almanac 1987* (Washington, DC: Congressional Quarterly, 1988).

139. "Lead in Drinking Water," CQ Press Electronic Library, CQ Almanac Online Edition, cqal88-1141456, http://library.cqpress.com/cqal88-1141456 (accessed August 23, 2010), originally published in *CQ Almanac 1988* (Washington, DC: Congressional Quarterly, 1989).

140. George H. W. Bush, "Message to the Congress Reporting on Environmental Quality," April 18, 1991, in *Public Papers of the President of the United States* (Washington, DC: U.S. Government Printing Office, 1992), 404–406.

141. George H. W. Bush, "Statement on Signing the Great Lakes Critical Programs Act of 1990," November 16, 1990, in *Public Papers of the President of the United States* (Washington, DC: U.S. Government Printing Office, 1991), 1609–1610.

142. "Oil Spill Liability, Prevention Bill Enacted," *Congressional Quarterly Almanac* 46 (1990): 283–286; George H. W. Bush, "Statement on Signing the Oil Pollution Act of 1990," August 18, 1990, in *Public Papers of the President of the United States* (Washington, DC: U.S. Government Printing Office, 1991), 1144–1145.

143. William J. Clinton, "The President's Radio Address," November 4, 1995, in *Public Papers of the Presidents of the United States*; John T. Woolley and Gerhard Peters, *The American Presidency Project,* http://www.presidency.ucsb.edu//ws/?pid=50736 (accessed May 5, 2011). William J. Clinton, "Proclamation 6664—Cancer Control Month, 1994," April 7, 1994, in *Public Papers of the Presidents of the United States*; John T. Woolley and Gerhard Peters, *The American Presidency Project,* http://www.presidency.ucsb.edu//ws/?pid=49924 (accessed May 5, 2011). William J. Clinton, "Remarks to the Community in Louisville," January 24, 1996, in *Public Papers of the Presidents of the United States*; John T. Woolley and Gerhard Peters, *The American Presidency Project,* http://www.presidency.ucsb.edu//ws/?pid=51630 (accessed May 5, 2011).

144. William J. Clinton, "Remarks on Clean Water Legislation, "May 30, 1995, in *Public Papers of the Presidents of the United States*; John T. Woolley and Gerhard Peters, *The American Presidency Project,* http://www.presidency.ucsb.edu//ws/?pid=51426 (accessed May 5, 2011).

145. William J. Clinton, "The President's Radio Address," August 14, 1999, in *Public Papers of the Presidents of the United States,* http://www.presidency.ucsb.edu//ws/?pid=51630 (accessed May 5, 2011).

146. William J. Clinton, "Statement on the Departments of Veterans Affairs and Housing and Urban Development Appropriations Legislation," July 18, 1995, in *Public Papers of the Presidents of the United States*; John T. Woolley and Gerhard Peters, *The American Presidency Project,* http://www.presidency.ucsb.edu//ws/?pid=51630 (accessed May 5, 2011). See also William J. Clinton, "Address Before a Joint Session of the Congress on the State of the Union," January 23, 1996, in *Public Papers of the Presidents of the United States*; John T. Woolley and Gerhard Peters, *The American Presidency Project,* http://www.presidency.ucsb.edu//ws/?pid=53091 (accessed May 5, 2011).

147. William J. Clinton, "Memorandum on Environmentally Beneficial Landscaping," April 26, 1994, in *Public Papers of the Presidents of the United States*; John T. Woolley and Gerhard Peters, *The American Presidency Project,* http://www.presidency.ucsb.edu//ws/?pid=51630 (accessed May 5, 2011).

148. William J. Clinton, "Proclamation 7250—America Recycles Day, 1999," November 15, 1999, in *Public Papers of the Presidents of the United States*; John T. Woolley and Gerhard Peters, *The American Presidency Project,* http://www.presidency.ucsb.edu//ws/?pid=56939 (accessed May 5, 2011).

149. "Hill Takes Up Wide Range of Environmental Issues," *Congressional Quarterly Almanac* 48 (1992): 289–295.

150. "Clean Water Rewrite Fails to Advance," *Congressional Quarterly Almanac* 49 (1993): 275–276; "No Update on Clean Water Law," *Congressional Quarterly Almanac* 50 (1994): 241; "Safe Drinking Water Overhaul Fails," *Congressional Quarterly Almanac* 48 (1992): 238–239.

151. "Clean Water Rewrite Stalls in Senate," *Congressional Quarterly Almanac* 51 (1995): 5-5–5-9.

152. "Drinking Water Act Wins Broad Support," *Congressional Quarterly Almanac* 52 (1996): 4-4–4-12; William J. Clinton, "Statement on Signing the Safe Drinking Water Act Amendments of 1996," August 6, 1996, in *Public Papers of the President of the United States* (Washington, DC: U.S. Government Printing Office, 1997), 1263–1264.

153. "Other Environmental Bills Considered in 1996: Batteries," *Congressional Quarterly Almanac* 52 (1996): 4–29.

154. William J. Clinton, "Statement on Signing the Coast Guard Authorization Act of 1996," October 19, 1996, in *Public Papers of the President of the United States* (Washington, DC: U.S. Government Printing Office, 1997), 1869–1870.

155. "Water Projects Bill Returned to Hill," *Congressional Quarterly Almanac* 63 (2007): D-12.

156. Shawn Zeller, "No Warmth for Airlines De-icing Limits," *CQ Weekly Online,* March 29, 2010: 730, http://library.cqpress.com/cqweekly/weeklyreport111-000003634195 (accessed September 2, 2010).

157. Michael Teitelbaum, "House-Passed Water Quality Package Would Fund Wastewater Treatment," *CQ Weekly Online,* March 16, 2009: 621, http://library.cqpress.com/cqweekly/weeklyreport111-000003075253 (accessed September 2, 2010).

158. Coral Davenport, "Hopes for Cleaner Fuel, Fears of Dirty Water," *CQ Weekly Online,* March 1, 2010: 475–476, http://library.cqpress.com/cqweekly/weeklyreport111-000003300845 (accessed September 2, 2010).

159. Jennifer Scholtes, "Bill Would Create Program to Thwart Oxygen Depletion, Algae in Waters," *CQ Weekly Online,* March 15, 2010: 635, http://library.cqpress.com/cqweekly/weeklyreport111-000003554116 (accessed September 2, 2010).

160. Colby Itkowitz, "Senate Committee OKs Protections for Bodies of Water," *CQ Weekly Online,* June 22, 2009: 1458, http://library.cqpress.com/cqweekly/weeklyreport111-000003149671 (accessed September 2, 2010).

161. Joanna Anderson, "Bill Would Require Stricter Guidelines for Chemical, Water Treatment Plants," *CQ Weekly Online,* November 9, 2009: 2600, http://library.cqpress.com/cqweekly/weeklyreport111-000003243408 (accessed September 2, 2010).

162. Kraft, *Environmental Policy:* 34.

Chapter 4 The Politics of Ocean Pollution

1. "No Final Action Taken on Oceanography Bill," CQ Press Electronic Library, CQ Almanac Online Edition, cqal65-1257801, http://library.cqpress.com/cqalmanac/cqal65-1257801 (accessed August 30, 2010), originally published in *CQ Almanac 1965* (Washington, DC: Congressional Quarterly, 1966).

2. "Oceanographic Research," cqal61-1371964.

3. "Oceanographic Research," CQ Press Electronic Library, CQ Almanac Online Edition, cqal61-1371964, http://library.cqpress.com/cqalmanac/cqal61-1371964 (accessed August 30, 2010), originally published in *CQ Almanac 1961* (Washington, DC: Congressional Quarterly, 1961); "President Kennedy's Message on Funds for Oceanographic Research," CQ Press Electronic Library, CQ Almanac Online Edition, cqal61-879-29200-1371388, http://library.cqpress.com/cqalmanac/cqal61-879-29200-1371388 (accessed August 30, 2010), originally published in *CQ Almanac 1961* (Washington, DC: Congressional Quarterly, 1961).

4. "Oceanographic Research," CQ Press Electronic Library, CQ Almanac Online Edition, cqal63-1315694, http://library.cqpress.com/cqalmanac/cqal63-13125694 (accessed August 30, 2010), originally published in *CQ Almanac 1963* (Washington, DC: Congressional Quarterly, 1964).

5. "National Aquarium," CQ Press Electronic Library, CQ Almanac Online Edition, cqal61-1373612, http://library.cqpress.com/cqalmanac/cqal61-1373612 (accessed August 30, 2010), originally published in *CQ Almanac 1961* (Washington, DC: Congressional Quarterly, 1961).

6. "National Aquarium," CQ Press Electronic Library, CQ Almanac Online Edition, cqal62-1324821, http://library.cqpress.com/cqalmanac/cqal62-1324821

(accessed August 30, 2010), originally published in *CQ Almanac 1962* (Washington, DC: Congressional Quarterly, 1963).

7. "National Aquarium," CQ Press Electronic Library, CQ Almanac Online Edition, cqal62-1324808, http://library.cqpress.com/cqalmanac/cqal62-1324808 (accessed August 30, 2010), originally published in *CQ Almanac 1962* (Washington, DC: Congressional Quarterly, 1963); citing "President Kennedy Vetoes Oceanographic Research Bill," CQ Press Electronic Library, CQ Almanac Online Edition, cqal61-1373612, http://library.cqpress.com/cqalmanac/cqal62-1324808 (accessed August 30, 2010), originally published in *CQ Almanac 1961* (Washington, DC: Congressional Quarterly, 1962).

8. "Oceanographic Research," cqal61-1371964.

9. "Oceanographic Research," cqal63-1315694.

10. "Message to Congress: Johnson's Conservation Message," CQ Press Electronic Library, CQ Almanac Online Edition, cqal68-1124522, http://library.cqpress.com/cqal68-1124522 (accessed August 23, 2010), originally published in *CQ Almanac 1968* (Washington, DC: Congressional Quarterly, 1969).

11. "No Final Action on Oceanography Bill."

12. "Sea Grant Colleges," CQ Press Electronic Library, CQ Almanac Online Edition, cqal66-1300138, http://library.cqpress.com/cqalmanac/cqal66-1300138 (accessed August 30, 2010), originally published in *CQ Almanac 1966* (Washington, DC: Congressional Quarterly, 1967).

13. Richard M. Nixon, "Special Message to the Congress on Marine Pollution from Oil Spills," May 20, 1970, in *Public Papers of the President of the United States* (Washington, DC: U.S. Government Printing Office, 1971), 443–447.

14. "Presidential Statement to Congress: Nixon's 1970 Environmental Message," CQ Press Electronic Library, CQ Almanac Online Edition, cqal71-869-26707-1254975, http://library.cqpress.com/cqal71-869-26707-1254975 (accessed August 23, 2010), originally published in *CQ Almanac 1970* (Washington, DC: Congressional Quarterly, 1972).

15. Richard M. Nixon, "Special Message to the Congress about Reorganization Plans to Establish the Environmental Protection Agency and the National Oceanic and Atmospheric Administration," July 9, 1970, in *Public Papers of the President of the United States* (Washington, DC: U.S. Government Printing Office, 1971), 578–586.

16. "Presidential Statement to Congress: Nixon's Message on Ocean Pollution," CQ Press Electronic Library, CQ Almanac Online Edition, cqal70-1290688, http://library.cqpress.com/cqalmanac/cqal70-1290688 (accessed August 30, 2010), originally published in *CQ Almanac 1970* (Washington, DC: Congressional Quarterly, 1971).

17. "Presidential Statement to Congress: Nixon's Environmental Reorganization Plan," CQ Press Electronic Library, CQ Almanac Online Edition, cqal70-1290928, http://library.cqpress.com/cqalmanac/cqal70-1290928 (accessed August 23, 2010), originally published in *CQ Almanac 1970* (Washington, DC: Congressional Quarterly, 1971).

18. Richard M. Nixon, "Message to the Congress Transmitting Annual Report on the Federal Ocean Program," September 28, 1973, in *Public Papers of the*

President of the United States (Washington, DC: U.S. Government Printing Office, 1974), 832–834.

19. Richard M. Nixon, "Message to the Senate Transmitting the Convention on Ocean Dumping," February 28, 1973, in *Public Papers of the President of the United States* (Washington, DC: U.S. Government Printing Office, 1974), 133.

20. Richard M. Nixon, "Message to the Senate Transmitting the Convention on International Trade in Endangered Species of Wild Fauna and Flora," April 13, 1973, in *Public Papers of the President of the United States* (Washington, DC: U.S. Government Printing Office, 1974), 285–286.

21. "Natural Resources," CQ Press Electronic Library, CQ Almanac Online Edition, cqal69-1248497, http://library.cqpress.com/cqalmanac/cqal69-1248497 (accessed August 30, 2010), originally published in *CQ Almanac 1969* (Washington, DC: Congressional Quarterly, 1970).

22. Ibid.

23. "Major Anti-Pollution Measures Clear 91st Congress," *Congressional Quarterly Almanac* 26 (1970), 127–129.

24. "Environment: Action Not Completed," *Congressional Quarterly Almanac* 27 (1971): 29.

25. "Environment and Resources: Action Completed," *Congressional Quarterly Almanac* 28 (1972): 17.

26. "Most Productive Environmental Session in History," CQ Press Electronic Library, CQ Almanac Online Edition, cqal72-1250140, http://library.cqpress.com/cqalmanac/cqal72-1250140 (accessed August 31, 2010), originally published in *CQ Almanac 1972* (Washington, DC: Congressional Quarterly, 1973).

27. "Ocean Dumping: House and Senate Pass Controls," *Congressional Quarterly Almanac* 27 (1971): 720–727; "Ocean Pollution," *Congressional Quarterly Almanac* 28 (1972): 700–701; "Environment and Resources: Action Completed": 17.

28. "Marine Safety," *Congressional Quarterly Almanac* 28 (1972): 234.

29. Ibid.

30. "Oil Pollution Act," CQ Press Electronic Library, CQ Almanac Online Edition, cqal72-1249169, http://library.cqpress.com/cqal72-1249169 (accessed August 23, 2010), originally published in *CQ Almanac* (Washington, DC: Congressional Quarterly, 1973).

31. "Coastal Management," CQ Press Electronic Library, CQ Almanac Online Edition, cqal71-1254664, http://library.cqpress.com/cqalmanac/cqal71-1254664 (accessed August 30, 2010), originally published in *CQ Almanac 1971* (Washington, DC: Congressional Quarterly, 1972).

32. "Treaties on Ocean Pollution," CQ Press Electronic Library, CQ Almanac Online Edition, cqal71-1254369, http://library.cqpress.com/cqalmanac/cqal71-1254369 (accessed August 30, 2010), originally published in *CQ Almanac 1971* (Washington, DC: Congressional Quarterly, 1972).

33. "Presidential Statement: Ford on Oil Pollution," CQ Press Electronic Library, CQ Almanac Online Edition, cqal75-865-26297-1210678, http://library.cqpress.com/cqal75-865-26297-1210678 (accessed August 31, 2010), originally published in *CQ Almanac 1975* (Washington, DC: Congressional Quarterly, 1976).

34. "Marine Protection," *Congressional Quarterly Almanac* 30 (1974): 744.

35. "High Seas Oil Pollution," CQ Press Electronic Library, CQ Almanac Online Edition, cqal74-1222874, http://library.cqpress.com/cqal74-1222874 (accessed August 31, 2010), originally published in *CQ Almanac* (Washington, DC: Congressional Quarterly, 1974).

36. "Pollution Prevention Treaty," CQ Press Electronic Library, CQ Almanac Online Edition, cqal73-1227767, http://library.cqpress.com/cqal73-1227767 (accessed August 31, 2010), originally published in *CQ Almanac* (Washington, DC: Congressional Quarterly, 1974).

37. "Marine Protection," *Congressional Quarterly Almanac* 31 (1975): 190.

38. Jimmy Carter, "Remarks and a Question-and-Answer Session at the Clinton Town Meeting," March 16, 1977, in *P Public Papers of the President of the United States* (Washington, DC: U.S. Government Printing Office, 1978), 387–388.

39. Jimmy Carter, "Message to the Congress on Oil Pollution of the Oceans," March 17, 1977, in *Public Papers of the President of the United States* (Washington, DC: U.S. Government Printing Office, 1978), 458–459.

40. Ibid.; Jimmy Carter, "Message to the Senate Transmitting the Convention on Pollution from Ships," March 22, 1977, in *Public Papers of the President of the United States* (Washington, DC: U.S. Government Printing Office, 1978), 476.

41. "Presidential Statement: Carter on Oil Pollution," CQ Press Electronic Library, CQ Almanac Online Edition, cqal77-863-26256-1200425, http://library.cqpress.com/cqal77-863-26256-1200425 (accessed August 31, 2010), originally published in *CQ Almanac 1977* (Washington, DC: Congressional Quarterly, 1978); Jimmy Carter, "Convention on Pollution from Ships," January 19, 1979, in *Public Papers of the President of the United States* (Washington, DC: U.S. Government Printing Office, 1980), 85–86.

42. "Oil Spill Liability," *Congressional Quarterly Almanac* 34 (1978): 71; "No Final Action Taken on Oceanography Bill," cqal63–1257801.

43. "Marine Mammals," *Congressional Quarterly Almanac* 33 (1977): 673; "Ocean Dumping," *Congressional Quarterly Almanac* 33 (1977): 675.

44. "Marine Mammal Protection," *Congressional Quarterly Almanac* 34 (1978): 7165.

45. "Seabed Mining," CQ Press Electronic Library, CQ Almanac Online Edition, cqal78-1236886, http://library.cqpress.com/cqalmanac/cqal78-1236886 (accessed August 30, 2010), originally published in *CQ Almanac 1978* (Washington, DC: Congressional Quarterly, 1979).

46. "Marine Mammals," *Congressional Quarterly Almanac* 33 (1977): 673; "Ocean Dumping," *Congressional Quarterly Almanac* 33 (1977): 675.

47. "Ocean Pollution Research," *Congressional Quarterly Almanac* 34 (1978): 715.

48. "Marine Protection Act," *Congressional Quarterly Almanac* 36 (1980): 609–610.

49. Ibid.

50. Ibid.

51. "Campeche Bay Oil Spill," CQ Press Electronic Library, CQ Almanac Online Edition, cqal80-1175094, http://library.cqpress.com/cqal80-1175094 (accessed

August 31, 2010), originally published in *CQ Almanac 1980* (Washington, DC: Congressional Quarterly, 1981).

52. Ronald W. Reagan, "Statement on Signing the Medical Waste Tracking Act of 1988," November 2, 1988, in *Public Papers of the President of the United States* (Washington, DC: U.S. Government Printing Office, 1989), 1430–1431.

53. Ronald W. Reagan, "Statement on Signing a Bill Terminating Ocean Dumping of Sewage, Sludge, and Industrial Waste," November 18, 1988, in *Public Papers of the President of the United States* (Washington, DC: U.S. Government Printing Office, 1989), 1558.

54. "Ocean Dumping Sites," *Congressional Quarterly Almanac* 37 (1981): 526.

55. "Ocean Dumping Legislation," *Congressional Quarterly Almanac* 38 (1982): 459.

56. Ibid.

57. "Oil Spill Liability Limits," CQ Press Electronic Library, CQ Almanac Online Edition, cqal82-1163134, http://library.cqpress.com/cqal82-1163134 (accessed August 31, 2010), originally published in *CQ Almanac 1982* (Washington, DC: Congressional Quarterly, 1983).

58. "Marine Sanctuaries Act," CQ Press Electronic Library, CQ Almanac Online Edition, cqal81-1171588, http://library.cqpress.com/cqalmanac/cqal81-1171588 (accessed August 30, 2010), originally published in *CQ Almanac 1981* (Washington, DC: Congressional Quarterly, 1982).

59. "NOAA Authorization Fails," CQ Press Electronic Library, CQ Almanac Online Edition, cqal82-1163440, http://library.cqpress.com/cqalmanac/cqal82-1163440 (accessed August 30, 2010), originally published in *CQ Almanac 1982* (Washington, DC: Congressional Quarterly, 1983).

60. "Ocean Dumping Legislation," *Congressional Quarterly Almanac* 39 (1983): 360; "Ocean Dumping Bill Dies," *Congressional Quarterly Almanac* 40 (1984): 333.

61. Ibid.; "Ocean Dumping Bill Dies," *Congressional Quarterly Almanac* 40 (1984): 333.

62. "Marine Mammal Protection," *Congressional Quarterly Almanac* 40 (1984): 334.

63. "Marine Sanctuaries," CQ Press Electronic Library, CQ Almanac Online Edition, cqal83-1199307, http://library.cqpress.com/cqalmanac/cqal83-1199307 (accessed August 30, 2010), originally published in *CQ Almanac 1983* (Washington, DC: Congressional Quarterly, 1984).

64. "Reagan Vetoes First Comprehensive NOAA Bill," CQ Press Electronic Library, CQ Almanac Online Edition, cqal84-1153412, http://library.cqpress.com/cqalmanac/cqal84-1153412 (accessed August 30, 2010), originally published in *CQ Almanac 1984* (Washington, DC: Congressional Quarterly, 1985).

65. "Oil Spill Liability," CQ Press Electronic Library, CQ Almanac Online Edition, cqal86-1149921, http://library.cqpress.com/cqal86-1149921 (accessed August 31, 2010), originally published in *CQ Almanac 1986* (Washington, DC: Congressional Quarterly, 1987).

66. "Ocean Pollution Controlled," CQ Press Electronic Library, CQ Almanac Online Edition, cqal88-1141433, http://library.cqpress.com/cqalmanac/cqal88-1141433 (accessed August 30, 2010), originally published in *CQ Almanac 1988*

(Washington, DC: Congressional Quarterly, 1989); "Ocean Dumping of Wastes," *Congressional Quarterly Almanac* 41 (1985): 202.

67. Daniel J. Fiorino, *Making Environmental Policy* (Berkeley: University of California Press, 1995): 154–155.

68. "Medical Waste Bill Enacted," *Congressional Quarterly Almanac* 44 (1988): 161–162.

69. Ibid.

70. Reagan, "Medical Waste Tracking Act": 1430–1431.

71. "Environment: Plastic Pollution," *Congressional Quarterly Almanac* 43 (1987): 17; "Congress Prohibits Dumping Plastics at Sea," *Congressional Quarterly Almanac* 43 (1987): 303–304.

72. "Ocean Pollution Controlled," *CQ Press Electronic Library*, CQ Almanac Online Edition, cqa 88-1141433, http://library.cqpress.com/cqalmanac/cqal88-1141433 (accessed May 6, 2011).

73. George H. W. Bush, "White House Fact Sheet on Environmental Initiatives," September 18, 1989, in *Public Papers of the President of the United States* (Washington, DC: U.S. Government Printing Office, 1990), 1212–1215: 1213.

74. George H. W. Bush, "Statement on Signing the Oil Pollution Act of 1990," August 18, 1990, in *Public Papers of the President of the United States* (Washington, DC: U.S. Government Printing Office, 1991), 1144–1145.

75. George H. W. Bush, "Message to the Senate Transmitting the International Cooperation on Oil Pollution Preparedness, Response and Cooperation," August 1, 1991, in *Public Papers of the President of the United States* (Washington, DC: U.S. Government Printing Office, 1992), 1011.

76. "Approval of Liability Bills Spurred by Alaska Spill," CQ Press Electronic Library, CQ Almanac Online Edition, cqal89-1139842, http://library.cqpress.com/cqal89-1139842 (accessed August 31, 2010), originally published in *CQ Almanac 1989* (Washington, DC: Congressional Quarterly, 1990).

77. "National Oceanic and Atmospheric Administration's (NOAA) Reauthorization," CQ Press Electronic Library, CQ Almanac Online Edition, cqal89-1139914, http://library.cqpress.com/cqalmanac/cqal89-1139914 (accessed August 30, 2010), originally published in *CQ Almanac 1989* (Washington, DC: Congressional Quarterly, 1990).

78. "Beach Pollution Bill Dies," CQ Press Electronic Library, CQ Almanac Online Edition, cqal90-1112580, http://library.cqpress.com/cqalmanac/cqal90-1112580 (accessed August 30, 2010), originally published in *CQ Almanac 1990* (Washington, DC: Congressional Quarterly, 1991).

79. "Florida Keys Declared Marine Sanctuary," CQ Press Electronic Library, CQ Almanac Online Edition, cqal90-1112634, http://library.cqpress.com/cqalmanac/cqal90-1112634 (accessed August 30, 2010), originally published in *CQ Almanac 1990* (Washington, DC: Congressional Quarterly, 1991).

80. "Research Bill for National Oceanographic and Atmospheric Administration (NOAA) Secured House Nod, but Didn't Reach Senate Floor," CQ Press Electronic Library, CQ Almanac Online Edition, cqal91-1110474, http://library.cqpress.com/cqalmanac/cqal91-1110474 (accessed August 30, 2010), originally

published in *CQ Almanac 1991* (Washington, DC: Congressional Quarterly, 1992); "National Oceanic and Atmospheric Administration (NOAA) Program Funded," CQ Press Electronic Library, CQ Almanac Online Edition, cqal92-1107948, http://library.cqpress.com/cqalmanac/cqal92-1107948 (accessed August 30, 2010), originally published in *CQ Almanac 1992* (Washington, DC: Congressional Quarterly, 1993).

81. "House Committee Passed Beach-Water Testing Bill," CQ Press Electronic Library, CQ Almanac Online Edition, cqal91-111463, http://library.cqpress.com/cqalmanac/cqal91-111463 (accessed August 30, 2010), originally published in *CQ Almanac 1991* (Washington, DC: Congressional Quarterly, 1992).

82. William J. Clinton, "Remarks on Earth Day," April 21, 1993, in *Public Papers of the Presidents of the United States*; John T. Woolley and Gerhard Peters, *The American Presidency Project,* http://www.presidency.ucsb.edu//ws/?pid=46460 (accessed May 5, 2011).

83. William J. Clinton, "Remarks to the National Oceans Conference in Monterey, California," June 12, 1998, in *Public Papers of the Presidents of the United States*; John T. Woolley and Gerhard Peters, *The American Presidency Project,* http://www.presidency.ucsb.edu//ws/?pid=56132 (accessed May 5, 2011).

84. William J. Clinton, "Background Briefing by Senior Administration Officials," February 11, 1994, in *Public Papers of the Presidents of the United States*; John T. Woolley and Gerhard Peters, *The American Presidency Project,* http://www.presidency.ucsb.edu//ws/?pid=59718 (accessed May 5, 2011). See also William J. Clinton, "Remarks to the National Oceans Conference in Monterey, California," June 12, 1998, in *Public Papers of the Presidents of the United States*; John T. Woolley and Gerhard Peters, *The American Presidency Project,* http://www.presidency.ucsb.edu//ws/?pid=56132 (accessed May 5, 2011).

85. "Hill Revamps Marine Mammal Act," *Congressional Quarterly Almanac* 50 (1994): 257–258.

86. "Bill Aims at Limiting Dumping of Radioactive Waste," CQ Press Electronic Library, CQ Almanac Online Edition, cqal94-1103417, http://library.cqpress.com/cqal94-1103417 (accessed August 23, 2010), originally published in *CQ Almanac 1994* (Washington, DC: Congressional Quarterly, 1995).

87. "Environmental Legislation Highlights Coastal Zones, Waste Management, and Antarctica Protections," CQ Press Electronic Library, CQ Almanac Online Edition, cqal96-1092246, http://library.cqpress.com/cqal96-1092246 (accessed August 23, 2010), originally published in *CQ Almanac 1996* (Washington, DC: Congressional Quarterly, 1997).

88. William J. Clinton, "Remarks to the National Oceans Conference in Monterey, California," June 12, 1998, in *Public Papers of the Presidents of the United States*; John T. Woolley and Gerhard Peters, *The American Presidency Project,* http://www.presidency.ucsb.edu//ws/?pid=56132 (accessed May 5, 2011).

89. George W. Bush, "Message to the Senate Transmitting the Protocol of 1997 to Amend the International Convention for the Prevention of Pollution from Ships," May 15, 2003, in *Public Papers of the President of the United States* (Washington, DC: U.S. Government Printing Office, 2004), 496–497.

90. Shawn Zeller, "Island Hopping on Bush's Legacy," *CQ Weekly Online*, July 27, 2009: 1754, http://library.cqpress.com/cqweekly/weeklyreport111-000003177533 (accessed September 2, 2010).

91. "New Pipeline Safety Bill Cleared," CQ Press Electronic Library, CQ Almanac Online Edition, cqal06-1421515, http://library.cqpress.com/cqal06-1421515 (accessed August 31, 2010), originally published in *CQ Almanac 2006* (Washington, DC: Congressional Quarterly, 2007).

92. Barack Obama, "Proclamation 8392—National Oceans Month, 2009," June 12, 2009 *The Federal Register*, June 17, 2009.

93. Ibid.; Barack Obama, "Proclamation 8531—National Oceans Month, 2010," May 28, 2010 *The Federal Register*, June 7, 2010.

94. Barack Obama, "Memorandum on National Policy for the Oceans, Our Coasts, and the Great Lakes," June 12, 2009 *The Federal Register*, June 17, 2009.

95. Barack Obama, "Executive Order 13547—Stewardship of the Ocean, Our Coasts, and the Great Lakes," July 19, 2010 *The Federal Register*, July 22, 2010.

Chapter 5 The Politics of Pesticides

1. Daniel J. Fiorino, *Making Environmental Policy* (Berkeley: University of California Press, 1995), 154.

2. Ibid., 32.

3. "Congress Weighs Stronger Controls on Pesticides," CQ Press Electronic Library, CQ Almanac Online Edition, cqal64-1303995, http://library.cqpress.com/cqalmanac/cqal64-1303995 (accessed August 30, 2010), originally published in *CQ Almanac 1964* (Washington, DC: Congressional Quarterly, 1965).

4. "Presidential Statement to Congress: Nixon's 1970 Environmental Message," CQ Press Electronic Library, CQ Almanac Online Edition, cqal71-869-26707-1254975, http://library.cqpress.com/cqal71-869-26707-1254975 (accessed August 23, 2010), originally published in *CQ Almanac 1971* (Washington, DC: Congressional Quarterly, 1972).

5. Ibid.

6. Ibid.

7. Ibid.

8. "Pollution Problems," CQ Press Electronic Library, CQ Almanac Online Edition, cqal69-1248491, http://library.cqpress.com/cqal69-1248491 (accessed August 23, 2010), originally published in *CQ Almanac 1969* (Washington, DC: Congressional Quarterly, 1970).

9. "Environment: Action Not Completed," *Congressional Quarterly Almanac* 27 (1971): 29; "Most Productive Environmental Session in History," *Congressional Quarterly Almanac* 28 (1972): 115–116; Richard M. Nixon, "Statement on Signing the Federal Environmental Pesticide Control Act of 1972," October 21, 1972, in *Public Papers of the President of the United States* (Washington, DC: U.S. Government Printing Office, 1973), 1005; "Nixon Signs Comprehensive Bill to Regulate

Pesticides," *Congressional Quarterly Almanac* 28 (1972): 934–944; Richard N. L. Andrews, *Managing the Environment, Managing Ourselves* (New Haven, CT: Yale University Press, 1999), 243.

10. Mary Clifford, "A Review of Federal Environmental Legislation," in Mary Clifford, ed., *Environmental Crime* (Gaithersburg, MD: Aspen, 1998), 95–109: 99.

11. Fiorino, *Making Environmental Policy:* 32.

12. "Pesticide Control: House Passes Measure," CQ Press Electronic Library, CQ Almanac Online Edition, cqal71-1254364, http://library.cqpress.com/cqalmanac/cqal71-1254364 (accessed August 30, 2010), originally published in *CQ Almanac 1971* (Washington, DC: Congressional Quarterly, 1972).

13. Gerald Ford, "Statement on Signing the Toxic Substances Control Act," October 12, 1976 in *Public Papers of the Presidents of the United States*; John T. Woolley and Gerhard Peters, *The American Presidency Project,* http://www.presidency.ucsb.edu//ws/?pid=6445 (accessed May 5, 2011).

14. "Energy and Environment: Action Completed," *Congressional Quarterly Almanac* 31 (1975): 14.

15. "Congress Clears Pesticide Regulation Bill," *Congressional Quarterly Almanac* 31 (1975): 201–206.

16. Jimmy Carter, "The Environment," May 23, 1977, in *Public Papers of the President of the United States* (Washington, DC: U.S. Government Printing Office, 1978), 967–986.

17. Jimmy Carter, "Statement on Signing S. 1678 into Law," October 2, 1978, in *Public Papers of the President of the United States* (Washington, DC: U.S. Government Printing Office, 1979), 1696.

18. Jimmy Carter, "Message to the Congress on Environmental Priorities and Programs," August 2, 1979, in *Public Papers of the President of the United States* (Washington, DC: U.S. Government Printing Office, 1980), 1353–1373; Jimmy Carter, "Memorandum from the President on Integrated Pest Management," August 2, 1979, in *Public Papers of the President of the United States* (Washington, DC: U.S. Government Printing Office, 1980), 1383.

19. "Pesticides Registration," *Congressional Quarterly Almanac* 33 (1977): 680–681.

20. "Pesticides Marketing Bill," CQ Press Electronic Library, CQ Almanac Online Edition, cqal78-1236872, http://library.cqpress.com/cqalmanac/cqal78-1236872 (accessed August 30, 2010), originally published in *CQ Almanac 1978* (Washington, DC: Congressional Quarterly, 1979).

21. "Pesticide Control Extension," *Congressional Quarterly Almanac* 35 (1979): 681–682.

22. "Federal Pesticides Rules," *Congressional Quarterly Almanac* 36 (1980): 610.

23. Ronald W. Reagan, "Remarks to the National Campers and Hikers Association in Bowling Green, Kentucky," July 12, 1984 in *Public Papers of the Presidents of the United States*; John T. Woolley and Gerhard Peters, *The American*

Presidency Project, http://www.presidency.ucsb.edu//ws/?pid=40151 (accessed May 5, 2011).

24. Ronald W. Reagan, "Executive Order 12420—Incentive Pay for Hazardous Duty," in May 11, 1983, in *Public Papers of the Presidents of the United States*; John T. Woolley and Gerhard Peters, *The American Presidency Project,* http://www.presidency.ucsb.edu//ws/?pid=41308 (accessed May 5, 2011).

25. Ronald W. Reagan, "Remarks in Atlanta, Georgia, at the Annual Convention of the National Conference of State Legislatures," July 30, 1981, in *Public Papers of the Presidents of the United States*; John T. Woolley and Gerhard Peters, *The American Presidency Project,* http://www.presidency.ucsb.edu//ws/?pid=6445 (accessed May 5, 2011).

26. Ronald W. Reagan, "Proclamation 4893—Bicentennial Year of the American Bald Eagle Day," January 28, 1982 in *Public Papers of the Presidents of the United States*; John T. Woolley and Gerhard Peters, *The American Presidency Project,* http://www.presidency.ucsb.edu//ws/?pid=42831 (accessed May 5, 2011).

27. "Conflicts Doom Passage of Pesticide Bill," *Congressional Quarterly Almanac* 38 (1982): 363.

28. "Pesticide Reauthorization," CQ Press Electronic Library, CQ Almanac Online Edition, cqal83-1199440, http://library.cqpress.com/cqalmanac/cqal83-1199440 (accessed August 30, 2010), originally published in *CQ Almanac 1983* (Washington, DC: Congressional Quarterly, 1984).

29. "Legislative Summary: Pesticide Control," *Congressional Quarterly Almanac* 42 (1986): 21; "Pesticide Bill Stalls at Session's End," *Congressional Quarterly Almanac* 42 (1986): 120–126.

30. "Pesticide Bill Stalls at Session's End," *Congressional Quarterly Almanac* 42 (1986): 120–126.

31. "Energy/Environment: Pesticides," *Congressional Quarterly Almanac* 44 (1988): 18; "Congress Speeds Up Pesticide Testing," *Congressional Quarterly Almanac* 44 (1988): 139–142.

32. "Pesticides Proposal," CQ Press Electronic Library, CQ Almanac Online Edition, cqal89-1139133, http://library.cqpress.com/cqalmanac/cqal89-1139133 (accessed August 30, 2010), originally published in *CQ Almanac 1989* (Washington, DC: Congressional Quarterly, 1990).

33. "Pesticide Rules Remain Unchanged," *Congressional Quarterly Almanac* 48 (1992): 212–213.

34. William J. Clinton, "Statement on Senate Action on Food Quality Protection Legislation," July 24, 1996 in *Public Papers of the Presidents of the United States*; John T. Woolley and Gerhard Peters, *The American Presidency Project,* http://www.presidency.ucsb.edu//ws/?pid=53110 (accessed May 5, 2011).

35. William J. Clinton, "Proclamation 6664—Cancer Control Month, 1994" April 7, 1994, in *Public Papers of the Presidents of the United States*; John T. Woolley and Gerhard Peters, *The American Presidency Project,* http://www.presidency.ucsb.edu//ws/?pid=49924 (accessed May 5, 2011).

36. William J. Clinton, "The President's Radio Address," August 3, 1996, in *Public Papers of the Presidents of the United States*; John T. Woolley and Gerhard Peters,

The American Presidency Project, http://www.presidency.ucsb.edu//ws/?pid=53154 (accessed May 5, 2011).

37. William J. Clinton, "Memorandum on Environmentally Beneficial Landscaping," April 26, 1994 in *Public Papers of the Presidents of the United States*; John T. Woolley and Gerhard Peters, *The American Presidency Project,* http://www.presidency.ucsb.edu//ws/?pid=50051 (accessed May 5, 2011).

38. William J. Clinton, "Address Before a Joint Session of the Congress on the State of the Union," January 23, 1986, in *Public Papers of the Presidents of the United States*; John T. Woolley and Gerhard Peters, *The American Presidency Project,* http://www.presidency.ucsb.edu//ws/?pid=53091 (accessed May 5, 2011).

39. William J. Clinton, "The President's Radio Address," August 3, 1996, in *Public Papers of the Presidents of the United States*; John T. Woolley and Gerhard Peters, *The American Presidency Project,* http://www.presidency.ucsb.edu//ws/?pid=53154 (accessed May 5, 2011).

40. "No Action Taken on Pesticide Regulation," *Congressional Quarterly Almanac* 49 (1993): 229–230.

41. "FIFRA Delay," *Congressional Quarterly Almanac* 50 (1994): 199.

42. "No Action Taken on Pesticide Regulation."

43. "Pesticides Rewrite Draws Wide Support," CQ Press Electronic Library, CQ Almanac Online Edition, cqal96-1092048, http://library.cqpress.com/cqalmanac/cqal96-1092048 (accessed August 30, 2010), originally published in *CQ Almanac 1996* (Washington, DC: Congressional Quarterly, 1997).

44. Michael E. Kraft, "Environmental Policy in Congress," in Norman J. Vig and Michael E. Kraft, eds., *Environmental Policy: New Directions for the Twenty-First Century* (Washington, DC: CQ Press, 2006), 124–147: 139.

45. George W. Bush, "Press Briefing by Ari Fleischer," April 19, 2001, in *Public Papers of the Presidents of the United States*; John T. Woolley and Gerhard Peters, *The American Presidency Project,* http://www.presidency.ucsb.edu//ws/?pid=47512(accessed May 5, 2011).

46. George W. Bush, "The President's Radio Address," April 28, 2001, in *Public Papers of the Presidents of the United States*; John T. Woolley and Gerhard Peters, *The American Presidency Project,* http://www.presidency.ucsb.edu//ws/?pid=45621 (accessed May 5, 2011).

47. George W. Bush, "Message to the Senate Transmitting the Stockholm Convention on Persistent Organic Pollutants," May 6, 2002, in *Public Papers of the Presidents of the United States*; John T. Woolley and Gerhard Peters, *The American Presidency Project,* http://www.presidency.ucsb.edu//ws/?pid=73294 (accessed May 5, 2011).

48. George W. Bush, "Remarks at the World Bank," July 17, 2001, in *Public Papers of the Presidents of the United States*; John T. Woolley and Gerhard Peters, *The American Presidency Project,* http://www.presidency.ucsb.edu//ws/?pid=73621 (accessed May 5, 2011).

49. Ellyn Ferguson, "Agribusiness Unhappy that EPA Is Studying Week Killer Again," *CQ Weekly Online,* May 10, 2010: 1130, http://library.cqpress.com/cqweekly/weeklyreport111-000003658681 (accessed September 2, 2010).

Chapter 6 The Politics of Solid Waste

1. Daniel D. Chiras, *Beyond the Fray* (Boulder, CO: Johnson Books, 1990), 4–5; Michael E. Kraft, *Environmental Policy and Politics* (New York: Longman, 2001), 43.

2. "Natural Beauty Sought as U.S. Goal," *Congressional Quarterly Almanac* 21 (1965): 722–723.

3. Lyndon B. Johnson, "Special Message to the Congress on Conservation: 'To Renew a Nation,'" Public Papers of the Presidents of the United States, John T. Woolley and Gerhard Peters, *The American Presidency Project*, http://www.presidency.ucsb.edu/ws/?pid=28719 (accessed May 6, 2011).

4. "Message to Congress: Johnson's Conservation Message," CQ Press Electronic Library, CQ Almanac Online Edition, cqal68-1124522, http://library.cqpress.com/cqal68-1124522 (accessed August 23, 2010), originally published in *CQ Almanac 1968* (Washington, DC: Congressional Quarterly, 1969).

5. Janet Schaeffer, "Solid Wastes," in James Rathlesberger, ed., *Nixon and the Environment* (New York: Village Voice Book, 1972), 239–256: 241.

6. "Solid Waste: Bill Stresses Recovery and Recycling," CQ Press Electronic Library, CQ Almanac Online Edition, cqal70-1293864, http://library.cqpress.com/cqal70-1293864 (accessed August 23, 2010), originally published in *CQ Almanac 1970* (Washington, DC: Congressional Quarterly, 1971).

7. "Pollution Problems," CQ Press Electronic Library, CQ Almanac Online Edition, cqal69-1248491, http://library.cqpress.com/cqal69-1248491 (accessed August 23, 2010), originally published in *CQ Almanac 1969* (Washington, DC: Congressional Quarterly, 1970).

8. "Bill Stresses Recovery."

9. Ibid.

10. Ibid.; "Presidential Statement to Congress: Nixon on the Environment 1970," CQ Press Electronic Library, CQ Almanac Online Edition, cqal70-1290732, http://library.cqpress.com/cqal70-1290732 (accessed August 23, 2010), originally published in *CQ Almanac 1970* (Washington, DC: Congressional Quarterly, 1971).

11. "Presidential Statement to Congress: Nixon's 1970 Environmental Message," CQ Press Electronic Library, CQ Almanac Online Edition, cqal71-869-26707-1254975, http://library.cqpress.com/cqal71-869-26707-1254975 (accessed August 23, 2010), originally published in *CQ Almanac 1970* (Washington, DC: Congressional Quarterly, 1972).

12. "Pollution Problems."

13. "Bill Stresses Recovery."

14. "Clean Air, Waste Disposal," CQ Press Electronic Library, CQ Almanac Online Edition, cqal73-1227314, http://library.cqpress.com/cqal73-1227314 (accessed August 23, 2010), originally published in *CQ Almanac 1971* (Washington, DC: Congressional Quarterly, 1972).

15. "Nonreturnable Containers," *Congressional Quarterly Almanac* 30 (1974): 337–338.

16. "Solid Waste Programs," CQ Press Electronic Library, CQ Almanac Online Edition, cqal74-1222884, http://library.cqpress.com/cqal74-1222884 (accessed August 23, 2010), originally published in *CQ Almanac 1974* (Washington: Congressional Quarterly, 1975).

17. Gerald R. Ford, "Statement on Signing the Bill Providing for Loan Guarantees for Construction of Municipal Waste Water Treatment Plants," October 20, 1976, in *Public Papers of the President of the United States* (Washington, DC: U.S. Government Printing Office, 1975), 2584.

18. "Solid Waste Bill," *Congressional Quarterly Almanac* 32 (1976): 199–201.

19. Jimmy Carter, "The Environment," May 23, 1977, in *Public Papers of the President of the United States* (Washington, DC: U.S. Government Printing Office, 1978), 967–986.

20. "Recycled Motor Oil," CQ Press Electronic Library, CQ Almanac Online Edition, cqal80-1174715, http://library.cqpress.com/cqal80-1174715 (accessed August 23, 2010), originally published in *CQ Almanac 1980* (Washington, DC: Congressional Quarterly, 1981).

21. "Interstate Waste Compact," CQ Press Electronic Library, CQ Almanac Online Edition, cqal82-1163094, http://library.cqpress.com/cqal82-1163094 (accessed August 23, 2010), originally published in *CQ Almanac 1982* (Washington, DC: Congressional Quarterly, 1983).

22. RonaldW.Reagan,"Proclamation5830—NationalRecyclingMonth"June14, 1988, in *Public Papers of the Presidents of the United States;* John T. Woolley and Gerhard Peters, *The American Presidency Project,* http://www.presidency.ucsb.edu//ws/?pid=35971 (accessed May 6, 2011).

23. Ronald W. Reagan, "Question-and-Answer Session with Farmers in State Center, Iowa," August 2, 1982, in *Public Papers of the Presidents of the United States;* John T. Woolley and Gerhard Peters, *The American Presidency Project,* http://www.presidency.ucsb.edu//ws/?pid=42812 (accessed May 6, 2011).

24. "Amtrak Told to Curb Waste Disposal," CQ Press Electronic Library, CQ Almanac Online Edition, cqal90-1112889, http://library.cqpress.com/cqal90-1112889 (accessed August 23, 2010), originally published in *CQ Almanac 1989* Washington, DC: Congressional Quarterly, 1990).

25. George H. W. Bush, "Message to the Congress Transmitting the Report of the Council on Environmental Quality," June 23, 1989, in *Public Papers of the Presidents of the United States;* John T. Woolley and Gerhard Peters, *The American Presidency Project,* http://www.presidency.ucsb.edu//ws/?pid=17204 (accessed May 6, 2011).

26. George H. W. Bush, "Remarks at the Washington Centennial Celebration in Spokane," September 19, 1989, in *Public Papers of the Presidents of the United States;* John T. Woolley and Gerhard Peters, *The American Presidency Project,* http://www.presidency.ucsb.edu//ws/?pid=17542 (accessed May 6, 2011).

27. George H. W. Bush, "Proclamation 5957—National Recycling Month, 1989," April 19, 1989 in *Public Papers of the Presidents of the United States;* John T. Woolley and Gerhard Peters, *The American Presidency Project,* http://www.presidency.ucsb.edu//ws/?pid=20446 (accessed May 6, 2011).

28. "Moves on Waste Laws for U.S. Sites Stall," CQ Press Electronic Library, CQ Almanac Online Edition, cqal90-1112651, http://library.cqpress.com/cqal90-1112651 (accessed August 23, 2010), originally published in *CQ Almanac 1990* (Washington, DC: Congressional Quarterly, 1991).

29. "Federal Facilities Cleanup," *Congressional Quarterly Almanac* 47 (1991): 222–223.

30. "Federal Agencies Liable for Waste Violations," CQ Press Electronic Library, CQ Almanac Online Edition, cqal92-1107831, http://library.cqpress.com/cqal92-1107831 (accessed August 23, 2010), originally published in *CQ Almanac 1992* (Washington, DC: Congressional Quarterly, 1993).

31. "Solid Waste Problem Remains Insoluble," CQ Press Electronic Library, CQ Almanac Online Edition, cqal92-1107854, http://library.cqpress.com/cqal92-1107854 (accessed August 23, 2010), originally published in *CQ Almanac 1991* (Washington, DC: Congressional Quarterly, 1992).

32. "Senate Passes Waste Disposal Bill," *Congressional Quarterly Almanac* 51 (1995): 5–17; "Other Environmental Bills Considered in 1996: Municipal Waste," *Congressional Quarterly Almanac* 52 (1996): 4–28.

33. William J. Clinton, "Executive Order 12837—Federal Acquisition and Waste Prevention," October 20, 1993, in *Public Papers of the Presidents of the United States;* John T. Woolley and Gerhard Peters, *The American Presidency Project,* http://www.presidency.ucsb.edu//ws/?pid=61566 (accessed May 6, 2011).

34. William J. Clinton, "Statement on Government Use of Recycled Products," September 14, 1998, in *Public Papers of the Presidents of the United States;* John T. Woolley and Gerhard Peters, *The American Presidency Project,* http://www.presidency.ucsb.edu//ws/?pid=54901 (accessed May 6, 2011); William J. Clinton, "Executive Order 13101—Greening the Government Through Waste Prevention, Recycling, and Federal Acquisitions," September 14, 1998, in *Public Papers of the Presidents of the United States;* John T. Woolley and Gerhard Peters, *The American Presidency Project,* http://www.presidency.ucsb.edu//ws/?pid=54902 (accessed May 6, 2011).

35. William J. Clinton, "Proclamation7377—America Recycles Day, 2000," November 15, 2000 in *Public Papers of the Presidents of the United States;* John T. Woolley and Gerhard Peters, *The American Presidency Project,* http://www.presidency.ucsb.edu//ws/?pid=62384 (accessed May 6, 2011).

36. "Senate Passes Waste Disposal Bill," *Congressional Quarterly Almanac* 51 (1995): 5–17.

37. "Environmental Legislation Highlights Coastal Zones, Waste Management, and Antarctica Protections," CQ Press Electronic Library, CQ Almanac Online Edition, cqal96-1092246, http://library.cqpress.com/cqal96-1092246 (accessed August 23, 2010), originally published in *CQ Almanac 1996* (Washington, DC: Congressional Quarterly, 1997).

38. George W. Bush, "Proclamation 7627—America Recycles Day, 2002," November 14, 2002, in *Public Papers of the Presidents of the United States;* John T. Woolley and Gerhard Peters, *The American Presidency Project,* http://www.presidency.ucsb.edu//ws/?pid=61918 (accessed May 6, 2011); George W. Bush, "Proclamation

7846—America Recycles Day, 2004," November 15, 2004, in *Public Papers of the Presidents of the United States;* John T. Woolley and Gerhard Peters, *The American Presidency Project,* http://www.presidency.ucsb.edu//ws/?pid=62242 (accessed May 6, 2011).

39. "GOP Balks at Wage Plan in Water Bill," CQ Press Electronic Library, CQ Almanac Online Edition, cqal07-1006-44902-2047652, http://library.cqpress.com/cqal07-1112651 (accessed August 23, 2010), originally published in *CQ Almanac 2007* (Washington, DC: Congressional Quarterly, 2008).

Chapter 7 The Politics of Toxic and Hazardous Waste

1. "Three Mile Island Cleanup," CQ Press Electronic Library, CQ Almanac Online Edition, cqal82-1164403, http://library.cqpress.com/cqal82-1164403 (accessed August 24, 2010), originally published in *CQ Almanac 1982* (Washington, DC: Congressional Quarterly, 1983).

2. Walter A. Rosenbaum, *Environmental Politics and Policy,* 6th ed. (Washington, DC: CQ Press, 2005), 214.

3. Michael E. Kraft, *Environmental Policy and Politics* (New York: Longman, 2001), 39.

4. Ibid., 41.

5. John H. Baldwin, *Environmental Planning and Management* (Boulder, CO: Westview Press, 1985), 184.

6. Kraft, *Environmental Policy:* 43.

7. Daniel J. Fiorino, *Making Environmental Policy* (Berkeley: University of California Press, 1995), 30.

8. Samuel P. Hays, *Environmental Politics since 1945* (Pittsburgh: University of Pittsburgh Press, 2000), 117.

9. Richard N. L. Andrews, *Managing the Environment, Managing Ourselves* (New Haven, CT: Yale University Press, 1999), 184.

10. Lyndon B. Johnson, "Letter Assigning to HEW Responsibility for Developing a Computer-Based File on Toxic Chemicals," June 20, 1966, in *Public Papers of the President of the United States* (Washington, DC: U.S. Government Printing Office, 1967), 637.

11. Richard Nixon, "Special Message to the Congress Proposing the 1971 Environment Program," February 8, 1971 in *Public Papers of the Presidents of the United States* (Washington, DC: U.S. Government Printing Office, 1972), 125–142:132.

12. "Presidential Statement to Congress: Nixon's 1970 Environmental Message," CQ Press Electronic Library, CQ Almanac Online Edition, cqal71-869-26707-1254975, http://library.cqpress.com/cqal71-869-26707-1254975 (accessed August 23, 2010), originally published in *CQ Almanac 1971* (Washington, DC: Congressional Quarterly, 1972).

13. "Toxic Substances Control Measure Dies in House," *Congressional Quarterly Almanac* 28 (1972): 993–996.

14. "Toxic Substances Control," CQ Press Electronic Library, CQ Almanac Online Edition, cqal73-1227405, http://library.cqpress.com/cqalmanac/

cqal73-1227405 (accessed August 30, 2010), originally published in *CQ Almanac 1973* (Washington, DC: Congressional Quarterly, 1974).

15. Gerald R. Ford, "Statement on Signing the Toxic Substances Control Act," October 12, 1976, in *Public Papers of the President of the United States* (Washington, DC: U.S. Government Printing Office, 1977), 2486–2487.

16. "Toxic Substances Control Bill Cleared," *Congressional Quarterly Almanac* 32 (1976): 120–125; "Energy and Environment: Action Completed," *Congressional Quarterly Almanac* 32 (1976): 11.

17. Fiorino, *Making Environmental Policy:* 32; Baldwin, *Environmental Planning and Management:* 189.

18. Mary Clifford, "A Review of Federal Environmental Legislation," in Mary Clifford, ed., *Environmental Crime* (Gaithersburg, MD: Aspen, 1998), 95–109: 103.

19. "Toxic Substances Control Bill Cleared."

20. Gerald R. Ford, "Statement on Signing the Resource Conservation and Recovery Act of 1976," October 22, 1976, in *Public Papers of the President of the United States* (Washington, DC: U.S. Government Printing Office, 1977), 2610–2611.

21. Clifford, "Review of Federal Environmental Legislation": 102.

22. Baldwin, *Environmental Planning and Management:* 190.

23. Andrews, *Managing the Environment:* 247.

24. "Small Business Amendments," *Congressional Quarterly Almanac* 32 (1976): 89.

25. Gerald R. Ford, "Statement on Signing the Small Business Omnibus Bill," June 4, 1976, in *Public Papers of the President of the United States* (Washington, DC: U.S. Government Printing Office, 1977), 1790–1792.

26. Jimmy Carter, "The Environment," May 23, 1977, in *Public Papers of the President of the United States* (Washington, DC: U.S. Government Printing Office, 1978), 967–986.

27. Ibid.

28. "Message to Congress: State of Union," *Congressional Quarterly Almanac* 36 (1980): 20E.

29. Jimmy Carter, "Environmental Priorities and Programs," August 2, 1979, in *Public Papers of the President of the United States* (Washington, DC: U.S. Government Printing Office, 1980), 1353–1373: 1357.

30. Ibid.

31. Jimmy Carter, "State of Union Address," January 25, 1979, in *Public Papers of the President of the United States* (Washington, DC: U.S. Government Printing Office, 1980), 149.

32. "Congress Fails to Set Nuclear Waste Policy," CQ Press Electronic Library, CQ Almanac Online Edition, cqal80-1174736, http://library.cqpress.com/cqal80-1174736 (accessed August 23, 2010), originally published in *CQ Almanac 1980* (Washington, DC: Congressional Quarterly, 1981).

33. "Carter's Second Session Agenda," January 21, 1980, CQ Press Electronic Library, CQ Almanac Online Edition, cqal80-860-25882-1174115, http://library.cqpress.com/cqalmanac/cqal80-860-25882-1174115 (accessed August 30, 2010), originally published in *CQ Almanac 1980* (Washington, DC: Congressional Quarterly, 1981).

34. "Presidential Statement: Carter's Nuclear Waste Program," CQ Press Electronic Library, CQ Almanac Online Edition, cqal80-860-25882-1174371, http://library.cqpress.com/cqal80-860-25882-1174371 (accessed August 23, 2010), originally published in *CQ Almanac 1980* (Washington, DC: Congressional Quarterly, 1981).

35. "Toxic Substances," *Congressional Quarterly Almanac* 33 (1977): 676–677.

36. "Uranium Mill Waste Control," *Congressional Quarterly Almanac* 34 (1978): 750.

37. "Three Mile Island Cleanup," CQ Press Electronic Library, CQ Almanac Online Edition, cqal82-1164403, http://library.cqpress.com/cqal82-1164403 (accessed August 24, 2010), originally published in *CQ Almanac 1982* (Washington, DC: Congressional Quarterly, 1983).

38. Donald J. Rebovich, *Dangerous Ground* (New Brunswick, NJ: Transaction, 1992), 4–5.

39. "Environment: Toxic Waste 'Superfund,'" *Congressional Quarterly Almanac* 36 (1980): 24.

40. Andrews, *Managing the Environment:* 248.

41. Baldwin, *Environmental Planning and Management:* 191.

42. "Congress Clears 'Superfund' Legislation," *Congressional Quarterly Almanac* 36 (1980): 584–587.

43. Ibid.

44. Fiorino, *Making Environmental Policy:* 31.

45. "Congress Clears 'Superfund' Legislation."

46. "EPA Powers over Illegal Dumping Increased," *Congressional Quarterly Almanac* 36 (1980): 605–609.

47. "Toxic Substances," *Congressional Quarterly Almanac* 35 (1979): 682.

48. "Nuclear Waste Disposal," CQ Press Electronic Library, CQ Almanac Online Edition, cqal79-1184356, http://library.cqpress.com/cqal79-1184356 (accessed August 23, 2010), originally published in *CQ Almanac 1978* (Washington, DC: Congressional Quarterly, 1979).

49. "Congress Fails to Set Nuclear Waste Policy," CQ Press Electronic Library, CQ Almanac Online Edition, cqal80-1174736, http://library.cqpress.com/cqal80-1174736 (accessed August 23, 2010), originally published in *CQ Almanac 1980* (Washington, DC: Congressional Quarterly, 1981).

50. Ronald W. Reagan, "Message to the Congress Transmitting the Annual Report of the Council on Environmental Quality," June 27, 1983, in *Public Papers of the President of the United States* (Washington, DC: U.S. Government Printing Office, 1984): 919–920.

51. Ronald W. Reagan, "Remarks at the Swearing in Ceremony of William D. Ruckelshaus as Administrator of the Environmental Protection Agency," May 18, 1983, in *Public Papers of the President of the United States* (Washington, DC: U.S. Government Printing Office, 1984), 733–735.

52. Ronald W. Reagan, "State of Union," January 18, 1984, in *Public Papers of the President of the United States* (Washington, DC: U.S. Government Printing Office, 1985), 90.

53. Ronald W. Reagan, "Statement on Proposed Superfund Reauthorization Legislation," February 22, 1985, in *Public Papers of the President of the United States* (Washington, DC: U.S. Government Printing Office, 1986), 205.

54. "Comprehensive Nuclear Waste Plan Enacted," *Congressional Quarterly Almanac* 38 (1982): 304–308.

55. "Nuclear Waste Disposal Bills Reported," CQ Press Electronic Library, CQ Almanac Online Edition, cqal81-1173250, http://library.cqpress.com/cqal81-1173250 (accessed August 23, 2010), originally published in *CQ Almanac 1981* (Washington, DC: Congressional Quarterly, 1982).

56. "RCRA Reauthorization Dies," CQ Press Electronic Library, CQ Almanac Online Edition, cqal82-1163149, http://library.cqpress.com/cqal82-1163149 (accessed August 23, 2010), originally published in *CQ Almanac 1982* (Washington, DC: Congressional Quarterly, 1983).

57. "Congress Tightens Hazardous Waste Controls," *Congressional Quarterly Almanac* 40 (1984): 303–304; "House Votes to Tighten Hazardous Waste Law," *Congressional Quarterly Almanac* 39 (1983): 335.

58. "Congress Tightens Hazardous Waste Controls."

59. "Environment: Hazardous Waste Control," *Congressional Quarterly Almanac* 40 (1984): 20.

60. Fiorino, *Making Environmental Policy:* 31.

61. Ibid., 152.

62. "Environment: Superfund," *Congressional Quarterly Almanac* 40 (1984): 20; "Superfund Reauthorization Passed by House," *Congressional Quarterly Almanac* 40 (1984): 309–313.

63. "Nuclear Waste Sites," CQ Press Electronic Library, CQ Almanac Online Edition, cqal84-1153190, http://library.cqpress.com/cqal84-1153190 (accessed August 23, 2010), originally published in *CQ Almanac 1984* (Washington, DC: Congressional Quarterly, 1985).

64. "Hazardous Materials," CQ Press Electronic Library, CQ Almanac Online Edition, cqal84-1152902, http://library.cqpress.com/cqal84-1152902 (accessed August 23, 2010), originally published in *CQ Almanac 1984* (Washington, DC: Congressional Quarterly, 1985).

65. "Low-Level Nuclear Waste Bill Clears Congress," *Congressional Quarterly Almanac* 41 (1985): 214–216.

66. Ibid.

67. "Reagan Signs 'Superfund' Waste-Cleanup Bill," *Congressional Quarterly Almanac* 42 (1986): 111–120; "Environment: Superfund Reauthorization," *Congressional Quarterly Almanac* 41 (1985): 17; "House, Senate Pass Superfund Authorization," *Congressional Quarterly Almanac* 41 (1985): 191–198; Ronald W. Reagan, "Statement on Signing the Superfund Amendments and Reauthorization Act of 1986," October 17, 1986, in *Public Papers of the President of the United States* (Washington, DC: U.S. Government Printing Office, 1987), 1401–1402.

68. Fiorino, *Making Environmental Policy:* 31.

69. Ibid.

70. "Hazardous-Waste Liability," CQ Press Electronic Library, CQ Almanac Online Edition, cqal85-1147705, http://library.cqpress.com/cqal85-1147705

(accessed August 23, 2010), originally published in *CQ Almanac 1985* (Washington, DC: Congressional Quarterly, 1986).

71. "Nevada Chosen to Receive Nuclear Waste," CQ Press Electronic Library, CQ Almanac Online Edition, cqal87-1145032, http://library.cqpress.com/cqal87-1145032 (accessed August 23, 2010), originally published in *CQ Almanac 1987* (Washington, DC: Congressional Quarterly, 1988).

72. "House Tightens Regulations for Disposal of Polychlorinated Biphenyls (PCBs)," CQ Press Electronic Library, CQ Almanac Online Edition, cqal88-1141449, http://library.cqpress.com/cqal88-1141449 (accessed August 23, 2010), originally published in *CQ Almanac 1988* (Washington, DC: Congressional Quarterly, 1989).

73. George H. W. Bush, "White House Fact Sheet on Environmental Initiatives," September 18, 1989, in *Public Papers of the President of the United States* (Washington, DC: U.S. Government Printing Office, 1990), 1212–1215.

74. George H. W. Bush, "Statement on Signing Legislation Waiving Federal Immunity Relating to Solid and Hazardous Waste," October 6, 1992, in *Public Papers of the President of the United States* (Washington, DC: U.S. Government Printing Office, 1993), 1760–1761.

75. "Resource Conservation and Recovery Act (RCRA) Extended to Federal Sites," CQ Press Electronic Library, CQ Almanac Online Edition, cqal89-1139830, http://library.cqpress.com/cqal89-1139830 (accessed August 23, 2010), originally published in *CQ Almanac 1989* (Washington, DC: Congressional Quarterly, 1990).

76. "Yucca Mountain Dump Site Still Debated," CQ Press Electronic Library, CQ Almanac Online Edition, cqal91-1110430, http://library.cqpress.com/cqal91-1110430 (accessed August 23, 2010), originally published in *CQ Almanac 1991* (Washington, DC: Congressional Quarterly, 1992).

77. "Nuclear Waste Dump Gets Cautious Nod," CQ Press Electronic Library, CQ Almanac Online Edition, cqal92-1107825, http://library.cqpress.com/cqal92-1107825 (accessed August 23, 2010), originally published in *CQ Almanac 1992* (Washington, DC: Congressional Quarterly, 1993).

78. Fiorino, *Making Environmental Policy:* 148.

79. "Federal Facilities Cleanup," CQ Press Electronic Library, CQ Almanac Online Edition, cqal91-1110417, http://library.cqpress.com/cqal91-1110417 (accessed August 23, 2010), originally published in *CQ Almanac 1991* (Washington, DC: Congressional Quarterly, 1992).

80. William J. Clinton, "Remarks to the Community in Louisville," January 24, 1996, in *Public Papers of the Presidents of the United States*; John T. Woolley and Gerhard Peters, *The American Presidency Project,* http://www.presidency.ucsb.edu//ws/?pid=53125 (accessed May 5, 2011); William J. Clinton, "Remarks to the Community In Hackensack, New Jersey," March 11, 1996, in *Public Papers of the Presidents of the United States*; John T. Woolley and Gerhard Peters, *The American Presidency Project,* http://www.presidency.ucsb.edu//ws/?pid=52526 (accessed May 5, 2011).

81. William J. Clinton, "Proclamation 6664—Cancer Control Month, 1994,"April 7, 1994, in *Public Papers of the Presidents of the United States*; John

T. Woolley and Gerhard Peters, *The American Presidency Project,* http://www.presidency.ucsb.edu//ws/?pid=49924 (accessed May 5, 2011).

82. William J. Clinton, "Remarks in Freehold Borough, New Jersey," September 24, 1996, in *Public Papers of the Presidents of the United States*; John T. Woolley and Gerhard Peters, *The American Presidency Project,* http://www.presidency.ucsb.edu//ws/?pid=51979 (accessed May 5, 2011).

83. William J. Clinton, "Remarks on the Observance of Earth Day," April 21, 1994, in *Public Papers of the Presidents of the United States*; John T. Woolley and Gerhard Peters, *The American Presidency Project,* http://www.presidency.ucsb.edu//ws/?pid=50012 (accessed May 5, 2011).

84. William J. Clinton, "Interview with Mark Riley and Laura Blackburne of WLIB Radio New York City," October 18, 1994, in *Public Papers of the Presidents of the United States*; John T. Woolley and Gerhard Peters, *The American Presidency Project,* http://www.presidency.ucsb.edu//ws/?pid=49317 (accessed May 5, 2011).

85. William J. Clinton, "Remarks to the Governor's Leadership Conference in New York City," October 19, 1994, in *Public Papers of the Presidents of the United States*; John T. Woolley and Gerhard Peters, *The American Presidency Project,* http://www.presidency.ucsb.edu//ws/?pid=49326 (accessed May 5, 2011); William J. Clinton, "Remarks to the Community in Stratford, Connecticut," October 15, 1994, in *Public Papers of the Presidents of the United States*; John T. Woolley and Gerhard Peters, *The American Presidency Project,* http://www.presidency.ucsb.edu//ws/?pid=49307 (accessed May 5, 2011).

86. William J. Clinton, "Remarks on Environmental Protection in Baltimore, Maryland," August 8, 1995, in *Public Papers of the Presidents of the United States*; John T. Woolley and Gerhard Peters, *The American Presidency Project,* http://www.presidency.ucsb.edu//ws/?pid=51718 (accessed May 5, 2011).

87. William J. Clinton, "The President's News Conference in Moscow," April 20, 1996, in *Public Papers of the Presidents of the United States*; John T. Woolley and Gerhard Peters, *The American Presidency Project,* http://www.presidency.ucsb.edu//ws/?pid=65089 (accessed May 5, 2011).

88. William J. Clinton, "Background Briefing by Senior Administration Officials," February 11, 1994, in *Public Papers of the Presidents of the United States*; John T. Woolley and Gerhard Peters, *The American Presidency Project,* http://www.presidency.ucsb.edu//ws/?pid=59718 (accessed May 5, 2011).

89. "No Floor Action on Superfund Bill," CQ Press Electronic Library, CQ Almanac Online Edition, cqal94-1103302, http://library.cqpress.com/cqal94-1103302 (accessed August 23, 2010), originally published in *CQ Almanac 1994* (Washington, DC: Congressional Quarterly, 1995).

90. "No Progress on Superfund Overhaul," *Congressional Quarterly Almanac* 51 (1995): 5-11–5-13; "GOP Scraps Superfund Plans," *Congressional Quarterly Almanac* 52 (1996): 4-13–4-14.

91. "$76 Billion Cut from 1995 Spending," *Congressional Quarterly Almanac* 51 (1995): 11-96–11-105: 11-104.

92. "House Puts Stop to Texas Plans for Radioactive Waste Dumps," CQ Press Electronic Library, CQ Almanac Online Edition, cqal95-1100418, http://library.cqpress.com/cqal95-1100418 (accessed August 23, 2010), originally published in *CQ Almanac 1994* (Washington, DC: Congressional Quarterly, 1995).

93. "Plans to Store Nuclear Waste Stalls," *Congressional Quarterly Almanac* 51 (1995): 5–27; "Nevadans Postpone Dump Site Bill," *Congressional Quarterly Almanac* 52 (1996): 4-30–4-32.

94. Ibid.

95. "Waste Isolation Pilot Plant Stalls in House," CQ Press Electronic Library, CQ Almanac Online Edition, cqal96-1092262, http://library.cqpress.com/cqal96-1092262 (accessed August 23, 2010), originally published in *CQ Almanac 1996* (Washington, DC: Congressional Quarterly, 1997).

96. "Environmental Legislation Highlights Coastal Zones, Waste Management, and Antarctica Protections," CQ Press Electronic Library, CQ Almanac Online Edition, cqal96-1092246, http://library.cqpress.com/cqal96-1092246 (accessed August 23, 2010), originally published in *CQ Almanac 1996* (Washington, DC: Congressional Quarterly, 1997).

97. "Partisan Disputes Stop Superfund Bill," *Congressional Quarterly Almanac* 53 (1997): 4–11.

98. "Congress Unable to Resolve Partisan Differences on Superfund Overhaul," *Congressional Quarterly Almanac* 54 (1998): 11-5.

99. "Nevada Waste Site Plan Again Stalls," CQ Press Electronic Library, CQ Almanac Online Edition, cqal97-0000181023, http://library.cqpress.com/cqal97-0000181023 (accessed August 23, 2010), originally published in *CQ Almanac 1997* (Washington, DC: Congressional Quarterly, 1998).

100. "Tri-State Radioactive Waste Pact Gets House Endorsement," CQ Press Electronic Library, CQ Almanac Online Edition, cqal97-000018011, http://library.cqpress.com/cqal97-000018011 (accessed August 23, 2010), originally published in *CQ Almanac 1997* (Washington, DC: Congressional Quarterly, 1998).

101. "Congress Clears 'Superfund' Legislation," CQ Press Electronic Library, CQ Almanac Online Edition, cqal97-0000181031, http://library.cqpress.com/cqal97-0000181031 (accessed August 23, 2010), originally published in *CQ Almanac 1997* (Washington, DC: Congressional Quarterly, 1998).

102. "Senate Postpones Action on Nuclear Waste Storage until Second Session," *Congressional Quarterly Almanac* 55 (1999): 12-3–12-7.

103. "Clinton Rejects Legislation to Establish Temporary Nuclear Waste Dump Site," *Congressional Quarterly Almanac* 55 (1999): 10-3–10-4.

104. "Presidential Veto Message: Clinton Claims Nuclear Waste Bill Would Halt New EPA Radiation Standards," CQ Press Electronic Library, CQ Almanac Online Edition, cqal00-834-24292-1081635, http://library.cqpress.com/cqal00-834-24292-1081635 (accessed August 23, 2010), originally published in *CQ Almanac 2000* (Washington, DC: Congressional Quarterly, 2001).

105. "Superfund Overhauls Stalls: Narrow Exemption Included in Omnibus Spending Bill," *Congressional Quarterly Almanac* 55 (1999): 12-11.

106. "Congress Unable to Satisfy Gas Industry and Farers over Possible Ban of MTBE," *Congressional Quarterly Almanac 55* (1999): 10-19–10-21.

107. George W. Bush, "The President's Radio Address," April 24, 2004, in *Public Papers of the President of the United States* (Washington, DC: U.S. Government Printing Office, 2005), 665–667.

108. Michael E. Kraft, "Environmental Policy in Congress," in Norman J. Vig and Michael E. Kraft, eds., *Environmental Policy: New Directions for the Twenty-First Century* (Washington, DC: CQ Press, 2006), 124–147: 139.

109. George W. Bush, "Letter to Congressional Leaders Recommending the Yucca Mountain Site for the Disposal of Spent Nuclear Fuel and Nuclear Waste," February 15, 2002, in *Public Papers of the President of the United States* (Washington, DC: U.S. Government Printing Office, 2003); From John T. Woolley and Gerhard Peters, *The American Presidency Project,* http://www.presidency.ucsb.edu/ws/?pid=72967 (accessed May 18, 2011).

110. "House OKs Nuclear Liability Limit," *Congressional Quarterly Almanac* 57 (2001): 9–10.

111. "Bush, Hill Back Nuclear Waste Sites," CQ Press Electronic Library, CQ Almanac Online Edition, cqal02-236-10379-664529, http://library.cqpress.com/cqal02-236-10379-664529 (accessed August 23, 2010), originally published in *CQ Almanac 2002* (Washington, DC: Congressional Quarterly, 2003).

112. "Congress Clears Brownfields Bill," CQ Press Electronic Library, CQ Almanac Online Edition, cqal01-106-6390-328777, http://library.cqpress.com/cqal01-106-6390-328777 (accessed August 23, 2010), originally published in *CQ Almanac 2001* (Washington, DC: Congressional Quarterly, 2002).

113. "Bill to Aid Brownfields Cleanup Passes House, Dies in Senate," CQ Press Electronic Library, CQ Almanac Online Edition, cqal02-236-10379-664535, http://library.cqpress.com/cqal02-236-10379-664535 (accessed August 23, 2010), originally published in *CQ Almanac 2002* (Washington, DC: Congressional Quarterly, 2003).

114. Coral Davenport, "Beyond Yucca, a Blurry View," *CQ Weekly Online,* April 27, 2009: 966–972, http://library.cqpress.com/cqweekly/weeklyreport111-000003103264 (accessed September 2, 2010).

115. Joanna Anderson, "Energy-Water Bill Heeds Obama Wish: Zero for Yucca Mountain Nuclear Site," *CQ Weekly Online,* September 2, 2010: 1810, http://library.cqpress.com/cqweekly/weeklyreport111-000003709314 (accessed September 2, 2010).

116. Eugene Mulero, "Bill to Tighten Nuclear Waste Imports Angers GOP," *CQ Weekly Online,* November 9, 2009: 2601, http://library.cqpress.com/cqweekly/weeklyreport111-000003243417 (accessed September 2, 2010).

117. Joanna Anderson, "Bill Would Require Stricter Guidelines for Chemical, Water Treatment Plants," *CQ Weekly Online,* November 9, 2009: 2600, http://library.cqpress.com/cqweekly/weeklyreport111-000003243408 (accessed September 2, 2010).

118. Eugene Mulero, "Senate Panel OKs Grants to Help with E-Waste," *CQ Weekly Online,* December 14, 2009: 2894, http://library.cqpress.com/cqweekly/weeklyreport111-000003265937 (accessed September 2, 2010).

119. Kraft, *Environmental Policy:* 38.

Chapter 8 The Politics of Land

1. Richard N. L. Andrews, *Managing the Environment, Managing Ourselves* (New Haven, CT: Yale University Press, 1999), ix, 309.

2. Philip F. Cramer, *Deep Environmental Politics* (Westport, CT: Praeger, 1998), 81.

3. "Soil Conservation and Fertilizer," CQ Press Electronic Library, CQ Almanac Online Edition, cqal47-1398378, http://library.cqpress.com/cqalmanac/cqal47-1398378 (accessed August 31, 2010), originally published in *CQ Almanac 1947* (Washington, DC: Congressional Quarterly, 1948).

4. "Forest Survey," CQ Press Electronic Library, CQ Almanac Online Edition, cqal49-1399230, http://library.cqpress.com/cqalmanac/cqal49-1399230 (accessed August 31, 2010), originally published in *CQ Almanac 1949* (Washington, DC: Congressional Quarterly, 1950).

5. Dwight Eisenhower, "Special Message to the Congress on the Nation's Natural Resources," July 31, 1953 in John T. Woolley and Gerhard Peters, *The American Presidency Project*, http://www.presidency.ucsb.edu/ws/?pid=9660 (accessed May 18, 2011.

6. "President Seeks Conservation Program," CQ Press Electronic Library, CQ Almanac Online Edition, cqal62-878-28126-1324166, http://library.cqpress.com/cqal62-878-28126-1324166 (accessed August 23, 2010), originally published in CQ *Almanac 1962* (Washington, DC: Congressional Quarterly, 1963).

7. "Senate Approves Wilderness System, 78–8," CQ Press Electronic Library, CQ Almanac Online Edition, cqal61-1371932, http://library.cqpress.com/cqal61-1371932 (accessed August 23, 2010), originally published in *CQ Almanac 1961* (Washington, DC: Congressional Quarterly, 1962).

8. "Wetlands Program," CQ Press Electronic Library, CQ Almanac Online Edition, cqal61-1372672, http://library.cqpress.com/cqal61-1372672 (accessed August 23, 2010), originally published in *CQ Almanac 1961* (Washington, DC: Congressional Quarterly, 1961).

9. "Wilderness System," CQ Press Electronic Library, CQ Almanac Online Edition, cqal63-1315690, http://library.cqpress.com/cqal63-1315690 (accessed August 23, 2010), originally published in *CQ Almanac 1963* (Washington, DC: Congressional Quarterly, 1964).

10. "Congress Passes Wilderness Act," CQ Press Electronic Library, CQ Almanac Online Edition, cqal64-1303184, http://library.cqpress.com/cqal64-1303184 (accessed August 23, 2010), originally published in *CQ Almanac 1964* (Washington, DC: Congressional Quarterly, 1965).

11. "Land and Water Conservation Act Passed," CQ Press Electronic Library, CQ Almanac Online Edition, cqal64-1303165, http://library.cqpress.com/cqal64-1303165 (accessed August 30, 2010), originally published in *CQ Almanac 1964* (Washington, DC: Congressional Quarterly, 1965).

12. "Major Study of Public Land Laws Authorized," CQ Press Electronic Library, CQ Almanac Online Edition, cqal64-1303218, http://library.cqpress.com/cqal64-1303218 (accessed August 30, 2010), originally published in *CQ Almanac 1964* (Washington, DC: Congressional Quarterly, 1965).

13. "Message to Congress: Johnson's Conservation Message," CQ Press Electronic Library, CQ Almanac Online Edition, cqal68-1124522, http://library.cqpress.com/cqal68-1124522 (accessed August 23, 2010), originally published in *CQ Almanac 1968* (Washington, DC: Congressional Quarterly, 1969).

14. "Surface Mining," CQ Press Electronic Library, CQ Almanac Online Edition, cqal67-1314394, http://library.cqpress.com/cqal67-1314394 (accessed August 30, 2010), originally published in *CQ Almanac 1967* (Washington, DC: Congressional Quarterly, 1968).

15. "Johnson's Conservation Message."

16. "Natural Beauty Sought as U.S. Goal," CQ Press Electronic Library, CQ Almanac Online Edition, cqal65-1257710, http://library.cqpress.com/cqal65-1257710 (accessed August 23, 2010), originally published in *CQ Almanac 1965* (Washington, DC: Congressional Quarterly, 1966).

17. Henry P. Caulfield, "The Conservation and Environmental Movements: An Historical Analysis," in James P. Lester, ed., *Environmental Politics and Policy* (Durham, NC: Duke University Press, 1989), 37.

18. "San Rafael Wilderness," CQ Press Electronic Library, CQ Almanac Online Edition, cqal67-1312822, http://library.cqpress.com/cqalmanac/cqal67-1312822 (accessed August 30, 2010), originally published in *CQ Almanac 1967* (Washington, DC: Congressional Quarterly, 1968).

19. "Estuary Preservation," CQ Press Electronic Library, CQ Almanac Online Edition, cqal68-1283827, http://library.cqpress.com/cqalmanac/cqal68-1283827 (accessed August 30, 2010), originally published in *CQ Almanac 1968* (Washington, DC: Congressional Quarterly, 1969).

20. "Surface Mining," CQ Press Electronic Library, CQ Almanac Online Edition, cqal68-1284344, http://library.cqpress.com/cqal68-1284344 (accessed August 30, 2010), originally published in *CQ Almanac 1968* (Washington, DC: Congressional Quarterly, 1969).

21. "Major Anti-Pollution Measures Clear 91st Congress," CQ Press Electronic Library, CQ Almanac Online Edition, cqal70-1292699, http://library.cqpress.com/cqal70-1292699 (accessed August 23, 2010), originally published in *CQ Almanac 1970* (Washington, DC: Congressional Quarterly, 1971).

22. "Presidential Statement to Congress: Nixon's 1970 Environmental Message," CQ Press Electronic Library, CQ Almanac Online Edition, cqal71-869-26707-1254975, http://library.cqpress.com/cqal71-869-26707-1254975 (accessed August 31, 2010), originally published in *CQ Almanac 1971* (Washington, DC: Congressional Quarterly, 1972).

23. "Presidential Statement to Congress: Nixon on the Environment 1970," CQ Press Electronic Library, CQ Almanac Online Edition, cqal70-1290732, http://library.cqpress.com/cqal70-1290732 (accessed August 23, 2010), originally published in *CQ Almanac 1970* (Washington, DC: Congressional Quarterly, 1971).

24. "Presidential Statement to Congress: Nixon Proposes New Wilderness Protection Areas," CQ Press Electronic Library, CQ Almanac Online Edition, cqal 72-868-26690-1251948, http://library.cqpress.com/cqalmanac/cqal72-868-26690-1251948 (accessed August 30, 2010), originally published in *CQ Almanac 1972* (Washington, DC: Congressional Quarterly, 1973).

25. “Presidential Statement to Congress: Nixon Renews Call for Environmental Legislation,” CQ Press Electronic Library, CQ Almanac Online Edition, cqal 72-868-26690-1251768, http://library.cqpress.com/cqalmanac/cqal72-868-26690-1251768 (accessed August 30, 2010), originally published in *CQ Almanac 1972* (Washington, DC: Congressional Quarterly, 1973).

26. “New Wilderness Areas,” CQ Press Electronic Library, CQ Almanac Online Edition, cqal70-1293873, http://library.cqpress.com/cqalmanac/cqal70-1293873 (accessed August 30, 2010), originally published in *CQ Almanac 1970* (Washington, DC: Congressional Quarterly, 1971).

27. “National Land Use Policy: A Beginning in 1970,” CQ Press Electronic Library, CQ Almanac Online Edition, cqal70-1293822, http://library.cqpress.com/cqal70-1293822 (accessed August 30, 2010), originally published in *CQ Almanac 1970* (Washington, DC: Congressional Quarterly, 1971).

28. “Natural Resources,” CQ Press Electronic Library, CQ Almanac Online Edition, cqal69-1248497, http://library.cqpress.com/cqalmanac/cqal69-1248497 (accessed August 30, 2010), originally published in *CQ Almanac 1969* (Washington, DC: Congressional Quarterly, 1970).

29. “Parks and Conservation,” CQ Press Electronic Library, CQ Almanac Online Edition, cqal69-1246891, http://library.cqpress.com/cqalmanac/cqal69-1246891 (accessed August 30, 2010), originally published in *CQ Almanac 1969* (Washington, DC: Congressional Quarterly, 1970).

30. “Alaska Land Claims,” CQ Press Electronic Library, CQ Almanac Online Edition, cqal70-1294963, http://library.cqpress.com/cqal70-1294963 (accessed August 30, 2010), originally published in *CQ Almanac 1970* (Washington, DC: Congressional Quarterly, 1971).

31. “Cedar Keys,” CQ Press Electronic Library, CQ Almanac Online Edition, cqal72-1249026, http://library.cqpress.com/cqalmanac/cqal72-1249026 (accessed August 30, 2010), originally published in *CQ Almanac 1972* (Washington, DC: Congressional Quarterly, 1973).

32. “Lassen Volcanic Wilderness Area,” CQ Press Electronic Library, CQ Almanac Online Edition, cqal72-1249346, http://library.cqpress.com/cqalmanac/cqal72-1249346 (accessed August 30, 2010), originally published in *CQ Almanac 1972* (Washington, DC: Congressional Quarterly, 1973).

33. “National Preservation Sites,” CQ Press Electronic Library, CQ Almanac Online Edition, cqal72-1249399, http://library.cqpress.com/cqalmanac/cqal72-1249399 (accessed August 230, 2010), originally published in *CQ Almanac 1972* (Washington, DC: Congressional Quarterly, 1973).

34. “Sawtooth Recreation Area,” CQ Press Electronic Library, CQ Almanac Online Edition, cqal72-1250887, http://library.cqpress.com/cqalmanac/cqal72-1250887 (accessed August 23, 2010), originally published in *CQ Almanac 1972* (Washington, DC: Congressional Quarterly, 1973).

35. “Montana Wilderness Tracts,” CQ Press Electronic Library, CQ Almanac Online Edition, cqal71-1254395, http://library.cqpress.com/cqalmanac/cqal71-1254395 (accessed August 30, 2010), originally published in *CQ Almanac 1971* (Washington, DC: Congressional Quarterly, 1972); “Sawtooth Recreation Area”; “Sycamore Canyon Wilderness,” CQ Press Electronic Library, CQ Almanac Online

Edition, cqal72-1250436, http://library.cqpress.com/cqalmanac/cqal72-1250436 (accessed August 23, 2010), originally published in *CQ Almanac 1972* (Washington, DC: Congressional Quarterly, 1973).

36. "Wild Areas," CQ Press Electronic Library, CQ Almanac Online Edition, cqal72-1250573, http://library.cqpress.com/cqalmanac/cqal72-1250573 (accessed August 23, 2010), originally published in *CQ Almanac 1972* (Washington, DC: Congressional Quarterly, 1973).

37. "Bills Regulating Strip Mining Die in Senate," CQ Press Electronic Library, CQ Almanac Online Edition, cqal72-1249599, http://library.cqpress.com/cqal72-1249599 (accessed August 30, 2010), originally published in *CQ Almanac 1972* (Washington, DC: Congressional Quarterly, 1973).

38. "Senate Passes Strong Land-Use Planning Bill," CQ Press Electronic Library, CQ Almanac Online Edition, cqal73-1227374, http://library.cqpress.com/cqal73-1227374 (accessed August 30, 2010), originally published in *CQ Almanac 1973* (Washington, DC: Congressional Quarterly, 1974); "House Kills Land Use Bill on Procedural Vote," CQ Press Electronic Library, CQ Almanac Online Edition, cqal74-1222595, http://library.cqpress.com/cqal74-1222595 (accessed August 30, 2010), originally published in *CQ Almanac 1974* (Washington, DC: Congressional Quarterly, 1975).

39. "Land Management: Land and Water Conservation Act Passed," CQ Press Electronic Library, CQ Almanac Online Edition, cqal74-1222680, http://library.cqpress.com/cqal74-1222680 (accessed August 30, 2010), originally published in *CQ Almanac 1974* (Washington, DC: Congressional Quarterly, 1975).

40. "Senate Passes Strip Mining Bill; House Fails to Act," CQ Press Electronic Library, CQ Almanac Online Edition, cqal73-1227276, http://library.cqpress.com/cqal73-1227276 (accessed August 30, 2010), originally published in *CQ Almanac 1973* (Washington, DC: Congressional Quarterly, 1974).

41. "Strip Mining Bill Pocket Vetoed," CQ Press Electronic Library, CQ Almanac Online Edition, cqal74-1222443, http://library.cqpress.com/cqal74-1222443 (accessed August 30, 2010), originally published in *CQ Almanac 1974* (Washington, DC: Congressional Quarterly, 1975).

42. Andrews, *Managing the Environment:* ix, 310.

43. "Eastern Wilderness Areas," CQ Press Electronic Library, CQ Almanac Online Edition, cqal73-1227416, http://library.cqpress.com/cqalmanac/cqal73-1227416 (accessed August 23, 2010), originally published in *CQ Almanac 1973* (Washington, DC: Congressional Quarterly, 1974).

44. "Eastern Wilderness Areas," CQ Press Electronic Library, CQ Almanac Online Edition, cqal74-1222738, http://library.cqpress.com/cqalmanac/cqal74-1222738 (accessed August 23, 2010), originally published in *CQ Almanac 1974* (Washington, DC: Congressional Quarterly, 1975).

45. "Okefenokee Wilderness Area," CQ Press Electronic Library, CQ Almanac Online Edition, cqal74-1222281, http://library.cqpress.com/cqalmanac/cqal74-1222281 (accessed August 23, 2010), originally published in *CQ Almanac 1974* (Washington, DC: Congressional Quarterly, 1975).

46. "Cascade Head Forest Area," CQ Press Electronic Library, CQ Almanac Online Edition, cqal74-1222864, http://library.cqpress.com/cqalmanac/cqal74-1222864 (accessed August 31, 2010), originally published in *CQ Almanac 1974* (Washington, DC: Congressional Quarterly, 1975).

47. Gerald R. Ford, "Letter to the President of the Senate and the Speaker of the House Transmitting Proposals to Establish New National Wilderness Areas," December 4, 1974, in *Public Papers of the President of the United States* (Washington, DC: U.S. Government Printing Office, 1975), 703–704.

48. Gerald R. Ford, "Message to the Congress Proposing Establishment of New National Wilderness Areas," December 4, 1974, in *Public Papers of the President of the United States* (Washington, DC: U.S. Government Printing Office, 1975), 704–711.

49. "Presidential Statement: Ford Proposes Creation of Wilderness Areas," CQ Press Electronic Library, CQ Almanac Online Edition, cqal74-866-26343-1220574, http://library.cqpress.com/cqalmanac/cqal74-866-26343-1220574 (accessed August 23, 2010), originally published in *CQ Almanac 1974* (Washington, DC: Congressional Quarterly, 1975).

50. Gerald R. Ford, "Statement on Signing a Bill Designating a National Forest Area in Colorado as the Flat Tops Wilderness," December 13, 1975, in *Public Papers of the President of the United States* (Washington, DC: U.S. Government Printing Office, 1976), 1968–1969.

51. Gerald R. Ford, "Special Message to the Congress Transmitting Proposed Bicentennial Land Heritage Legislation," August 31, 1976, in *Public Papers of the President of the United States* (Washington, DC: U.S. Government Printing Office, 1977), 2193–2195.

52. "Energy and Environment 1974: Overview," CQ Press Electronic Library, CQ Almanac Online Edition, cqal74-1222249, http://library.cqpress.com/cqal74-1222249 (accessed August 31, 2010), originally published in *CQ Almanac 1974* (Washington, DC: Congressional Quarterly, 1975).

53. "Land Management," CQ Press Electronic Library, CQ Almanac Online Edition, cqal75-1213906, http://library.cqpress.com/cqal75-1213906 (accessed August 30, 2010), originally published in *CQ Almanac 1975* (Washington, DC: Congressional Quarterly, 1976).

54. "Public Land Management," CQ Press Electronic Library, CQ Almanac Online Edition, cqal76-1189303, http://library.cqpress.com/cqal76-1189303 (accessed August 30, 2010), originally published in *CQ Almanac 1976* (Washington, DC: Congressional Quarterly, 1977).

55. "House Sustains Strip Mining Veto," CQ Press Electronic Library, CQ Almanac Online Edition, cqal75-1213516, http://library.cqpress.com/cqalmanac/cqal75-1213516 (accessed August 31, 2010), originally published in *CQ Almanac 1975* (Washington, DC: Congressional Quarterly, 1976).

56. "Alpine Lakes Wilderness," CQ Press Electronic Library, CQ Almanac Online Edition, cqal76-1189059, http://library.cqpress.com/cqalmanac/cqal76-1189059 (accessed August 23, 2010), originally published in *CQ Almanac 1976* (Washington, DC: Congressional Quarterly, 1977).

57. Andrews, *Managing the Environment:* 311.

58. Tom Arrandale, "Access to Federal Lands," in Hoyt Gimlin, ed., *Environmental Issues: Prospects and Problems* (Washington, DC: Congressional Quarterly, 1982), 3–20: 13.

59. "Environmental Impact," CQ Press Electronic Library, CQ Almanac Online Edition, cqal75-1213592, http://library.cqpress.com/cqal75-1213592 (accessed August 30, 2010), originally published in *CQ Almanac 1975* (Washington, DC: Congressional Quarterly, 1976).

60. Jimmy Carter, "The Environment," May 23, 1977, in *Public Papers of the President of the United States* (Washington, DC: U.S. Government Printing Office, 1978), 967–986.

61. Jimmy Carter, "Remarks on Signing H.R. 3454 into Law," February 24, 1978, in *Public Papers of the President of the United States* (Washington, DC: U.S. Government Printing Office, 1979), 409–410.

62. "Carter Acts to Preserve Alaskan Wilderness," CQ Press Electronic Library, CQ Almanac Online Edition, cqal78-1236992, http://library.cqpress.com/cqalmanac/cqal78-1236992 (accessed August 23, 2010), originally published in *CQ Almanac 1978* (Washington, DC: Congressional Quarterly, 1979).

63. "Carter's Second Session Agenda, January 21, 1980," CQ Press Electronic Library, CQ Almanac Online Edition, cqal80-860-25882-1174115, http://library.cqpress.com/cqalmanac/cqal80-860-25882-1174115 (accessed August 30, 2010), originally published in *CQ Almanac 1980* (Washington, DC: Congressional Quarterly, 1981).

64. "Strip Mining Control Bill Signed," CQ Press Electronic Library, CQ Almanac Online Edition, cqal77-1203778, http://library.cqpress.com/cqal77-1203778 (accessed August 30, 2010), originally published in *CQ Almanac 1977* (Washington, DC: Congressional Quarterly, 1978).

65. "New Wilderness Areas," CQ Press Electronic Library, CQ Almanac Online Edition, cqal77-1204233, http://library.cqpress.com/cqalmanac/cqal77-1204233 (accessed August 23, 2010), originally published in *CQ Almanac 1977* (Washington, DC: Congressional Quarterly, 1978).

66. "EPA FY 80 Research Funds," CQ Press Electronic Library, CQ Almanac Online Edition, cqal80-1175137, http://library.cqpress.com/cqal80-1175137 (accessed August 31, 2010), originally published in *CQ Almanac 1980* (Washington, DC: Congressional Quarterly, 1981).

67. "House Passes Alaska Bill but Senate Stymied," CQ Press Electronic Library, CQ Almanac Online Edition, cqal79-1184209, http://library.cqpress.com/cqalmanac/cqal79-1184209 (accessed August 23, 2010), originally published in *CQ Almanac 1979* (Washington, DC: Congressional Quarterly, 1980).

68. "Congress Clears Alaska Lands Legislation," CQ Press Electronic Library, CQ Almanac Online Edition, cqal80-1175027, http://library.cqpress.com/cqal80-1175027 (accessed August 30, 2010), originally published in *CQ Almanac 1980* (Washington, DC: Congressional Quarterly, 1981).

69. "Rare II Wilderness Bills," CQ Press Electronic Library, CQ Almanac Online Edition, cqal80-1175204, http://library.cqpress.com/cqalmanac/cqal80-1175204

(accessed August 23, 2010), originally published in *CQ Almanac 1980* (Washington, DC: Congressional Quarterly, 1981).

70. "Strip Mine Exemptions," CQ Press Electronic Library, CQ Almanac Online Edition, cqal79-1184293, http://library.cqpress.com/cqal79-1184293 (accessed August 30, 2010), originally published in *CQ Almanac 1979* (Washington, DC: Congressional Quarterly, 1980).

71. "Strip Mining Amendments," CQ Press Electronic Library, CQ Almanac Online Edition, cqal80-1175172, http://library.cqpress.com/cqal80-1175172 (accessed August 30, 2010), originally published in *CQ Almanac 1980* (Washington, DC: Congressional Quarterly, 1981).

72. Ronald W. Reagan, "Message to the Congress Proposing Additions to the National Wild and Scenic Rivers and National Wilderness Preservation Systems," April 26, 1985, in *Public Papers of the President of the United States* (Washington, DC: U.S. Government Printing Office, 1986), 507–508.

73. Ronald W. Reagan, "Radio Address to the Nation on Environmental and Natural Resources Management," June 11, 1983, in *Public Papers of the President of the United States* (Washington, DC: U.S. Government Printing Office, 1984), 852–853.

74. "Presidential Veto Message: Reagan Vetoes Wilderness Bill," CQ Press Electronic Library, CQ Almanac Online Edition, cqal82-858-25803-1162810, http://library.cqpress.com/cqalmanac/cqal82-858-25803-1162810 (accessed August 23, 2010), originally published in *CQ Almanac 1982* (Washington, DC: Congressional Quarterly, 1983).

75. "Georgia Wilderness Bill," CQ Press Electronic Library, CQ Almanac Online Edition, cqal82-1163091, http://library.cqpress.com/cqalmanac/cqal82-1163091 (accessed August 23, 2010), originally published in *CQ Almanac 1982* (Washington, DC: Congressional Quarterly, 1983).

76. "California Wilderness Bill," CQ Press Electronic Library, CQ Almanac Online Edition, cqal81-1171613, http://library.cqpress.com/cqalmanac/cqal81-1171613 (accessed August 23, 2010), originally published in *CQ Almanac 1981* (Washington, DC: Congressional Quarterly, 1982).

77. "Florida Wilderness Measure," CQ Press Electronic Library, CQ Almanac Online Edition, cqal81-1171647, http://library.cqpress.com/cqalmanac/cqal81-1171647 (accessed August 23, 2010), originally published in *CQ Almanac 1981* (Washington, DC: Congressional Quarterly, 1982).

78. "Congress Clears RARE II Wilderness Bills," CQ Press Electronic Library, CQ Almanac Online Edition, cqal82-1163195, http://library.cqpress.com/cqalmanac/cqal82-1163195 (accessed August 23, 2010), originally published in *CQ Almanac 1982* (Washington, DC: Congressional Quarterly, 1983).

79. "National Trails Legislation," CQ Press Electronic Library, CQ Almanac Online Edition, cqal82-1163101, http://library.cqpress.com/cqalmanac/cqal82-1163101 (accessed August 23, 2010), originally published in *CQ Almanac 1982* (Washington, DC: Congressional Quarterly, 1983).

80. "Arizona Wilderness," CQ Press Electronic Library, CQ Almanac Online Edition, cqal83-1199220, http://library.cqpress.com/cqalmanac/cqal83-1199220

(accessed August 23, 2010), originally published in *CQ Almanac 1983* (Washington, DC: Congressional Quarterly, 1984).

81. "Millions of Acres Win Wilderness Protection," CQ Press Electronic Library, CQ Almanac Online Edition, cqal84-1152976, http://library.cqpress.com/cqalmanac/cqal84-1152976 (accessed August 23, 2010), originally published in *CQ Almanac 1984* (Washington, DC: Congressional Quarterly, 1985).

82. "Wilderness Protection," CQ Press Electronic Library, CQ Almanac Online Edition, cqal86-1149889, http://library.cqpress.com/cqalmanac/cqal86-1149889 (accessed August 23, 2010), originally published in *CQ Almanac 1986* (Washington, DC: Congressional Quarterly, 1987).

83. Ibid.

84. Ibid.

85. Ibid.

86. "Montana Wilderness Vetoed," CQ Press Electronic Library, CQ Almanac Online Edition, cqal88-1141422, http://library.cqpress.com/cqalmanac/cqal88-1141422 (accessed August 23, 2010, originally published in *CQ Almanac 1988* (Washington, DC: Congressional Quarterly, 1989); "Presidential Veto Message: Reagan Vetoes Montana Wilderness Legislation," CQ Press Electronic Library, CQ Almanac Online Edition, cqal88-852-25655-1140398, http://library.cqpress.com/cqalmanac/cqal88-852-25655-1140398 (accessed August 23, 2010), originally published in *CQ Almanac 1988* (Washington, DC: Congressional Quarterly, 1989).

87. "Polluted-Lands Curtailed," CQ Press Electronic Library, CQ Almanac Online Edition, cqal88-1141448, http://library.cqpress.com/cqal88-1141448 (accessed August 30, 2010), originally published in *CQ Almanac 1988* (Washington, DC: Congressional Quarterly, 1989).

88. "Strip-Mining Restrictions," CQ Press Electronic Library, CQ Almanac Online Edition, cqal87-1144999, http://library.cqpress.com/cqal87-11--4433999 (accessed August 30, 2010), originally published in *CQ Almanac 1987* (Washington, DC: Congressional Quarterly, 1988).

89. George H. W. Bush, "White House Fact Sheet on the President's Proposal for a Global Forest Convention," July 11, 1990, in *Public Papers of the President of the United States* (Washington, DC: U.S. Government Printing Office, 1991), 1002–1004.

90. George H. W. Bush, "Remarks to Members of Ducks Unlimited," June 8, 1989, from John T. Woolley and Gerhard Peters, *The American Presidency Project,* http://www.presidency.ucsb.edu/ws/?pid=17125 (accessed May 4, 2011); George H. W. Bush, "White House Fact Sheet on the President's Proposal for a Global Forest Convention," July 11, 1980 in John T. Woolley and Gerhard Peters, *The American Presidency Project,* http://www.presidency.ucsb.edu/ws/?pid=18671 (accessed May 4, 2011).

91. George H. W. Bush, "Message to the Congress Transmitting Proposed Legislation on Wilderness Designation of California Public Lands," July 26, 1991, in *Public Papers of the President of the United States* (Washington, DC: U.S. Government Printing Office, 1992), 966.

92. George H. W. Bush, "Statement on Wetlands Preservation," August 9, 1991, in *Public Papers of the President of the United States* (Washington, DC: U.S. Government Printing Office, 1992), 1032–1033.

93. George H. W. Bush, "Statement on Signing the Marsh-Billings National Historical Park Establishment Act," August 26, 1992, in *Public Papers of the President of the United States* (Washington, DC: U.S. Government Printing Office, 1993), 1433.

94. George H. W. Bush, "Statement on Signing the Rocky Mountain Arsenal National Wildlife Refuge Act of 1992," October 9, 1992, in *Public Papers of the President of the United States* (Washington, DC: U.S. Government Printing Office, 1993), 1784; George H. W. Bush, "Statement on Signing Legislation Establishing the Keweenaw National Historical Park," October 27, 1992, in *Public Papers of the President of the United States* (Washington, DC: U.S. Government Printing Office, 1993), 2032.

95. George H. W. Bush, "Letter to Congressional Leaders Transmitting Proposed Legislation on Nevada Public Lands Wilderness Designation," September 3, 1992, in *Public Papers of the President of the United States* (Washington, DC: U.S. Government Printing Office, 1993), 1472–1473; George H. W. Bush, "Letter to Congressional Leaders Transmitting Proposed Legislation on Idaho Public Lands Wilderness Designation," September 4, 1992, in *Public Papers of the President of the United States* (Washington, DC: U.S. Government Printing Office, 1993), 1478; George H. W. Bush, "Letter to Congressional Leaders Transmitting Proposed Legislation on Colorado Public Lands Wilderness Designation," January 7, 1993, in *Public Papers of the President of the United States* (Washington, DC: U.S. Government Printing Office, 1994), 2236–2237; George H. W. Bush, "Letter to Congressional Leaders Transmitting Proposed Legislation on Montana Public Lands Wilderness Designation," January 7, 1993, in *Public Papers of the President of the United States* (Washington, DC: U.S. Government Printing Office, 1994), 2237–2238.

96. "Nevada Wilderness Bill," CQ Press Electronic Library, CQ Almanac Online Edition, cqal89-1139837, http://library.cqpress.com/cqalmanac/cqal89-1139837 (accessed August 23, 2010), originally published in *CQ Almanac 1989* (Washington, DC: Congressional Quarterly, 1990).

97. "Arizona Wilderness Gains Protection," CQ Press Electronic Library, CQ Almanac Online Edition, cqal90-1112636, http://library.cqpress.com/cqalmanac/cqal90-1112636 (accessed August 23, 2010), originally published in *CQ Almanac 1990* (Washington, DC: Congressional Quarterly, 1991).

98. "Compromise to Protect Tongass Forest," CQ Press Electronic Library, CQ Almanac Online Edition, cqal90-1112597, http://library.cqpress.com/cqalmanac/cqal90-1112597 (accessed August 31, 2010), originally published in *CQ Almanac 1990* (Washington, DC: Congressional Quarterly, 1991).

99. "California Desert Protection Stalled in Senate," CQ Press Electronic Library, CQ Almanac Online Edition, cqal91-1110440, http://library.cqpress.com/cqalmanac/cqal91-1110440 (accessed August 23, 2010), originally published in *CQ Almanac 1991* (Washington, DC: Congressional Quarterly, 1992).

100. "National Monument and Wilderness Protection," CQ Press Electronic Library, CQ Almanac Online Edition, cqal91-1110453, http://library.cqpress.com/cqalmanac/cqal91-1110453 (accessed August 23, 2010), originally published in *CQ Almanac 1991* (Washington, DC: Congressional Quarterly, 1992).

101. "Chambers Çan't Agree on Montana Wilderness Bill," CQ Press Electronic Library, CQ Almanac Online Edition, cqal92-1107909, http://library.cqpress.com/cqalmanac/cqal92-1107909 (accessed August 23, 2010), originally published in *CQ Almanac 1992* (Washington, DC: Congressional Quarterly, 1993).

102. "Mining Law Overhaul Is Stymied Again," CQ Press Electronic Library, CQ Almanac Online Edition, cqal92-1107886, http://library.cqpress.com/cqalmanac/cqal92-1107886 (accessed August 31, 2010), originally published in *CQ Almanac 1992* (Washington, DC: Congressional Quarterly, 1993).

103. William J. Clinton, "Remarks on Earth Day in Great Falls, Maryland," Aoruk 22m 1996, in *Public Papers of the Presidents of the United States;* John T. Woolley and Gerhard Peters, *The American Presidency Project,* http://www.presidency.ucsb.edu//ws/?pid=52705 (accessed May 5, 2011).

104. William J. Clinton, "Remarks to the Community in Hackensack, New Jersey," March 11, 1996, in *Public Papers of the Presidents of the United States;* John T. Woolley and Gerhard Peters, *The American Presidency Project,* http://www.presidency.ucsb.edu//ws/?pid=52526 (accessed May 5, 2011).

105. William J. Clinton, "Remarks on Signing the Colorado Wilderness At of 1993 in Denver," August 13, 1993, in *Public Papers of the Presidents of the United States;* John T. Woolley and Gerhard Peters, *The American Presidency Project,* http://www.presidency.ucsb.edu//ws/?pid=46989 (accessed May 5, 2011); William J. Clinton, "The President's Radio Address," October 30, 1999, in *Public Papers of the Presidents of the United States;* John T. Woolley and Gerhard Peters, *The American Presidency Project,* http://www.presidency.ucsb.edu//ws/?pid=56830 (accessed May 5, 2011).

106. William J. Clinton, "Statement on Signing the Mollie Beattie Wilderness Area Act," July 29, 1996, in *Public Papers of the Presidents of the United States;* John T. Woolley and Gerhard Peters, *The American Presidency Project,* http://www.presidency.ucsb.edu//ws/?pid=53132 (accessed May 5, 2011).

107. William J. Clinton, "Statement on Signing the Black Canyon of the Gunnison National Park and Gunnison Gorge National Conservation Area Act of 1999," October 21, 1999, in *Public Papers of the Presidents of the United States;* John T. Woolley and Gerhard Peters, *The American Presidency Project,* http://www.presidency.ucsb.edu//ws/?pid=56778 (accessed May 5, 2011).

108. William J. Clinton, "Remarks In Sunrise, Florida," September 5, 1996, in *Public Papers of the Presidents of the United States;* John T. Woolley and Gerhard Peters, *The American Presidency Project,* http://www.presidency.ucsb.edu//ws/?pid=53290 (accessed May 5, 2011).

109. William J. Clinton, "Proclamation 6775—National Park Week, 1995," March 10, 1995, in *Public Papers of the Presidents of the United States;* John T. Woolley and Gerhard Peters, *The American Presidency Project,* http://www.presidency.ucsb.edu//ws/?pid=51085 (accessed May 5, 2011).

110. William J. Clinton, "Memorandum on Transportation Planning to Address Impacts of Transportation on National Parks," April 22, 1996 in *Public Papers of the Presidents of the United States;* John T. Woolley and Gerhard Peters, *The American Presidency Project,* http://www.presidency.ucsb.edu//ws/?pid=52707 (accessed May 5, 2011).

111. "California Desert Bill Advances," CQ Press Electronic Library, CQ Almanac Online Edition, cqal93-1105649, http://library.cqpress.com/cqalmanac/cqal93-1105649 (accessed August 23, 2010), originally published in *CQ Almanac 1993* (Washington, DC: Congressional Quarterly, 1994); "Fragile California Desert Bill Blooms Late in Session," CQ Press Electronic Library, CQ Almanac Online Edition, cqal94-1103287, http://library.cqpress.com/cqalmanac/cqal94-1103287 (accessed August 23, 2010), originally published in *CQ Almanac 1994* (Washington, DC: Congressional Quarterly, 1995).

112. "Congress Clears Protection of Colorado Wilderness," CQ Press Electronic Library, CQ Almanac Online Edition, cqal93-1105658, http://library.cqpress.com/cqalmanac/cqal93-1105658 (accessed August 23, 2010), originally published in *CQ Almanac 1993* (Washington, DC: Congressional Quarterly, 1994).

113. "Various Wilderness Bills Considered by Congress," CQ Press Electronic Library, CQ Almanac Online Edition, cqal93-1105662, http://library.cqpress.com/cqalmanac/cqal93-1105662 (accessed August 23, 2010), originally published in *CQ Almanac 1993* (Washington, DC: Congressional Quarterly, 1994).

114. "Montana Wilderness Bill Stalls," CQ Press Electronic Library, CQ Almanac Online Edition, cqal94-1103343, http://library.cqpress.com/cqalmanac/cqal94-1103343 (accessed August 23, 2010), originally published in *CQ Almanac 1994* (Washington, DC: Congressional Quarterly, 1995).

115. "Split in Idaho Delegation Sinks Wilderness Bill," CQ Press Electronic Library, CQ Almanac Online Edition, cqal94-1103351, http://library.cqpress.com/cqalmanac/cqal94-1103351 (accessed August 23, 2010), originally published in *CQ Almanac 1994* (Washington, DC: Congressional Quarterly, 1995).

116. "Committees OK Utah Wilderness Bill," CQ Press Electronic Library, CQ Almanac Online Edition, cqal95-1100422, http://library.cqpress.com/cqalmanac/cqal95-1100422 (accessed August 23, 2010), originally published in *CQ Almanac 1995* (Washington, DC: Congressional Quarterly, 1996).

117. "Boundary Wilderness," CQ Press Electronic Library, CQ Almanac Online Edition, cqal97-0000181027, http://library.cqpress.com/cqalmanac/cqal97-0000181027 (accessed August 23, 2010), originally published in *CQ Almanac 1997* (Washington, DC: Congressional Quarterly, 1998).

118. "House Panel Takes Step toward Establishing National Lands Protection Program," CQ Press Electronic Library, CQ Almanac Online Edition, cqal99-0000201148, http://library.cqpress.com/cqal99-0000201148 (accessed August 30, 2010), originally published in *CQ Almanac 1999* (Washington, DC: Congressional Quarterly, 2000).

119. George W. Bush, "Remarks at Sequoia National Park, California," May 30, 2001, in *Public Papers of the President of the United States* (Washington, DC: U.S. Government Printing Office, 2002), 598–601; George W. Bush, "Statement on the

Centennial Anniversary of the National Wildlife Refuge System," March 14, 2003, in *Public Papers of the President of the United States* (Washington, DC: U.S. Government Printing Office, 2004), 265.

120. George W. Bush, "Remarks on the Healthy Forests Initiative," May 20, 2003, in *Public Papers of the President of the United States* (Washington, DC: U.S. Government Printing Office, 2004), 514–518; George W. Bush, "Remarks in Redmond, Oregon," August 21, 2003, in *Public Papers of the President of the United States* (Washington, DC: U.S. Government Printing Office, 2004), 1039–1045; George W. Bush, "Remarks on Signing the Healthy Forests Restoration Act of 2003," December 3, 2003, in *Public Papers of the President of the United States* (Washington, DC: U.S. Government Printing Office, 2004), 1669–1671.

121. George W. Bush, "Remarks on Earth Day in Wells, Maine," April 22, 2004, in *Public Papers of the President of the United States* (Washington, DC: U.S. Government Printing Office, 2005), 649–653.

122. George W. Bush, "Remarks at Rookery Bay National Estuarine Research Reserve in Naples, Florida," April 23, 2004, in *Public Papers of the President of the United States* (Washington, DC: U.S. Government Printing Office, 2005), 655–659.

123. Michael E. Kraft, "Environmental Policy in Congress," in Norman J. Vig and Michael E. Kraft, eds., *Environmental Policy: New Directions for the Twenty-First Century* (Washington, DC: CQ Press, 2006), 124–147: 139–140.

124. Barack Obama, "Remarks on Signing the Omnibus Public Land Management Act of 2009," March 30, 2009 *The Federal Register,* April 3, 2009.

125. Barack Obama, "Proclamation 8409—National Wilderness Month, 2009," September 3, 2009 (Washington, DC: *The Federal Register,* September 8, 2009); Barack Obama, "Proclamation 8553—National Wilderness Month, 2010," August 31, 2010 (Washington, DC: *The Federal Register,* September 7, 2010).

126. Barack Obama, "Proclamation 8362—National Park Week, 2009," April 17, 2009 (Washington, DC: *The Federal Register,* April 22, 2009).

127. Avery Palmer, "Long-Stalled Lands Bill Gets Nod from Senate," *CQ Weekly Online,* January 19, 2009: 128, http://library.cqpress.com/cqweekly/weeklyreport111-000003012957 (accessed September 2, 2010); Avery Palmer, "Lands Omnibus Set to Clear after Senate Passage," *CQ Weekly Online,* March 23, 2009: 675, http://library.cqpress.com/cqweekly/weeklyreport111-000003081110 (accessed September 2, 2010).

128. Palmer, "Lands Omnibus Set to Clear"; Palmer, "Long-Stalled Lands Bill."

129. Avery Palmer, "Long-Delayed Public Lands Bill Sent to Obama," *CQ Weekly Online,* March 30, 2009: 727, http://library.cqpress.com/cqweekly/weeklyreport111-000003087457 (accessed September 2, 2010).

130. Jennifer Scholtes, "House Panel Rejects Attempt to Get Land Documents," *CQ Weekly Online,* May 10, 2010: 1155, http://library.cqpress.com/cqweekly/weeklyreport111-000003658708 (accessed September 2, 2010).

Chapter 9 The Politics of Endangered Species and Wildlife

1. Samuel P. Hays, *Environmental Politics since 1945* (Pittsburgh: University of Pittsburgh Press, 2000), 42.

2. Phillip F. Cramer, *Deep Environmental Politics* (Westport, CT: Praeger, 1998), 84.

3. "Threatened Wildlife Species," CQ Press Electronic Library, CQ Almanac Online Edition, cqal66-1300127, http://library.cqpress.com/cqalmanac/cqal66-1300127 (accessed August 31, 2010), originally published in *CQ Almanac 1966* (Washington, DC: Congressional Quarterly, 1967).

4. Tom Garrett, "Wildlife," in James Rathlesberger, ed., *Nixon and the Environment* (New York: Village Voice Book, 1972), 129–145: 131.

5. "Predator Controls and Endangered Species," *Congressional Quarterly Almanac* 28 (1972): 739–744: 742.

6. Jeffrey J. Pompe and James R. Rinehart, *Environmental Conflict* (Albany: State University of New York, 2002), 86.

7. "Wildlife Restoration Acts," CQ Press Electronic Library, CQ Almanac Online Edition, cqal70-1293888, http://library.cqpress.com/cqalmanac/cqal70-1293888 (accessed August 31, 2010), originally published in *CQ Almanac 1970* (Washington, DC: Congressional Quarterly, 1971).

8. "Marine Mammals," CQ Press Electronic Library, CQ Almanac Online Edition, cqal71-1254656, http://library.cqpress.com/cqalmanac/cqal71-1254656 (accessed August 31, 2010), originally published in *CQ Almanac 1971* (Washington, DC: Congressional Quarterly, 1972); "Congress Approves Moratorium on Sea Mammal Killing," CQ Press Electronic Library, CQ Almanac Online Edition, cqal72-1249699, http://library.cqpress.com/cqalmanac/cqal72-1249699 (accessed August 31, 2010), originally published in *CQ Almanac 1972* (Washington, DC: Congressional Quarterly, 1973).

9. "Predator Controls and Endangered Species."

10. Ibid.

11. "Wildlife Restoration," CQ Press Electronic Library, CQ Almanac Online Edition, cqal72-1249134, http://library.cqpress.com/cqalmanac/cqal72-1249134 (accessed August 31, 2010), originally published in *CQ Almanac 1972* (Washington, DC: Congressional Quarterly, 1973).

12. "San Francisco Bay Wildlife," CQ Press Electronic Library, CQ Almanac Online Edition, cqal72-1250876, http://library.cqpress.com/cqalmanac/cqal72-1250876 (accessed August 31, 2010), originally published in *CQ Almanac 1972* (Washington, DC: Congressional Quarterly, 1973).

13. "Hunting Animals from Aircraft," *Congressional Quarterly Almanac* 27 (1971): 727.

14. "Environment: Action Completed," *Congressional Quarterly Almanac* 27 (1971): 29.

15. "Predator Controls and Endangered Species": 743; "Bald Eagle Protection," CQ Press Electronic Library, CQ Almanac Online Edition, cqal72-1249131,

http://library.cqpress.com/cqalmanac/cqal72-1249131 (accessed August 31, 2010), originally published in *CQ Almanac 1972* (Washington, DC: Congressional Quarterly, 1973).

16. "Major Congressional Action," *Congressional Quarterly Almanac* 28 (1972): 743–745.

17. "Endangered Species," *Congressional Quarterly Almanac* 29 (1973): 670–673.

18. Cramer, *Deep Environmental Politics:* 94, citing Endangered Species Act, U.S. Code, title 16 (Conservation), chapter 35, section 1531.

19. Ibid.

20. Pompe and Rinehart, *Environmental Conflict:* 86.

21. Richard N.L. Andrews, *Managing the Environment, Managing Ourselves* (New Haven, CT: Yale University Press, 1999), 292.

22. Cramer, *Deep Environmental Politics:* 94.

23. Ibid., 93.

24. Richard M. Nixon, "Statement on Signing the Endangered Species Act of 1973," December 28, 1973, in *Public Papers of the President of the United States* (Washington, DC: U.S. Government Printing Office, 1974), 1027–1028.

25. Rosemary O'Leary, "Environmental Policy in the Courts," in Norman J. Vig and Michael E. Kraft, eds., *Environmental Policy: New Directions for the Twenty-First Century* (Washington, DC: CQ Press, 2006), 148–168: 157–158.

26. Ibid.

27. Ibid.

28. Gerald R. Ford, "Message to the Congress Reporting on International Whaling Operations and Conservation Programs," January 16, 1975, in *Public Papers of the President of the United States* (Washington, DC: U.S. Government Printing Office, 1976), 47–49.

29. Gerald R. Ford, "Message to the Senate Transmitting the Agreement on the Conservation of Polar Bears," November 29, 1975, in *Public Papers of the President of the United States* (Washington, DC: U.S. Government Printing Office, 1976), 1919–1920.

30. Gerald R. Ford, "Message to the Senate Transmitting the Convention for the Conservation of Antarctic Seals," December 17, 1975, in *Public Papers of the President of the United States* (Washington, DC: U.S. Government Printing Office, 1976), 1971–1972.

31. "Wildlife Refuges," CQ Press Electronic Library, CQ Almanac Online Edition, cqal76-1189022, http://library.cqpress.com/cqalmanac/cqal76-1189022 (accessed August 31, 2010), originally published in *CQ Almanac 1976* (Washington, DC: Congressional Quarterly, 1977).

32. Pompe and Rinehart, *Environmental Conflict:* 92.

33. Jimmy Carter, "The Environment," May 23, 1977, in *Public Papers of the President of the United States* (Washington, DC: U.S. Government Printing Office, 1978), 967–986.

34. Ibid.

35. Ibid.

36. Jimmy Carter, "Statement on Signing S. 2899 into Law," November 10, 1978, in *Public Papers of the President of the United States* (Washington, DC: U.S. Government Printing Office, 1979), 2002–2003.

37. "Endangered Species," *Congressional Quarterly Almanac* 33 (1977): 707.

38. "Endangered Species Curbs," *Congressional Quarterly Almanac* 34 (1978): 707–709.

39. Ibid.

40. "Marine Mammal Protection," CQ Press Electronic Library, CQ Almanac Online Edition, cqal78-1236952, http://library.cqpress.com/cqalmanac/cqal78-1236952 (accessed August 31, 2010), originally published in *CQ Almanac 1978* (Washington, DC: Congressional Quarterly, 1979).

41. "Nongame Wildlife Bill," CQ Press Electronic Library, CQ Almanac Online Edition, cqal78-1236928, http://library.cqpress.com/cqalmanac/cqal78-1236928 (accessed August 31, 2010), originally published in *CQ Almanac 1978* (Washington, DC: Congressional Quarterly, 1979).

42. "Environment: Endangered Species, Tellico Dam," *Congressional Quarterly Almanac* 35 (1979): 23.

43. Jimmy Carter, "Endangered Species Act Amendments," December 28, 1979, in *Public Papers of the President of the United States* (Washington, DC: U.S. Government Printing Office, 1980), 2288.

44. Ibid.; "Fish and Wildlife Grants," CQ Press Electronic Library, CQ Almanac Online Edition, cqal80-1175182, http://library.cqpress.com/cqalmanac/cqal80-1175182 (accessed August 31, 2010), originally published in *CQ Almanac 1980* (Washington, DC: Congressional Quarterly, 1981).

45. "Endangered Species Funds," CQ Press Electronic Library, CQ Almanac Online Edition, cqal80-1175185, http://library.cqpress.com/cqalmanac/cqal80-1175185 (accessed August 31, 2010), originally published in *CQ Almanac 1980* (Washington, DC: Congressional Quarterly, 1981).

46. "Fish and Wildlife Protection," CQ Press Electronic Library, CQ Almanac Online Edition, cqal80-1175178, http://library.cqpress.com/cqalmanac/cqal80-1175178 (accessed August 31, 2010), originally published in *CQ Almanac 1980* (Washington, DC: Congressional Quarterly, 1981).

47. Ronald W. Reagan, "Message to the Congress Reporting on the Whaling Activities of the Soviet Union," May 31, 1985, in *Public Papers of the President of the United States* (Washington, DC: U.S. Government Printing Office, 1986), 704–705.

48. "Illegal Wildlife Trafficking," *Congressional Quarterly Almanac* 37 (1981): 524.

49. "Endangered Species Act Reauthorized," *Congressional Quarterly Almanac* 38 (1982): 435–436.

50. "Marine Mammal Protection," CQ Press Electronic Library, CQ Almanac Online Edition, cqal81-1171607, http://library.cqpress.com/cqalmanac/cqal81-1171607 (accessed August 31, 2010), originally published in *CQ Almanac 1981* (Washington, DC: Congressional Quarterly, 1982).

51. "House Clears Wetlands Bill," CQ Press Electronic Library, CQ Almanac Online Edition, cqal83-1199269, http://library.cqpress.com/cqalmanac/

cqal83-1199269 (accessed August 31, 2010), originally published in *CQ Almanac 1983* (Washington, DC: Congressional Quarterly, 1984).

52. "Marine Mammal Protection," CQ Press Electronic Library, CQ Almanac Online Edition, cqal84-1153078, http://library.cqpress.com/cqalmanac/cqal84-1153078 (accessed August 31, 2010), originally published in *CQ Almanac 1984* (Washington, DC: Congressional Quarterly, 1985).

53. "Wetlands Protection Bill," CQ Press Electronic Library, CQ Almanac Online Edition, cqal84-1153095, http://library.cqpress.com/cqalmanac/cqal84-1153095 (accessed August 31, 2010), originally published in *CQ Almanac 1984* (Washington, DC: Congressional Quarterly, 1985).

54. "Wetlands Loan Extended," CQ Press Electronic Library, CQ Almanac Online Edition, cqal84-1153094, http://library.cqpress.com/cqalmanac/cqal84-1153094 (accessed August 31, 2010), originally published in *CQ Almanac 1984* (Washington, DC: Congressional Quarterly, 1985).

55. "Endangered Species Act," *Congressional Quarterly Almanac* 41 (1985): 201–202.

56. "Wildlife Refuge Veto," CQ Press Electronic Library, CQ Almanac Online Edition, cqal85-1147734, http://library.cqpress.com/cqalmanac/cqal85-1147734 (accessed August 31, 2010), originally published in *CQ Almanac 1985* (Washington, DC: Congressional Quarterly, 1986).

57. "Presidential Veto Message: Reagan Vetoes Wildlife Refuge," CQ Press Electronic Library, CQ Almanac Online Edition, cqal85-855-25694-1145851, http://library.cqpress.com/cqalmanac/cqal85-855-25694-1145851 (accessed August 31, 2010), originally published in *CQ Almanac 1985* (Washington, DC: Congressional Quarterly, 1986).

58. "Endangered Species Act," *Congressional Quarterly Almanac* 43 (1987): 305; "Energy/Environment: Endangered Species," *Congressional Quarterly Almanac* 44 (1988): 19; "Endangered Species Cleared," *Congressional Quarterly Almanac* 44 (1988): 156–157.

59. "Fish and Wildlife Bill," CQ Press Electronic Library, CQ Almanac Online Edition, cqal87-1145028, http://library.cqpress.com/cqalmanac/cqal87-1145028 (accessed August 31, 2010), originally published in *CQ Almanac 1987* (Washington, DC: Congressional Quarterly, 1988).

60. George H. W. Bush, "Letter to Congressional Leaders on Norwegian Whaling Activities," December 18, 1990, in *Public Papers of the President of the United States* (Washington, DC: U.S. Government Printing Office, 1991), 1181–1182.

61. George H. W. Bush, "Statement on Signing the Wild Bird Conservation Act of 1992," October 23, 1992, in *Public Papers of the President of the United States* (Washington, DC: U.S. Government Printing Office, 1993), 1938.

62. "Wetlands Bill Clears," CQ Press Electronic Library, CQ Almanac Online Edition, cqal89-1139833, http://library.cqpress.com/cqalmanac/cqal89-1139833 (accessed August 31, 2010), originally published in *CQ Almanac 1989* (Washington, DC: Congressional Quarterly, 1990).

63. "Endangered Species Act Revision Looms," CQ Press Electronic Library, CQ Almanac Online Edition, cqal91-1110365, http://library.cqpress.

com/cqalmanac/cqal91-1110365 (accessed August 31, 2010), originally published in *CQ Almanac 1991* (Washington, DC: Congressional Quarterly, 1992).

64. Ibid.

65. "Rocky Mountain Arsenal to Be Wildlife Refuge," CQ Press Electronic Library, CQ Almanac Online Edition, cqal92-1107834, http://library.cqpress.com/cqalmanac/cqal92-1107834 (accessed August 31, 2010), originally published in *CQ Almanac 1992* (Washington, DC: Congressional Quarterly, 1993).

66. William J. Clinton, "Message to the Congress Transmitting the Convention on Biological Diversity," November 19, 1993, in *Public Papers of the Presidents of the United States;* John T. Woolley and Gerhard Peters, *The American Presidency Project,* http://www.presidency.ucsb.edu//ws/?pid=46145 (accessed May 5, 2011).

67. William J. Clinton, "Statement of Administration Policy: HR 1420—National Wildlife Refuge System Improvement Act of 1997," June 2, 1997, in *Public Papers of the Presidents of the United States;* John T. Woolley and Gerhard Peters, *The American Presidency Project,* http://www.presidency.ucsb.edu//ws/?pid=74947 (accessed May 5, 2011).

68. William J. Clinton, "Proclamation 6993—National Wildlife Week, 1997," April 19, 1997, in *Public Papers of the Presidents of the United States;* John T. Woolley and Gerhard Peters, *The American Presidency Project,* http://www.presidency.ucsb.edu//ws/?pid=54024 (accessed May 5, 2011).

69. William J. Clinton, "Remarks on Steps to Remove the American Bald Eagle from the Endangered Species List," July 2, 1999, in *Public Papers of the Presidents of the United States;* John T. Woolley and Gerhard Peters, *The American Presidency Project,* http://www.presidency.ucsb.edu//ws/?pid=57822 (accessed May 5, 2011).

70. William J. Clinton, "Message to the Senate transmitting the Protocol to the Caribbean Environmental Convention," April 20, 1993, in *Public Papers of the Presidents of the United States;* John T. Woolley and Gerhard Peters, *The American Presidency Project,* http://www.presidency.ucsb.edu//ws/?pid=46456 (accessed May 5, 2011).

71. William J. Clinton, "Message to the Congress on Rhinoceros and Tiger Trade by China and Taiwan," November 8, 1993, in *Public Papers of the Presidents of the United States;* John T. Woolley and Gerhard Peters, *The American Presidency Project,* http://www.presidency.ucsb.edu//ws/?pid=46081 (accessed May 5, 2011).

72. William J. Clinton, "Statement on Trade Sanctions Against Taiwan," April 11, 1994, in *Public Papers of the Presidents of the United States;* John T. Woolley and Gerhard Peters, *The American Presidency Project,* http://www.presidency.ucsb.edu//ws/?pid=49947 (accessed May 5, 2011).

73. William J. Clinton, "Statement on Signing Wildlife and Wetlands Legislation," October 30, 1998, in *Public Papers of the Presidents of the United States;* John T. Woolley and Gerhard Peters, *The American Presidency Project,* http://www.presidency.ucsb.edu//ws/?pid=55195 (accessed May 5, 2011).

74. William J. Clinton, "Message to the Senate Transmitting the Inter-American Convention for the Protection and Conservation of Sea Turtles," May 22, 1998, in *Public Papers of the Presidents of the United States;* John T. Woolley

and Gerhard Peters, *The American Presidency Project,* http://www.presidency.ucsb.edu//ws/?pid=56018 (accessed May 5, 2011).

75. "Exemption Extended for Marine Mammal Kills," CQ Press Electronic Library, CQ Almanac Online Edition, cqal93-1105671, http://library.cqpress.com/cqalmanac/cqal93-1105671 (accessed August 31, 2010), originally published in *CQ Almanac 1993* (Washington, DC: Congressional Quarterly, 1994).

76. "Hill Revamps Marine Mammal Act," CQ Press Electronic Library, CQ Almanac Online Edition, cqal94-110395, http://library.cqpress.com/cqalmanac/cqal94-110395 (accessed August 31, 2010), originally published in *CQ Almanac 1994* (Washington, DC: Congressional Quarterly, 1995).

77. "House Approves Biological Inventory," CQ Press Electronic Library, CQ Almanac Online Edition, cqal93-1105626, http://library.cqpress.com/cqalmanac/cqal93-1105626 (accessed August 31, 2010), originally published in *CQ Almanac 1993* (Washington, DC: Congressional Quarterly, 1994).

78. "Neither Chamber Advances Endangered Species Bill," *Congressional Quarterly Almanac* 51 (1995): 5-13–5-15.

79. "Senate Starts on Species Act Rewrite," *Congressional Quarterly Almanac* 53 (1997): 4-7–4-10.

80. "Endangered Species Act," *Congressional Quarterly Almanac* 53 (1997): 5–25.

81. Ibid.

82. George W. Bush, "Statement on Signing Legislation Establishing the Detroit River International Wildlife Refuge," December 21, 2001, in *Public Papers of the President of the United States* (Washington, DC: U.S. Government Printing Office, 2002), 1542.

83. Barack Obama, "Remarks on the 160th Anniversary of the Department of the Interior," March 3, 2009 *The Federal Register,* March 6, 2009; Barack Obama, "Memorandum on the Endangered Species Act," March 3, 2009 (Washington, DC: *The Federal Register,* March 6, 2009).

84. Shawn Zeller, "A Spurt in Gun Sales Is Also Helping Wildlife," *CQ Weekly Online,* April 5, 2010: 806, http://library.cqpress.com/cqweekly/weeklyreport111-000003636579 (accessed September 2, 2010).

85. Stacey Skotzko, "Carpe That Carp, Seize That Snake," *CQ Weekly Online,* January 18, 2010: 158, http://library.cqpress.com/cqweekly/weeklyreport111-000003279786 (accessed September 2, 2010).

Chapter 10 Conclusion

1. Walter A. Rosenbaum, *Environmental Politics and Policy,* 6th ed. (Washington, DC: CQ Press, 2005), 15.

2. Stephan Schmidheiny, "Looking Forward: Our Common Enterprise," in Rao V. Kolluru, ed., *Environmental Strategies Handbook* (New York: McGraw-Hill, 1994), 1–8: 6.

3. Rosenbaum, *Environmental Politics:* 5.

4. Phillip F. Cramer, *Deep Environmental Politics* (Westport, CT: Praeger, 1998), 64.

5. George W. Bush, "Remarks on Proposed Clear Skies Legislation," September 16,2003" in John T. Woolley and Gerhard Peters, *The American Presidency Project*, http://www.presidency.ucsb.edu/ws/?pid=64401 (accessed May 4, 2011).

6. Rosenbaum, *Environmental Politics:* 5.

7. Sally M. Edwards, "Environmental Criminal Enforcement: Efforts by the States," in Sally M. Edwards, Terry D. Edwards, and Charles B. Fields, eds., *Environmental Crime and Criminality* (New York: Garland, 1996), 205–244: 205.

8. William R. Lowry, "A Return to Traditional Priorities in Natural Resource Policies," in Norman J. Vig and Michael E. Kraft, eds., *Environmental Policy: New Directions for the Twenty-First Century* (Washington, DC: CQ Press, 2006), 311 –332: 314.

9. Ibid., 118.

10. Norman J. Vig and Michael E. Kraft," Environmental Policy from the 1970s to 2000: An Overview," in Norman J. Vig and Michael E. Kraft, eds., *Environmental Policy: New Directions for the Twenty-First Century* (Washington, DC: CQ Press, 2006), 1–31: 2.

11. Michael E. Kraft, *Environmental Policy and Politics* (New York: Longman, 2001), 5.

12. Norman J. Vig and Michael E. Kraft, "Preface," in Norman J. Vig and Michael E. Kraft, *Environmental Policy: New Directions:* xii.

Index

About the Author

NANCY E. MARION is professor of political science at the University of Akron, Akron, Ohio. She holds a PhD in political science/public policy from the State University of New York at Binghamton. Dr. Marion's research interests revolve around federal public policy issues and the reaction of Congress and the president to different public concerns. She is the author of numerous books and articles that focus on congressional policy, including *Government vs. Organized Crime* (Prentice Hall, 2008). She is also the co-author of *The Public Policy of Crime and Criminal Justice* (Prentice Hall, 2011).